THE MALARIA GENOME PROJECTS
Promise, Progress, and Prospects

THE MALARIA GENOME PROJECTS
Promise, Progress, and Prospects

Irwin W Sherman
University of California, USA

Imperial College Press

Published by

Imperial College Press
57 Shelton Street
Covent Garden
London WC2H 9HE

Distributed by

World Scientific Publishing Co. Pte. Ltd.
5 Toh Tuck Link, Singapore 596224
USA office: 27 Warren Street, Suite 401-402, Hackensack, NJ 07601
UK office: 57 Shelton Street, Covent Garden, London WC2H 9HE

British Library Cataloguing-in-Publication Data
A catalogue record for this book is available from the British Library.

THE MALARIA GENOME PROJECTS
Promise, Progress, and Prospects

Copyright © 2012 by Imperial College Press

All rights reserved. This book, or parts thereof, may not be reproduced in any form or by any means, electronic or mechanical, including photocopying, recording or any information storage and retrieval system now known or to be invented, without written permission from the Publisher.

For photocopying of material in this volume, please pay a copying fee through the Copyright Clearance Center, Inc., 222 Rosewood Drive, Danvers, MA 01923, USA. In this case permission to photocopy is not required from the publisher.

ISBN-13 978-1-84816-903-6
ISBN-10 1-84816-903-5

Typeset by Stallion Press
Email: enquiries@stallionpress.com

Printed in Singapore by Mainland Press Pte Ltd.

Dedication

This book is dedicated to the memory of my wife
Vilia G. Sherman (1941–2009).
Her critical thinking and love of language,
people and life were the inspiration
for this work and so much more.

Contents

Preface   xi

Chapter 1. Introduction   1
   Plasmodium discovered   2
   Plasmodium transmitted   10
   Plasmodium in hiding   19

Chapter 2. Respice: Before the Genome Project   23

Chapter 3. The Nature of *Plasmodium falciparum* and its Genome   41
   A home for the genome   43
   Location, location, location   47
   The malaria genome capers   49

Chapter 4. Chipping Away at the Genome   63
   Turning "on," turning "off"   71
   Evasion   77

Chapter 5. The Importance of Import   85

Chapter 6. Remodeling the Genome's Home   103

Chapter 7. Getting on the Inside   111

Chapter 8. The Great Escape   125

Chapter 9.	The Neglected Malaria, *Plasmodium vivax*	133
	Resistance	135
	Secrets hidden in the genome	141
Chapter 10.	The *Anopheles* Genome and Transmission Control	145
	Making mosquitoes immune	147
	Making mosquitoes insecticide resistant	150
	Replacement and/or reduction	150
Chapter 11.	The Monkey's Tale	155
Chapter 12.	A Not So Sweet Solution	171
	Isoenzymes, the Achilles' heels of *Plasmodium*	178
	Tackling the Krebs "cycle"	181
Chapter 13.	To Search and Find	187
	Rapid diagnostic tests	188
	Nucleic acid-based tests	189
	Loop-mediated isothermal amplification	193
Chapter 14.	The Elusive Vaccines	195
	Antigens before the genome	196
	Two blood-stage vaccine candidates	217
	The road to merozoite surface protein	218
	The road to apical membrane antigen 1	222
	A failed promise, SPf66	224
	Transmission-blocking vaccines	227
	The sporozoite vaccine revisited	233
	Development of the pre-blood-stage vaccine, RTS,S	242
	Back to the future	249
	Learning from the liver	253

	Vaccine approaches before sequencing the genome	261
	Vaccine approaches after sequencing the genome	264
Chapter 15.	**New Medicines, Old Problems**	**269**
	The old problem of resistance	273
	Chloroquine	275
	Resistance to anti-folate drugs	281
	Mefloquine and multidrug resistance	284
	The fall of mefloquine	288
	Atovaquone	289
	Artemisia to artemisinins	293
Chapter 16.	**Prospice: Looking to the Future**	**301**
References		305
Index		359

Preface

To decode the genome — the entire sequence of genes — of the most deadly human malarial parasite *Plasmodium falciparum* took six years of effort by a multinational team of scientists, driven by lofty expectations and hubris. The feat was announced on 3 October 2002 in *Nature* magazine and was heralded by a *New York Times* headline: "Genetic decoding may bring advances in worldwide fight against malaria." The need for the Malaria Genome Project seemed obvious to many: each year the disease causes hundreds of millions of people to become sick and more than a million — mostly African children under the age of five — die. Few infectious diseases can rival malaria's ability to kill and cripple. Then, as today, "it has ducked every vaccine attempt and shaken off most of the drugs developed to knock out the disease" [512].

In 1996, the proponents of the Malaria Genome Project argued that it was more than an academic exercise; in their view it was a global public health imperative. Its grand promise was that a more complete understanding of the malaria parasite's genes would provide a rational basis for the development of antimalarial drugs and vaccines, allow a better understanding of the regulation of the complex life cycle in the red blood and liver cells of the human, identify those genes that the parasite uses to thwart the host immune response, and explain the manner by which it is possible for the parasite to evade cure after drug treatments. The hope of this new breed of "gene hunters" was that cracking the life code of *P. falciparum* would reveal the novel biochemical Achilles' heels of the parasite, lead to new drugs and enable pharmaceutical companies to manufacture protective vaccines through the discovery of the targets of protective immunity.

Support for the Malaria Genome Project was not wishful thinking. It was based on previous scientific successes beginning in 1995 when it was possible to "read" the genome of the free-living microbe *Haemophilus influenzae*, followed a year later by the pathogens *Mycoplasma genitalae*, *Borrelia burgdorfi*, the causative agent of Lyme disease, *Helicobacter pylori*, implicated in gastric ulcers, and *Mycobacterium tuberculosis*, responsible for tuberculosis. Further, there were cost-effective technologies to isolate DNA from individual chromosomes, machines to "read" the sequences of genes and computer programs for aligning the pieces to produce a linear map. The cost, the project promoters suggested, would be substantially less than that spent on the Human Genome Project. Deciphering all 23 million letters in the *P. falciparum* genome would be much cheaper than the much larger human genome — ~3 billion letters. Indeed, improvements in instrumentation and powerful computers would not require $28 billion, as did the Human Genome Project; nevertheless tens of millions of dollars would still be needed. As no single funding source was large enough to take on such an immense project, several agencies were tapped, including the U.S. Department of Defense, the Wellcome Trust and the National Institutes of Health (NIH). In order to get the sequencing work done, an international consortium of scientists was assembled in the United States at The Institute for Genomic Research (TIGR), Sloan Kettering Institute, Stanford University, Harvard University and University of Florida, in the UK at the National Institute for Medical Research (NIMR), the Wellcome Trust Sanger Institute, Edinburgh University and Oxford University, and in Australia at the Menzies School of Health Research and the Walter and Eliza Hall Institute.

My own involvement with *Plasmodium* DNA began several decades earlier and was by accident rather than by design. In the summer of 1960 I began post-doctoral studies in the laboratory of William Trager (1910–2005) at the Rockefeller University (formerly the Rockefeller Institute). This was several decades before the human malaria *P. falciparum* could be grown continuously in large quantities in laboratory dishes, so a convenient model for studying human malaria was the bird malaria *P. lophurae*. *Plasmodium lophurae* had been isolated in 1938 by Lowell T. Coggeshall (of the Rockefeller Foundation) from the blood of a Borneo fireback pheasant, *Lophurae igniti igniti*, kept captive in the Bronx Zoo. Thereafter the parasite was maintained by transferring infected blood into chickens and ducks. The growth of *P. lophurae* in young ducklings was phenomenal. At five days after inoculation (by hypodermic syringe) of suitably diluted infected blood almost 90% of the recipient's red blood cells were infected. It was also possible to obtain 100 ml of whole blood from a single duckling; after spinning in a centrifuge it was possible to obtain 10–20 ml of packed red cells with a wet weight of ~15 g — an amount equivalent to that from a luxuriant culture of bacteria. My research focus was biochemical, i.e. enzymes. At the time, purifying an enzyme required fairly large quantities of crude material. *Plasmodium lophurae* was perfectly suited for this, and I became quite adept at isolating substantial quantities of parasites removed from their hemoglobin-rich habitat — the red blood cell.

At the time of my arrival at Rockefeller, Trager had two small laboratories and an office on the third floor of Theobald Smith Hall. The equipment was minimal for biochemistry — a spectrophotometer placed in an abandoned broom closet, a few refrigerated centrifuges, incubators, glassware, cotton-plugged test tubes and pipettes, an assortment of balances for weighing out reagents, and a wood-framed glass hood and bunsen burners for sterile transfer. All of this was in support of Trager's attempts to grow *P. lophurae* in glass dishes both within and removed from red blood cells. Taking all this in I imagined myself in the microbiology laboratory of Robert Koch (1843–1910) or Paul Ehrlich (1854–1915), yet here I was a century later at the center of medical research, "the Rockefeller" in New York City, working with the doyen of malaria, William Trager. Once resident at Rockefeller, Trager and I had a conversation about a meeting he had attended in Florida where another participant was the Brandeis University molecular biologist Julius Marmur (1926–1996). Marmur, following hard on the discovery by the Rockefeller Institute physician Oswald Avery of pneumococcal DNA (called transforming principle) as the genetic material, had worked with Avery's colleague, Rollin Hotchkiss (1911–2004), at Rockefeller (above us on the fourth floor of Theobald Smith Hall!) and demonstrated that the DNA of pneumococcus coded for an enzyme. Later, after developing highly reproducible methods for DNA purification, Marmur was able to show that upon heating purified DNA the helical strands separated (called melting) and the melting temperature was dependent on the composition of the DNA, that is, the percent of guanine plus cytosine (G + C). Marmur was attempting to use the melting temperature of DNA from a variety of bacteria to be used "as a valuable asset in their classification." He told Trager he needed some one-celled microbes that were not bacteria for comparative purposes, and he asked whether Trager could help supply these. Trager had no intention of spending his valuable time in preparation of gram quantities of malaria parasites as this would require abandoning his beloved cultures for several days. So he "volunteered" me to send Marmur some of my precious samples of *P. lophurae*. I did so and when the results were published I was thanked for "useful discussions" but unexpectedly the data for the *P. lophurae* DNA were not included [584]. When I asked Marmur about the omission he simply replied, "It was so low in G + C that it didn't fit." Indeed, what Marmur had found was that the G + C content was 20%!

Eight years later, when I was on the faculty of the University of California at Riverside, Charles Walsh, a graduate student in my laboratory, confirmed Marmur's unpublished findings, but our publication received little attention. Others regarded it to have no real significance as this malaria parasite came from a bird with little in common with the human malaria, *P. falciparum*. It was gratifying many years later, when *P. falciparum* could be grown in culture so its DNA could be analyzed, to find that its G + C was 18–20% (similar to that of *P. lophurae*) and very different from the malarias infecting rodents and monkeys, which were 24% and 37%, respectively.

Although my venture with malaria DNA might have led me to join the ranks of the "gene hunters" it did not. Rather I continued to study the products of genes — protein enzymes — and only at various intervals did I direct my research toward DNA, RNA and protein synthesis. However, never in the succeeding years did I ever lose interest in the research being carried out by others on the genes of malaria parasites and culminating in the Malaria Genome Project.

For whom is this book intended? My aim is to tell a story for those who want to know more about the biology of malaria parasites and the mosquitoes that transmit the disease, how the *P. falciparum* Genome Project came into being, the people who created it and the cadre of scientists — biochemists, parasitologists, microbiologists, molecular geneticists, immunologists — who are attempting to see the promise of the 10-year-old project realized. I have approached the subject in an historic fashion so that those new to the field, or those with insufficient background, will have an easier entry point. Hopefully, through this approach even those already in the field may better appreciate how discoveries made in the past can impact the direction of future malaria research.

■ Chapter 1 ■

Introduction

Today there are approximately 3.3 billion people — ~50% of all the people on this planet — who are at risk of developing malaria each year, with at least 500 million cases, and nearly a million deaths annually [299]. This amounts, on average, to one person dying from malaria every 30 seconds.

Here's how one victim describes the disease: "I wanted to sit up, but felt that I didn't have the strength to, that I was paralyzed. The first signal of an imminent attack is a feeling of anxiety, which comes on suddenly and for no clear reason. Something has happened to you, something bad. If you believe in spirits, you know what it is: someone has pronounced a curse, and an evil spirit has entered you, disabling you and rooting you to the ground. Hence the dullness, the weakness, and the heaviness that comes over you. Everything is irritating. First and foremost, the light; you hate the light. And others are irritating — their loud voices, their revolting smell, their rough touch. But you don't have a lot of time for these repugnances and loathings. For the attack arrives quickly, sometimes quite abruptly, with few preliminaries. It is a sudden, violent onset of cold. A polar, arctic cold. Someone has taken you, naked, toasted in the hellish heat of the Sahel and the Sahara and has thrown you straight into the icy highlands of Greenland or Spitsbergen, amid the snows, winds, and blizzards. What a shock! You feel the cold in a split second, a terrifying, piercing, ghastly cold. You begin to tremble, to quake, to thrash about. You immediately recognize, however, that this is not a trembling you are familiar with from earlier experiences — when you caught cold one winter in a frost; these tremors and convulsions tossing you around are of a kind that any moment now will tear you to shreds. Trying to save yourself, you begin to beg for help. What can bring relief? The only thing that really helps is if someone covers you. But not simply throws a blanket or quilt over you. The thing you are being covered with must crush you with its weight, squeeze you, flatten you. You dream of being pulverized. You desperately long for a steamroller to pass over you" [360]. Indeed, "a man right after a strong attack...is a human rag. He lies in a puddle of sweat, he is still feverish, and he can move neither hand nor foot. Everything hurts; he is dizzy and nauseous. He is exhausted, weak, and limp. Carried by someone else, he gives the impression of having no bones and muscles. And many days must pass before he can get up on his feet again" [46].

Plasmodium discovered

For over 2,500 years the prevalent belief was that "the fever" (malaria) was due to miasmas — deadly vapors arising from wetlands and hence the French name *paludisme* from the Latin meaning swamp. In England in the 15th and 16th centuries the fevers were seasonal. They were called ague (meaning acute fever) and were common in low lying and marshy areas called the unwholesome fens in Kent, Cambridgeshire, Essex and the Thames estuary. The word malaria, literally "bad air," was unknown in the English language until the writer Horace Walpole in 1740, during a visit to Italy, wrote: "There is a horrid thing called malaria that comes to Rome every summer and kills one." Other diseases such as cholera, tuberculosis and diphtheria, in addition to malaria, were believed to be due to "bad air" but in the mid-19th century, after Louis Pasteur (1822–1895) and Robert Koch (1843–1910) identified microbes as the cause of anthrax, cholera, tuberculosis, diphtheria and tetanus, it became clear that germs, not miasmas, were the cause. Once germs (microbes) were shown to be responsible for a variety of infectious diseases it provoked intrepid physicians to find the germ responsible for malaria.

There were various reports of a sighting of the microbe causing malaria and of particular interest was one in 1879 by two investigators in Italy, Edwin Klebs and Corrado Tommasi-Crudeli, who found a rod-shaped bacterium (a bacillus) in the mud and waters of the Pontine marsh in the malarious Roman Campagna. When the bacteria were injected into rabbits, fever resulted, the spleen was enlarged, and the same bacteria could be re-isolated from the sick rabbits. They named the germ *Bacillus malariae*. This finding, quite naturally, attracted worldwide attention. Klebs and Tommasi-Crudeli were well-respected scientists. Klebs held the professorship of pathology in Prague and had conducted research on the relationship of bacteria to disease with Koch. Tommasi-Crudeli, who had studied with the eminent pathologist Rudolph Virchow in Berlin, was Director of the Pathological Institute in Rome with a specialty in malarial fevers. Initially, support for *B. malariae* as the agent of malaria was forthcoming. In 1880 Ettore Marchiafava, a loyal first assistant to Tommasi-Crudeli, and Giuseppe Cuboni reported isolation of the same bacillus from the bodies of patients who had died from malaria, and when studies in Klebs' laboratory found that the antimalarial drug quinine killed *B. malariae* there appeared little doubt that malaria was a bacterial disease [294].

In the United States, the U.S. Board of Health commissioned Major George Sternberg to try to repeat the experiments of Klebs and Tommasi-Crudeli in the malaria-affected area around New Orleans. He found bacteria similar to *B. malariae* in the mud from the Mississippi delta; however, after the bacteria were injected into rabbits the fevers produced were atypical of malaria in that they were not periodic. Sternberg was also able to produce the disease in rabbits by injecting them with his own saliva. He concluded that the disease was septicemia, not malaria, and suggested the bacteria were contaminants [294]. Others raised questions when they were unable to grow *B. malariae* outside the body using the blood of patients with malaria, and there was a suggestion that the killing effect of quinine was not specific. This lack of reproducibility certainly acted to undermine the belief that malaria was a bacterial infection; more telling, however, was the discovery made by an obscure French military physician, Charles-Louis Alphonse Laveran (1845–1922).

In 1875, after reading the literature on malaria and becoming acquainted with physicians who had worked in the French territory of Algeria where cases of malaria were numerous, Laveran wrote a treatise on military epidemiology. He noted that malaria could occur in countries where the climate was cold and could be absent from tropical climes, but that the fevers became more severe as one moved from the poles to the equator. Although swamps and low humid plains were the most favorable environment for malaria, he concluded that swamps themselves did not cause the fever, as even in hot countries not all swamps gave rise to fever. Laveran's treatise made a prescient statement: "Swamp fevers are due to a germ" [347].

After several military assignments in France, Laveran was transferred in 1878 to Bone, Algeria. The North African coast was rife with malaria and Laveran spent most of his time looking at autopsy specimens from those who died from the disease. One of the signal characteristics of malaria in cadavers is the enlarged and blackened spleen and liver, due to accumulation of a brownish-black pigment, called hemozoin. Laveran has been described as "bespectacled with sharp features and a small trim beard" [425]. He was reputed to be extraordinarily precise, meticulous, singularly sharp-minded, incisive and self-opinionated. In short, he did not suffer fools gladly. Laveran spent most of his time looking at preserved dead material from the deceased who had succumbed to "the fever," but he also examined fresh specimens. His microscope was not a good one but he was patient and determined. On 6 November 1880, while examining a drop of fresh liquid blood from a feverish artilleryman, he saw several transparent mobile filaments — flagella — emerging from a clear spherical body. He recognized that these bodies were alive, and that he was looking at an animal, not a bacterium or a fungus. Subsequently he examined blood samples from 192 malaria patients: in 148 of these he found the telltale crescents. Where there were no crescents, there were no symptoms of malaria. Laveran also found spherical bodies in or on the blood cells of those suffering with malaria. Remarkably, and in testimony to his expertise in microscopy, Laveran's discovery was made with a microscope having a magnification of only 400 diameters. He named the beast *Oscillaria malariae* and communicated his findings to the Société Médicale des Hôspitaux on 24 December 1880 [362].

Laveran was anxious to confirm his observations on malaria in other parts of the world, and so he traveled to the Santo Spirito Hospital in Rome where he met with two Italian physicians (one of whom was Marchiafava, Tommasi-Crudeli's assistant, and the other Angelo Celli, Professor of Hygiene) and showed them his slides. The Italians, whose chief interest was *B. malariae*, were unconvinced and told him that the spherical bodies he had seen were nothing more than degenerating red blood cells caused by *B. malariae* or some other cause. Consequently, few of those attending the sick and suffering with "the fever" paid attention to Laveran's observations.

In 1883, Marchiafava and Celli claimed to have seen the same bodies as described by Laveran but without any pigment granules. They also denied the visit by Laveran two years earlier. The Italians were unsuccessful in growing the bodies outside the body of the malaria patient. The lack of consensus on the causative agent was due not only to differences in interpretation of what was seen with the microscope, it was a matter of focus: the Italians emphasized the smallest forms in the blood, whereas Laveran concentrated on the whip-like filaments [294]. Although the search for the parasite itself occupied most research during this period, there was a serendipitous finding of great significance: malaria was an infectious disease, but not one that could be contracted by simply being exposed to a patient with a fever. In 1884, C. Gerhardt deliberately induced malaria for therapeutic purposes in two patients with tertiary syphilis by injection of blood from another patient with intermittent fever, and then cured them all by treatment with quinine. A year later Marchiafava and Celli, working on the wards of Rome's Santo Spirito Hospital, gave multiple injections of infected blood intravenously and sub-cutaneously to five healthy subjects. Parasites were recovered from three of the five who came down with malaria; all recovered after receiving quinine [294]. Clearly, it was the blood of a malaria patient that was infectious, not his breath.

In the 1880s, largely due to the studies of Paul Ehrlich (1854–1918) working in Koch's Berlin laboratory, staining of microbes and tissues became popular among pathologists and microbe hunters. Ehrlich introduced at least 12 aniline dyes as stains, and especially useful was methylene blue for coloring bacteria and live nerve fibers. Ten years later, after returning from Egypt, where he had gone to recuperate from tuberculosis, and knowing that methylene blue stained malaria parasites in a drop of blood after it was smeared out on a microscope slide, he administered the dye to two patients who had been admitted to the Moabite Hospital in Berlin suffering from mild malaria. Both recovered, and, although methylene blue was later found to be ineffective against the more severe forms of malaria, this was the first instance of a synthetic drug being used with success against a specific disease. Subsequent advances in staining technique allowed malaria parasites to be more easily seen with the microscope; however, it was not until late 1884 or early 1885 that Marchiafava and Celli abandoned their use of thin blood films containing killed and stained parasites and began to study, as had Laveran, fresh blood. Examining drops of liquid blood they observed the ameba-like movements of the parasite within the red blood cell and hence they called it *Plasmodium* (from the Latin "plasmo" meaning "mold"). They also witnessed emerging whip-like filaments (called flagella) from the clear spherical bodies within the red blood cell although they questioned the significance of the flagella in the disease [635]. The differences between the interpretations of Laveran and Marchiafava and Celli are now evident: for several years the Italians examined only preserved (killed) specimens and so they missed seeing the movements of the living parasite that had caused Laveran to give it the name *Oscillaria*.

Using both dried and liquid drops of infected blood Marchiafava and Celli were able to trace the development of the small non-pigmented bodies within the red blood cell; they also described the hemozoin as the by-product of the destruction of the red cell's hemoglobin by the growing parasite. In 1886, during a visit to Europe, Major George Sternberg visited Celli at the Santo Spirito Hospital. Celli drew a drop of blood from the fingertip of a malaria patient and was able to show Sternberg the ameba-like movement of the parasite and the emergence of flagella. Sternberg returned to the United States and, working with blood taken from a malaria patient in the Bay View Hospital in Baltimore, was able to find Laveran's parasite in Welch's laboratory at Johns Hopkins University. A year later Welch separated the two kinds of malarias with a 48-hour fever peak, called tertian (because the fever peak occurred every third day) malarias; one would be named *P. vivax* and the other he named *P. falciparum* because it had sickle-shaped crescents (and "falcip" in Latin means "sickle or scythe") [632].

Although Laveran saw another kind of malaria parasite in drops of fresh human blood in patients in Algeria in 1880, it wasn't until 1890 that Battista Grassi and Raimondo Feletti gave it the name *P. malariae*; in this parasite fever occurs every fourth day (because it has a 72-hour developmental cycle) and so it was called a quartan parasite. In 1918, John Stephens working in West Africa discovered in patients a fourth kind of malaria that resembled *P. vivax*; he described it in 1922 and because of its ovoid shape he named it *P. ovale* [158]. In 2004 it was claimed that there was a fifth kind of human malaria, *P. knowlesi*, previously believed to be restricted to wild primates [734]. It was identified in several hundred patients living in Malaysia with macaque and leaf monkeys serving as reservoirs [534]. "In fact, however, it is a simian malaria, whether or not it is occasionally (or, in some areas, more frequently) transmitted to humans. Until it is established that *P. knowlesi* is cyclically transmitted by mosquitoes from human to human, it should be considered a simian malaria and, hence, a zoonotic infection" [149]. (Previously these human infections were misidentified as *P. malariae*.)

Beginning in 1885, Camillo Golgi (1843–1926) of the University of Pavia, convinced by the Italian workers' confirmation of Laveran's observations, felt malaria parasites deserved further study [254]. In 22 patients with *P. malariae* he traced the tiny, non-pigmented bodies of Marchiafava over three days until they grew to fill the red cell, and on the day of the fever paroxysm he found the pigment to concentrate in the center of the parasite as it divided. Golgi discovered that the parasite reproduced asexually by fission and correlated the clinical course of fever with destruction of the red blood cell to release the parasite. In 1886, when he noted that in both the tertian and the quartan fevers there were no crescents, he effectively had distinguished *P. vivax* from *P. malariae* based on fever symptoms alone. Marchiafava and his student Amico Bignami took Golgi's studies of malaria further. Cases of tertian and quartan fever cycle malarias occurred throughout the year in Italy, whereas in the autumn and summer they were outnumbered by a much more severe tertian type, called aestivo-autumnal malaria or malignant malaria. Later, it would be recognized that the deadly tertian disease was caused by the malaria parasite with crescents, *P. falciparum*.

Clinicians in the United States, however, continued to be skeptical of the significance of Laveran's discovery to malaria as a disease. William Osler, the premier blood specialist of his day, on hearing a paper presented at the inaugural meeting of the American Association of Physicians (June 1886) by W.T. Councilman, in which he found flagellated parasites in 80 attempts, challenged his findings. Osler also questioned the role of the flagellated bodies of Laveran, finding them improbable and contrary to all past experience of flagellated organisms occurring in the blood. By late 1886, however, after verifying the existence of the parasites with his own eyes — and postponing his Canadian vacation to examine the blood of every malaria patient he could find — Osler became a convert to the doctrine of "Laveranity." Osler published his findings on 28 October 1886 and in his 1889 treatise *Hematozoa of Malaria*. Osler's paper was read by Henry Vandyke Carter, a pathologist working at the Grant Medical College in India. Previously, Vandyke Carter had been unable to find the malaria parasite in the blood, but with Osler's guidance he succeeded. Vandyke Carter published his findings of three kinds of malaria parasites in India but his report received little notice for a decade among his colleagues in the Indian Medical Service [294]. Despite this, malaria parasites were now being identified elsewhere: in Russia by Metchnikoff, by Morado and Coronado in Cuba, Anderson in Mauritius, and Atkinson in Hong Kong. By 1890 almost all the world believed in both the existence of Laveran's "beasts" as well as in their being the cause of the disease malaria. The significance of Laveran's observation of the release of motile filaments (flagella) or exflagellation would remain unappreciated, however, until William MacCallum and Eugene Opie made some seminal observations.

William MacCallum (1894–1944) and Eugene Opie (1893–1971), two medical interns at Johns Hopkins University, were given the task by their Professor William Thayer of following up Laveran's observations using the malaria-like parasites found in the blood of birds. In 1897, Opie described some of these parasites in wild-caught birds. During that same summer MacCallum, on vacation outside Toronto, Canada, studied one of these "malarias" named *Haemoproteus*, where the male and female sex cells (gametocytes) in the blood are clearly different from one another even in unstained preparations. (This is unlike human malarias where the gametocytes are very similar in appearance.) One type, the clear form, put out Laveran's flagella, whereas the granular forms freed themselves from the red cell and remained quiescent. Observing the two forms in the same field under the microscope, he found the released flagella to invade and unite with the clear form to produce a worm-like gliding form. MacCallum immediately recognized that the flagella were sperm-like, that the granular forms were egg-like, and that he had witnessed fertilization to form a vermicule (later called the ookinete). On his return to Baltimore MacCallum confirmed his discovery in a woman suffering from falciparum malaria. In his 1898 publication MacCallum described the ookinete: "The movement is slow and even…with the pointed end forward. It can move in any direction readily…Often it is seen to rotate continually along its long axis. The forward progression…occurs with considerable force…pushing directly through the obstacle. The ultimate fate and true significance of these forms is difficult to determine" [418].

Plasmodium transmitted

After the description of a variety of malaria parasites in the blood of patients the next great question was: How did people become infected? Ronald Ross (1857–1932), a Surgeon-Major in the Indian Medical Service serving in Bangalore, India, occasionally (when he was not painting watercolors, playing the violin, solving mathematical equations, or writing poetry) used his microscope to look at the blood of soldiers ill with malaria. Finding nothing he shouted to all who could hear: "Laveran is wrong. There is no germ of malaria" [167]. In 1894, Ross returned to England on leave. By that time, he had spent 13 years in India and had few scientific accomplishments: he actually wrote a few papers for the *Indian Medical Gazette* and claimed (without any real evidence) that malaria was primarily an intestinal infection. His hunt for the way malaria was transmitted from person to person began on 9 April, when the 37-year-old Ross visited with the 50-year-old Patrick Manson (1844–1922) at his home at 21 Queen Street in London. At the time of Ross's visit Manson was physician to the Seaman's Hospital Society at Greenwich and a lecturer on Tropical Diseases at St George's Hospital and Charing Cross Hospital Medical Schools, where he had access to malaria contracted by sailors and others in the tropical regions of West Africa and India. Manson was a dedicated and experienced clinician as well as an expert microscopist and he had been shown Laveran's "beast" by H.G. Plimmer (1856–1918) of the University of London who in turn had been able to see the parasite under the guidance of Marchiafava during a visit to Rome. In his primitive laboratory on the top story of his house, night after night Manson watched the release of Laveran's flagella from the crescent forms. At the Seaman's Hospital Manson took a drop of fresh blood from a sailor ill with malaria and showed Ross Laveran's parasite peppered with the black-brown malaria pigment as well as the release of flagella. One day, as the two were walking along Oxford Street, Manson said: "Do you know Ross, I have formed the theory that mosquitoes carry malaria…the mosquitoes suck the blood of people sick with malaria…the blood has those crescents in it...they get into the mosquito stomach, shoot out those whips...the whips shake themselves free and get into the mosquito's carcass...where they turn into a tough form like the spore of an anthrax bacillus...the mosquitoes die...they fall into water...people drink a soup of dead mosquitoes and they become infected" [424].

Manson's idea was based on several false assumptions and analogies. One was the earlier report (1870) of Alexei Fedchenko working in Central Asia, who had shown that the guinea worm, *Dracunculus*, when released into water, entered into a tiny one-eyed crustacean, *Cyclops*, which when swallowed in the drinking water resulted in infection. Second, Manson did not realize that mosquitoes could feed more than once in a lifetime and did not die in the water after their eggs were laid after a single feeding. Thus, a string of seemingly "logical" deductions led to Manson's presumption that malaria parasites must be transmitted by ingestion. Manson had concocted a romantic story that was mostly a guess on his part, but the younger and inexperienced Ross took it as fact.

Ross left England in March 1895, and reached Bombay a month later. In June 1895, encouraged by Manson's passionate plea, he captured various kinds of mosquitoes, although he hadn't a clue what kind they were. He set up an experiment using the water in which an infected mosquito had laid her eggs, her young had been observed swimming, and in which the offspring had been allowed to die. The water with its dead and decaying mosquitoes was then given to a volunteer to drink. The man came down with a fever but, after a few days, no crescents could be found in his blood. The same experiment was repeated with two other men who were paid for their services. Failure again. It thus appeared that drinking water with mosquito-infected material did not produce the disease. Ross began to think that, perhaps, the mosquitoes had the disease, but that they probably gave it to human beings by biting them and not by being eaten. He began to work with patients whose blood contained crescent-shaped malaria parasites, and with mosquitoes bred from larvae. The first task was to get the mosquitoes to bite the patients. There are more than 2,500 different kinds of mosquitoes and, at the time, there were no good means of identifying most of them. Initially, Ross worked mostly with the gray and striped-wing kind. Although when these mosquitoes were dissected there were the whip-like flagella in the mosquito stomach no further development occurred. This result was no more informative than what Laveran had seen in a drop of blood on a microscope slide nearly 20 years earlier. Today, we understand (as Ross did not) why this was so: the gray mosquitoes are *Culex* and those with striped wings are *Aedes*, and these do not carry human malaria. No, the one that he should have used was the brown, spotted winged mosquito, *Anopheles*, but Ross did not recognize this for the entire year he dissected mosquitoes. Each mosquito dissection had required hours of effort at the microscope and yet his prodigious labors had yielded nothing more than thousands of mosquito carcasses.

Then, at the age of 40 and having spent 17 years in the Indian Medical Service, Ross turned from the insusceptible kind of mosquito to the susceptible, brown, spotted winged one. On 16 August 1897, his assistant brought him a bottle in which mosquitoes were being hatched from larvae. It contained "about a dozen big brown fellows, with fine tapered bodies hungrily trying to escape through the gauze covering of the flask which the angel of fate had given my humble retainer!" [560]. He fed them on Husein Khan, a patient who had crescents in his blood. There had been some casualties among the mosquitoes, and only three were left on the morning of 20 August 1897. One of these had died and swelled up with decay. At about 1 p.m., he sacrificed the last mosquito. As Ross described it the dissection was excellent and he went carefully through the tissues, now so familiar, searching every micron with the same passion and care as one would have in searching some vast ruined palace for a little hidden treasure. Nothing. But the stomach tissues still remained to be examined — lying there, empty and flaccid. There he saw a clear and almost perfectly circular outline of about 12 microns in diameter. Looking further, there was another, and another exactly similar cell. In each of these, there was a cluster of small granules, black as jet, and exactly like the black pigment granules of the crescents.

Here was the critical clue to the means of transmission. Ross had shown that, four or five days after feeding on infected blood, the mosquito had wart-like bodies (oocysts) on the outer surface of the stomach. He did not know if these kept on growing, however, or how the mosquitoes became infective. He planned to answer these questions shortly but, before that work could begin, Ross reported his findings to the *British Medical Journal* in a paper entitled "On some peculiar pigmented cells found in two mosquitoes fed on malarial blood." It appeared on 18 December 1897.

Ross felt he could wrap up the unfinished work in a matter of a few weeks, but then was struck by a blow from the Indian Medical Service by being ordered to proceed to Calcutta immediately. As soon as he arrived in Calcutta, he set his hospital assistants the task of hunting for the larvae and pupae of the brown, spotted winged mosquitoes. Soon he had a stock of these, and set about getting them to bite patients who were suffering from malaria. By flooding the ground outside the laboratory, he hoped to imitate rain puddles and with this he began to learn about mosquito breeding. "If I am not on the pigmented cells again in a week or two," he wrote to Manson, "my language will be dreadful."

In Calcutta, Ross was given a small laboratory. There were two Indian assistants who had already been working there when he arrived, but they were old men and not very intelligent, so he engaged two younger men, Purboona and Mahomed Bux. Both of these he paid out of his own money. As there were not a large number of cases of human malaria in the Calcutta hospitals, Ross turned to something that Manson had suggested earlier: the study of mosquitoes and malaria, as seen in birds. Pigeons, crows, larks and sparrows were caught and placed in cages on two old hospital beds. Mosquito nets were put over the beds and then, at night, infected mosquitoes were put under the nets. Before much time had passed the crows and pigeons were found to harbor malaria parasites in their blood; also, he found the pigmented cells, which he previously had spotted on the stomachs of mosquitoes that had been fed on infected larks. Ross became certain of the whole life history in the mosquito, except he had not actually seen the ookinetes (described by MacCallum) turning into oocysts. This was the last stage in the study. He found that the oocysts depended, as regards their size, exactly on the length of time since the mosquitoes had been fed on infected blood. They grew to their maximum size about six days after the mosquito had fed on infected blood. They left the stomach after this time, but he did not know what happened to them then.

One day, while studying some sparrows, he found that one was quite healthy, another contained a few of the malaria parasites, and the third had a large number of parasites in its blood. Each bird was put under a separate mosquito net and exposed to a group of mosquitoes from a batch that had been hatched out from grubs in the same bottle. Fifteen mosquitoes were fed on the healthy sparrow; on their stomachs, not one parasite was found. Nineteen mosquitoes were fed on the second sparrow; every one of these contained some parasites, although, in some cases, not very many. Twenty insects were fed on the third, badly infected, sparrow; every one of these contained some parasites on their stomachs and some contained huge numbers.

Ross wrote in his *Memoirs* [560]: "This delighted me! I asked the medical service for assistance and a leave, but was denied this. I wanted to provide the final story of the malaria parasite for this meeting; but, I knew that time was very short. I still did not have the full details of...the change from the oocysts in the mosquito's stomach into the stages that could infect human beings and birds. Then, I found that some of the oocysts seemed to have stripes or ridges in them; this happened on the 7th or 8th day after the mosquito had been fed on infected blood." He continued, "I spent hours every day peering into the microscope. The constant strain on mind and eye at this temperature is making me thoroughly ill." He had no doubt these oocysts with the stripes or rods burst, but did not know what happened to them. He asked himself: "If they burst, did they produce the same stages that infected human blood?"

Then, on 4 July 1898, Ross got something of value. Near a mosquito's head, he found a large branching gland. It led into the head of the mosquito. Ross mused, "It is a thousand to one that it is a salivary gland. Did this gland infect healthy creatures? Did it mean that if an infected mosquito fed off the blood of an uninfected human being or bird, then this gland would pour some of the parasites...into the blood of the healthy creature?" Shortly thereafter he found that it was possible to infect 21 out of 28 healthy sparrows by the bites of infected (*Culex*) mosquitoes. On 9 July he wrote Manson "One single experiment with crescents will enable me to bring human malaria in line with [bird malaria] — they are sure to be just the same."

This was the proof — this showed that malaria was not conveyed by bad air. After all, men and birds don't go about eating dead mosquitoes! On 25 July, now sure, he sent off a triumphant telegram to Patrick Manson, reporting the complete solution; three days later, Manson spoke at the British Medical Association meeting in Edinburgh, describing the long and painstaking research Ross had been carrying out for years past. Ross's findings were communicated on 28 July 1898, and published in the 20 August issue of *The Lancet* and the 24 September issue of the *British Medical Journal*. In addition to his malaria work Ross was assigned by the Medical Service to investigate kala-azar, and consequently had little time to continue the work. Ross never did the single experiment with crescents (using *Anopheles* mosquitoes) and he left India for good on 16 February 1899.

Ross's discovery of infectious stages in the mosquito salivary glands in a malaria infecting birds appeared to be the critical element in understanding transmission of the disease in humans. Manson rightly cautioned, however: "One can object that the fact determined for birds does not hold, necessarily, for humans." Ross and Manson wanted to grab the glory of discovery for themselves and for England but in such a quest they were not alone [294]. During the period 1894–1898 — a time when Ross was working alone in India — the Italians published virtually nothing except for the 1896 paper by Bignami. But by the middle of July 1898 — a time when Ross's proof was complete and partly published — the Italians, led by Giovanni Battista Grassi, began to work in earnest on the transmission of malaria [294]. Grassi received a medical degree from the University of Padua (1875); however, he never practiced medicine and by the time the work on transmission began he held the Chair in Comparative Anatomy in Rome. He was world-renowned for his studies in zoology including unraveling the complex and mysterious life history of the eel (1887); he was able to diagnose hookworm disease by finding eggs in the feces, identified fleas as vectors of dog and mouse tapeworms, and wrote a monograph on the marine chaetognaths. As early as 1890, he had worked with bird malaria, and this led quite naturally to studies of human malaria; together with Amico Bignami and Antonio Dionisi at the Santo Spirito Hospital they attempted (in 1894) to determine whether mosquitoes from the malarious areas were transmitters of the disease. In this, they were unsuccessful.

Grassi was delicate and small in stature but with a serious and forceful will. He had long, black, curly hair, beetling eyebrows and an unkempt shaggy beard. He usually wore a battered felt hat, tattered clothes and dark, smoked glasses. He seemed to enjoy a good quarrel and was pleased when such an opportunity presented itself. He was also very proud of the research he did and was resentful of those who questioned his authority. Grassi recognized that, insofar as malaria was concerned, there remained two main tasks: to demonstrate the developmental cycle of the human parasite in the mosquito and to identify the kind of mosquito that transmitted human malaria. To this end he assembled a team of colleagues to make an all-out push. The team consisted of Dionisi, who had worked with bird malaria and was to test Ross's findings; Bignami, who would test his mosquito "bite theory;" Grassi, who knew the different kinds of mosquitoes, and who would survey the malarious and non-malarious areas and by comparing the mosquito populations try to deduce which species were possible transmitters; and Bignami together with Giovanni Bastianelli, a careful microscopist who knew his malaria parasites well, would follow Ross's trail to determine the parasite's development in the mosquito.

Beginning on 15 July 1898 Grassi started to examine the marshes and swamps of Italy. Where Ross was patient and perseverant and willing to carry out a seemingly endless series of trial and error experiments Grassi was methodical and analytical — he was also able to distinguish the different kinds of mosquitoes. Grassi observed "there was not a single place where there is malaria where there aren't mosquitoes too, and either malaria is carried by one particular blood sucking mosquito out of the forty different kinds of mosquitoes in Italy or it isn't carried by mosquitoes at all" [167]. Working in the Roman Campagna and the area surrounding it Grassi collected mosquitoes and at the same time recorded information on the incidence of malaria among the people. (In effect, Grassi was carrying out an epidemiologic study.) Soon it became apparent that most of the mosquitoes could be eliminated as carriers of the disease because they occurred where there was no malaria. There was, however, an exception. Where there were "zanzarone," as the Italians called the large brown, spotted winged mosquitoes, there was always malaria. Grassi recognized that the zanzarone were *Anopheles* and he wrote: "It is the *Anopheles* mosquito that carries malaria...." With this work Grassi was able to prove that "It is not the mosquito's children, but only the mosquito who herself bites a malaria sufferer — it is only that mosquito who can give malaria to healthy people" [167]. In September Grassi read a paper before the prestigious Accademia Nazionale dei Lincei, founded in 1603 and counting Galileo Galilei as one of its most famous members, in which he stated: "If mosquitoes do carry malaria, they are most certainly the zanzarone — *Anopheles* — not any of the 30–40 other species" [167]. In October 1898, Grassi announced that he and his colleagues were experimenting with *Culex* and *Anopheles* and expected a definitive solution to the question of transmission shortly. A month later *Culex* had been eliminated. In papers published in November and December Grassi and colleagues reported on the development of the parasite in *Anopheles*. And he declared, "the victory was completely ours" [294].

Grassi found that the development of human malaria in *Anopheles* was as had been predicted from Ross's studies of bird malaria. Grassi acknowledged the assistance of Bignami and

Bastinelli but stated that the credit for the final completion belonged to him alone. Grassi also angered Ross with faint praise, stating: "It [the life cycle of *Plasmodium*] finds confirmation in that observed by Ross with the malaria of birds in the grey [*Culex*] mosquito." Although the Italians attempted to place their work on an equal footing and parallel to Ross's, the fact is they had a copy of his official report and had followed his published procedures step by step. Ross accused them of piracy. Grassi, as egotistical and stubborn as Ross, fought back, stating that he had worked independently of Ross and had made his discoveries without any prior knowledge of what Ross had done. He stigmatized Ross's bird malaria research saying the figures and descriptions were incomprehensible, and he doubted whether they were of any value as a guide as to what happens in human malaria as similar parasites may have different life cycles. In 1902, when the Nobel committee was considering splitting the prize for Physiology or Medicine between Ross and Grassi, Koch stood opposed. Indeed, Grassi was considered by Koch to be an enemy, and he called him a charlatan with neither brains nor ethics. Ross alone received the Nobel Prize, but despite this award for "work on malaria, by which he has shown how it enters the organism and thereby has laid the foundation for successful research on this disease and methods of combating it," for the remainder of his life he was embittered. This was largely due to his easily taking offense and magnifying a petty slight out of all proportion. Even apologies from Bignami and Bastianelli and their dissociation from Grassi's increasingly arrogant contentions did not mollify Ross's contempt for the Italians.

Ross flagged the "dapple-winged" mosquito as involved in malaria transmission, and Grassi specifically identified that mosquito as *Anopheles*. Once *Anopheles* had been identified as the transmitter of malaria, methods for controlling it became possible. Of the 450 species of *Anopheles* only ~50 are capable of transmitting the disease, and of these only 30 are considered to be highly efficient in transmission. Generally speaking, the measures for the prevention of malaria are to keep infected mosquitoes from feeding on humans, to eliminate the breeding sites of mosquitoes, to kill mosquito larvae, and to reduce as well the lifespan of the blood-feeding adult. Contact with adult mosquitoes can be prevented by using insect repellents, wearing protective clothing, using impregnated mosquito netting, and installing window screens in houses. Breeding sites can be controled by draining water, changing its salinity, flushing, altering water levels, and clearing vegetation. Destruction of adult mosquitoes can be accomplished by using sprays, and larvae can be destroyed by larvicides. Prevention can also be accomplished by education and treatment of the human population. Employing all these strategies has helped to eradicate malaria from many temperate parts of the world, but in the tropics and in developing countries, especially those with limited budgets for mounting public health campaigns and where there are drug-resistant parasites and insecticide-resistant mosquitoes, malaria is on the rise.

Plasmodium in hiding

Ronald Ross, shortly after discovering the mosquito as a transmitter for *Plasmodium*, presumed the inoculated parasites (called sporozoites) burrowed straightaway into red blood cells after entry into the bloodstream. His rival Battista Grassi, however, suggested that the malaria sporozoite was so different from the beasts in the blood that a considerable degree of transformation would be necessary to convert one directly into the other. Pursuing this line of thought, Grassi hypothesized in 1901 that an intermediate stage occurred somewhere in the body — a pre-blood [exo-erythrocytic (EE)] form — and that this would carry out the necessary transformation. Grassi's hypothesis quickly fell apart when two years later Fritz Schaudinn (1871–1906), a pre-eminent microbe hunter, claimed the direct penetration of the red blood cell by the sporozoite. Schaudinn, it is said, took ripe and ruptured oocysts from a mosquito infected with *P. vivax*, placed it in a warmed, dilute drop of his own blood obtained from a blood blister he got from rowing and then peered through his microscope for six uninterrupted hours. So persuasive was Schaudinn's description — sporozoites pushed into the red blood cell, first making a dent, then penetrating with their pointed tail and lastly pulling themselves inside by peristaltic jerks — that even Grassi did not pursue the matter further. "Schaudinn's curious delusion lay like a spell over subsequent investigators" for decades [294]. However, "science is a study of errors slowly corrected" and soon indirect evidence questioned Schaudinn's observations and conclusions. First, there was the failure to confirm Schaudinn's findings and second, in the use of malaria to treat patients with tertiary syphilis, the effects of quinine treatment were found to be markedly different in blood-induced and sporozoite-induced infections. In this so-called "therapy by malaria" the common practice was to induce malaria (mostly *P. vivax*) by direct inoculation of blood or by inoculating sporozoites by mosquito bite or in isolated salivary glands or in entire ground-up mosquitoes. The blood-inoculated patients were cured with quinine, whereas the sporozoite-induced infections relapsed after the same quinine therapy. Even more telling were the observations of the Australian Neil Fairley. During World War II, with the help of Australian army volunteers, Fairley measured the incubation period, i.e. the time it took for parasites to appear in the blood after a mosquito-induced infection; in *P. vivax* it was eight days and in *P. falciparum* it was five days [599]. He also found that during the incubation period the blood was not infectious by transfusion. Malaria parasites must have been hiding somewhere in the body, but the question was where?

Beginning in the mid-1930s, Clay Huff and William Bloom at the University of Chicago and Colonel Sydney Pryce James and Parr Tate in the UK, and then others, observed malaria parasites developing in endothelial cells and macrophages prior to the appearance of parasites in the red blood cells in bird as well as in lizard malarias. In 1945 Colonel James told P. C. C. (Cyril) Garnham, a young Medical Officer in Kenya, not to return from East Africa until he found pre-blood forms in a mammalian malaria. James' gentle insistence proved stimulating to Garnham and, two years later, after James' death, in the Medical Research Laboratory, Nairobi, Garnham found pre-blood stages in the liver of an African monkey infected with *P. kochi* [252]. Shortly thereafter Garnham joined H.E. Shortt at the London School of Hygiene & Tropical Medicine where work began using a monkey malaria (*P. cynomolgi*) with the expectation that the findings would relate to *P. vivax*. There were many attempts and many failures; success was finally achieved, however, when 500 infected mosquitoes were allowed to bite a single rhesus monkey, then for good measure the infected mosquitoes were macerated in monkey serum, and this brew also injected. Seven days later the monkey was sacrificed and its organs taken for microscopic examination. Shortt expected that the pre-blood stages would be found in locations similar to those described for bird malarias. This, however, turned out not to be the case. Instead, the site of pre-blood stages for *P. cynomolgi* was the liver, as had been the case with *P. kochi*. Shortt and Garnham promptly reported their findings in a 1948 paper published in *Nature* [609]. From that time forward pre-blood or EE stages have been described for the non-human primate malarias, the human malarias, as well as in many of the rodent malarias.

In *P. falciparum*, the disappearance of infected red blood cells from the peripheral blood (as demonstrated by simple microscopic examination of a stained blood film) may be followed by a reappearance of parasites in the blood. This type of relapse, called recrudescence, results from an increase in the number of pre-existing blood parasites. *Plasmodium vivax* and *P. ovale* also relapse, although the reappearance of parasites in the blood is not from a pre-existing population of blood stages and occurs after cure of the primary attack. The source of these blood stages remained controversial for many years, but in 1980 the origin of such relapses was identified. In relapsing malarias, induced by sporozoites, in 1982 W. Krotoski, working with Garnham's team, found dormant parasites, called hypnozoites, within liver cells [382]. The hypnozoites, by an unknown mechanism, are able to "awake" and these become the seeds for full EE development and then go on to establish a blood infection.

Thus, a complete description of the life cycle of the malaria parasite took 70 years. It is as follows: humans are infected through the bite of a female anopheline mosquito when during blood feeding she injects sporozoites with her saliva. The inoculated sporozoites travel via the bloodstream to the liver where they enter liver cells. Within the liver cell the non-pigmented parasite, the pre-blood or EE stage, multiplies asexually to produce infective offspring. These do not return to their spawning ground, the liver, but instead invade erythrocytes. By asexual reproduction of parasites in red blood cells (called schizogony or merogony) infectious offspring (merozoites) are released from the red blood cells. These can invade other red blood cells to continue the cycle of parasite multiplication, with extensive red blood cell destruction and deposition of hemozoin. In some cases the merozoites enter red blood cells but do not divide. Instead, they differentiate into male or female gametocytes. When the female mosquito takes a blood meal from an infected individual, in its stomach the male gametocyte divides into flagellated microgametes that escape from the enclosing red blood cell, these swim to the macrogamete, one fertilizes it, and the resultant motile zygote, the ookinete, moves across the stomach wall. This encysted zygote, now on the outer surface of the mosquito stomach, is an oocyst. Through asexual multiplication, thread-like sporozoites are produced in the oocyst, which bursts to release sporozoites into the body cavity of the mosquito. The sporozoites find their way to the mosquito salivary glands where they mature, and, when this female mosquito feeds again, sporozoites are introduced into the body and the transmission cycle has been completed.

Chapter 2

Respice: Before the Genome Project

DNA is the stuff of which genes are made. DNA was discovered more than a century ago by an obscure Swiss physician, Friedrich Miescher (1844–1895) [162]. Shortly before completing his doctoral dissertation (1868), Miescher elected not to follow in his father's footsteps as a hospital physician and teacher of pathologic anatomy and instead pursued a career in physiological research. In part, it is believed, this was due to his partial deafness, the result of an earlier attack of typhus that certainly would have limited his abilities to use the stethoscope and perform percussion and auscultation [392]. He joined the laboratory of Felix Hoppe-Seyler (1825–1895) at the University of Tübingen, Germany, with the intention of studying what enables the cell to live and how cells are fashioned into tissues. Hoppe-Seyler, whose interests were in the chemistry of blood, was one of the first to crystallize hemoglobin and describe the interaction of oxygen with hemoglobin in the red blood cell. He was now turning his attention to a white blood cell, the lymphocyte, present in pus as well as blood. Hoppe-Seyler thought an understanding of the chemistry of the lymphocyte might lead to a better view of why pus was formed during infections. The collaboration was ideal: Miescher wanted to analyze the chemical composition of the cell and Hoppe-Seyler had the "perfect" cell for Miescher to analyze, the lymphocyte. Today a study of pus cells would be impractical inasmuch as infections are rare, but in 1869 when Miescher began his studies there were no antibiotics and antiseptic methods during surgery were non-existent, so human pus was readily available from the surgical wards of many hospitals [392]. In addition, he could take advantage of the newly invented clean, sterile, absorbent cotton used for dressing pus-filled wounds [162]. Hoppe-Seyler encouraged young Miescher; however, he also cautioned that there was no method to study pus chemistry! Miescher learned by trial and error and after many failures finally succeeded when he extracted pus-laden bandages with a weakly alkaline solution; a highly viscous, snot-like material that was impossible to handle as it would not dissolve in water, acetic acid or sodium chloride was obtained. (In hindsight we now know that this occurred because the solution he used extracted high molecular weight DNA.) He asked himself, from where in the cell did this material come? Was it the nucleus, the membrane, the cytoplasm? Examining the pus cells under the microscope he observed that the alkaline solution caused the nucleus in the pus cell to swell and break open. This suggested that the nucleus was the source of this material and hence he named the substance "nuclein." Using element analysis, one of the few chemical methods available at the time, he found the new substance to contain 14% nitrogen, 3% phosphorous, and 2% sulfur. This ratio of nitrogen to phosphorous was unique. Further, finding it to be resistant to digestion by pepsin (found in the stomach), he concluded the material could not be a protein. By August of 1869 Miescher reported that the same material was found not only in pus cells but also in other cells with nuclei such as yeast, kidney, liver and chicken and duck red blood cells. Miescher did not understand the importance of nuclein as the carrier of inheritance [392]. Indeed, he considered nuclein to be a storehouse of phosphorous for the cell and to the end of his life he rejected the idea that nuclein might have something to do with heredity.

Miescher was anxious to have his research work published; however, publication of the discovery of nuclein in Hoppe-Seyler's *Medical-Chemical Journal* was delayed until 1871 because of difficulties with reproducibility of his work by other laboratories (including Hoppe-Seyler, who had doubts about the correctness of

Miescher's work) as well as the confusion that accompanied the Franco–Prussian War (1870–1871), which affected publication of everything except the daily news. Although Miescher recognized the reasons for the delays he specifically asked Hoppe-Seyler to put the date of the submission of the manuscript at the end of the published paper. This would ensure his receiving the proper credit should someone else complete similar studies and publish on nuclein after October 1869. Clearly, establishing priority for a discovery was as important then as it is in today's competitive world.

Leaving Tübingen Miescher took up a post at the University of Basel where he continued his studies on nuclein using salmon sperm. At the time salmon fishing on the Rhine was a robust industry with its center in Basel. Further, because salmon sperm was readily available at practically no cost it helped Miescher's research as he had little space, almost no equipment and very little funding. Even after he became a professor the space for his research was the corridor between two rooms in an old building and help consisted of a technician working one-quarter time [392]. There was, however, another plus: a sperm cell is essentially a nucleus propelled by a filamentous tail.

The substance Miescher called nuclein, extracted from the cell nucleus, was actually a mixture of DNA and protein and it caused considerable confusion in the literature until in 1889 the German biochemist (and a student of Miescher) Richard Altmann was able to separate nuclein into two distinct entities; the one with phosphorus he called "nucleic acid" and other was a protein, protamine.

Miescher gave a preliminary account of his work on salmon sperm nuclein at a meeting of the Basel Society for Biological Research in 1872 and a full account was published in the *Transactions of the Society* in 1874. After this publication he abandoned further research on nuclein. Miescher was shy and an intensely private person who had few friends. He was, according to his students, an uninspiring and oftentimes awkward lecturer, but was an indefatigable worker in the laboratory. Perhaps as a result of his deafness he found it much easier to communicate in writing rather than face to face. His publications were met with indifference, doubt and in some cases with mistrust. Never in robust health, in 1885 he suffered from pleurisy and in 1894 was diagnosed with tuberculosis of the lung. He was sent to a sanatorium in Davos, Switzerland, where he remained until his death aged 51, unrecognized by his peers for his seminal discovery of DNA.

Miescher's discovery of DNA remained unappreciated by his chemical colleagues; however, this was not the view of the biologists who were studying cells using microscopes with lenses having greater powers of magnification and resolution. At the time of Miescher's discovery of nuclein, Germany was the center for the synthesis and production of aniline dyes. Although these dyes were intended primarily for use on textiles to create more colorful fabrics, the staining properties were also being explored by microbiologists and pathologists in order to more clearly reveal internal structures invisible in the mostly colorless cell. If Miescher had applied one of the available basic dyes to the pus cells, the nucleus would have stained, and with an acidic dye the cytoplasm would have stained preferentially. Ironically, Miescher resisted using staining techniques to characterize nuclein, mistrusting the "dyer's" art and believing that most of them used staining without understanding the underlying chemistry. Others had no such misgivings, and in 1882 Walther Flemming (1843–1905) summarized the uses of the new dye techniques in his book *Cell Substance, Nucleus and Cell Division* where he introduced the term "chromatin" for the stained material in the nucleus; he suggested chromatin was identical with Miescher's nuclein [524]. At the time the function of the nucleus was unknown; however, when rod-like bodies that stained with the basic dyes appeared in cells undergoing division, and these then split longitudinally and were segregated to the daughter cells, it was postulated that the dyed rods (called chromosomes after the Greek "chromo" meaning color and "soma" meaning body) were the bearers of heredity.

Further, Oscar Hertwig (1849–1922) in studies of sea urchin egg fertilization showed that the sperm nucleus fused with the egg nucleus and subsequent divisions of the fertilized egg gave rise to the nuclei of the following generation and the development of the embryo. He accurately concluded "nuclein is the substance not only responsible for fertilization but also for the transmission of hereditary characteristics." In 1896 the American biologist E.B. Wilson, working at Columbia University in New York, wrote: "Chromatin is to be regarded as the physical basis of inheritance... chromatin is known to be similar to, if not identical with a substance...nuclein, which analysis shows to be a definite chemical compound of nucleic acid (a complex organic acid rich in phosphorous) and albumin" [524]. And what did Miescher say to all this? He was simply not impressed by the evidence and wrote, "I must defend my skin against the guild dyers who insist there is nothing but chromatin [nuclein]."

To be sure, there was reason to doubt the role of chromatin as sometimes it disappeared from within the nucleus. This failure of staining was taken, by some, as an indication that the cell no longer contained chromatin. In actual fact, it was not absent from the nucleus but rather, by unraveling its configuration, had changed so that it was less compact and hence inaccessible to the stain. (Consider this: a DNA molecule in a human cell is an attenuated thread that if stretched out would be six feet in length.) A further argument against nuclein or DNA being the hereditary material came from chemical analysis: DNA was believed to be a comparatively small molecule with a monotonous structure of four nucleic acid bases in equal amounts. Thus, at the beginning of the 20th century DNA appeared to be rather unattractive as the structural basis of genetic information, whereas proteins, with their high molecular weight, complicated structure and composition of many different amino acids, seemed a more likely candidate [524].

It was studies of pneumonia, or rather the microbe causing pneumonia, that turned favor from protein to DNA. Before the advent of antibiotics pneumonia was a disease that claimed many lives each year. It was then and still can be a public health problem of enormous proportions. *Streptococcus pneumoniae* is the causative organism of bacterial pneumonia. Obtained from a patient with disease and grown in the laboratory on a plate of nutrient jelly or agar, *S. pneumoniae* multiplies rapidly to form colonies that appear smooth and shiny to the naked eye. The smooth kind is the normal, virulent disease producer, and the smooth appearance is due to a coating — a mucilaginous capsule — that probably prevents the bacteria from being engulfed by the white blood cells, allowing them to multiply and cause symptoms of disease. Some strains of *S. pneumoniae* have lost the ability to produce the mucilaginous capsule; when grown on agar the colonies appear rough. The rough forms are not virulent and do not cause disease because they are easily digested and destroyed by the white blood cells. Each strain of bacteria breeds true: rough strains give rise to rough strains and smooth strains give rise to smooth strains [603].

In 1928 Frederick Griffith (1879–1941), a medical officer in the Ministry of Health in England, reported a remarkable finding. If he injected living rough cells into a mouse, it lived. If he injected the smooth cells into a mouse it died. When he killed the bacteria by heating the broth in which they were growing to 60°C and then injected each of these strains into mice they lived. So far, there are no unexpected results. However, when Griffith injected mice with a mixture of living rough cells and some killed smooth cells the mouse died. The cause of death was pneumonia and the bacteria recovered from the body multiplied on agar to form smooth colonies. Obviously there had been a conversion of the rough cells into smooth cells, and Griffith postulated that an agent had been given off by the dead smooth cells that was taken up by the living rough cells; this agent was transmissible and conferred hereditary properties on the recipient. The agent was called transforming principle. Griffith was in his fifties, well established and known for his careful experimentation. However, he was a virtual recluse known to only a few of his associates, was modest and retiring and enjoyed working quietly on his own in the laboratory. His unusual findings were published in the *Journal of Hygiene* in 1928 and his work received little attention for more than a decade [524].

In 1901 John D. Rockefeller, the robber baron oil millionaire, became a philanthropist and provided a gift to establish the Rockefeller Institute in New York for the purposes of conducting biomedical research in the service of clinical medicine. In 1913, Oswald T. Avery (1877–1955), a physician who tired of the clinic, wanted to devote himself entirely to research. Avery joined the Institute, which at the time had as its primary focus pneumonia. He remained at the Institute working on pneumonia until his retirement at age 71. Avery was a bachelor, lived within walking distance of the Institute, and spent long hours in the laboratory. He was a small, slender individual with a high forehead. From his professorial demeanor he was given the nickname "Fess." "Avery was an extremely private person with a very gracious exterior, but he also had a 'brooding' forehead and could be a melancholy figure whistling gently to himself. He was a man of immense charm who always appeared to be interested in what you were interested in, but in fact, always turned the conversation around to what he was interested in — the chemical composition of the pneumonia-causing bacteria" [524].

Avery did not work alone. In 1934 a young physician, Colin McLeod (1909–1972), joined his laboratory and in 1941 another physician interested in infectious disease, Maclyn McCarty (1911–2005), joined the laboratory. Griffith's results caught the attention of Avery and he and his co-workers now dedicated themselves to isolating and chemically characterizing the transforming principle. In 1943 Avery, McCarty and McCleod wrote a paper describing their results. They identified the transforming principle as the chemical substance DNA. The manuscript was hand-delivered to Peyton Rous, the Editor of the *Journal of Experimental Medicine*, published by the Rockefeller University Press. Rous returned the manuscript in two weeks with hand-written comments and crossed out some of the evidence he considered insufficient. Before the revised paper was published Avery gave a seminar in which he was uncharacteristically effusive saying, "The belief is that DNA is the fundamental unit of the transforming principle" [524].

Avery, McCarty and McLeod's paper is now considered a classic of 20th century biochemistry. Sadly, Griffith was not alive to appreciate it as he was killed while working in his laboratory during the 1941 blitz of London. Indeed, it is unlikely that Griffith would have recognized that the virulence factor being supplied was the DNA molecule. Even Avery had been skeptical at first! As with Griffith's work, at the time of Avery's publication it received little attention. In fact it was mostly ignored.

Between 1947 and 1951 transformation experiments proved difficult to repeat and even Avery's group experienced difficulties. Many of these problems turned out to be technical ones in that some serum samples were ineffective for transformation in the test tube, and traits such as resistance to certain drugs were initially found not to be transferable. This lack of appreciation of Avery's work has been ascribed to publication in a journal devoted mostly to immunology and not widely read by geneticists who were skeptical about the applicability of bacteria to understanding inheritance [524]. Timing too was poor. The paper was published in 1944 near the height of US involvement in World War II. Another factor was the criticism leveled by a fellow member at Rockefeller, Alfred Mirsky, who was convinced Avery's preparations were contaminated with residual protein that he considered to be the transforming principle. One of Avery's collaborators, Maclyn McCarty, claimed that Mirsky's criticism so swayed the Nobel committee that Avery's nomination was postponed. Regrettably, acceptance of Avery's findings did not occur until 1952 when Alfred Hershey (1908–1997) and Martha Chase (1928–2003) showed that genetic information was carried on DNA not protein. (The key experiment involved shaking off the proteinaceous viral shell using a Waring blender, allowing the viruses to inject their DNA, and showing that the removal of protein did not diminish infection.) However, despite this confirmation of the work done at the Rockefeller Institute by Avery's group, the award of a Nobel Prize was denied Avery by his death in 1955.

Phoebus A. Levene (1869–1940), a contemporary of Avery at the Rockefeller Institute, was born in Russia, trained in medicine, and as a result of increasing anti-Semitism in Russia he and his family immigrated to the United States [524]. Levene practiced medicine in New York City until 1896 when he came down with tuberculosis. As a result he spent the next few years recuperating in sanatoria in Saranac Lake, New York, and Davos, Switzerland, as had Miescher before him. Upon recuperation he decided to devote himself to biochemical research rather than clinical medicine. In 1905 he joined the Rockefeller Institute and was put in charge of its Division of Chemistry. Levene has been described as wiry, thin, short in stature, with penetrating dark eyes and a shock of black hair. He had a stern, somewhat authoritarian demeanor, and was a glutton for laboratory work, mainly on nucleic acids. He and his collaborator Walter Jacobs (1883–1967) elucidated the way the building blocks were put together. The sugar in nucleic acid was identified as the five-carbon ribose (not the six-carbon glucose) and it was linked to purine or pyrimidine (each being referred to simply as a base) and the phosphate was linked to the sugar to form a nucleotide. A base without the phosphate was called a nucleoside. When Levene and Jacobs analyzed the nucleic acid from yeast they obtained four different nucleosides in roughly equal amounts and on further breakdown of the nucleoside obtained the bases adenine (abbreviated A), guanine (abbreviated G), cytosine (abbreviated C) and uracil (abbreviated U). They concluded that yeast nucleic acid was made of these bases in the form of a repeat of four, called a tetranucleotide. Levene and Jacobs did not recognize they were dealing with a mixture of both DNA and RNA and that the equal proportion of the four nucleotides was accidental. Indeed, Levene was unaware of this and until his death (in 1940) he believed DNA and RNA consisted of tetranucleotides, repeated only a few times. Despite this error, Levene's stature was so great — he published more than 700 scientific papers — his tetranucleotide hypothesis (formulated in 1910) was widely accepted. As a consequence the size of the DNA molecule was underestimated and the tetranucleotide hypothesis did further damage in that it diverted attention away from nucleic acids as carriers of genetic information.

As Levene had shown, each DNA molecule is made up of building blocks called nucleotides and, in turn, each is made up of three major constituents: phosphate, a sugar, deoxyribose, and a base. The phosphate and sugar are constant features, but the bases come in four varieties: the double-ringed purines, adenine (A) and guanine (G), and the single-ringed pyrimidines, thymine (T) and cytosine (C). (Levene did not know that the pyrimidine uracil [U] is not found in DNA but only in RNA; thus when he isolated uracil from his "DNA preparations" they were most certainly contaminated with RNA.)

Levene's tetranucleotide hypothesis was shattered in 1950 by the biochemist Erwin Chargaff (1905–2002) working at Columbia University, just a few miles away from the Rockefeller Institute where Levene worked. The two, however, never shared their findings. Indeed, even when Avery, while working in the same Institute, consulted Levene about the possibility that nucleic acids might be involved in the biological activity of transforming principle he discouraged him, citing the essential invariability of the nucleic acids on the basis of his tetranucleotide hypothesis.

Chargaff had immigrated to the United States from Austria in 1935 to escape the virulent anti-Semitism by the Nazis, to become Assistant Professor of Biochemistry at Columbia University. He was so moved by reading Avery's paper that he embarked on studies of DNA. Between 1945 and 1949 he used paper chromatography, a better method of analyzing the kinds of bases in DNA than had been available to Levene. Using highly purified DNA from a variety of sources he found the proportion of bases from tissues of any particular species was the same; however, the composition of the DNA varied from species to species, and the percentage of purines always equaled that of pyrimidines; in a shorthand way $A = T$ and $G = C$, and $A + G = T + C$. A review of the work, published in 1950 in *Experientia*, was titled "Chemical specificity of nucleic acids and mechanisms of their enzymatic degradation." In his 1978 memoir, *Heraclitean Fire. Sketches from Life before Nature* [125], Chargaff writes that in 1947 when he began getting results on base composition he "began to think about the way in which differences in composition, even slight differences, could influence the content of biological information…Then, one late afternoon [in 1950], while sitting at my desk in my narrow tube of an office on the fifth floor of the medical school, I asked myself, what would happen if I assume DNA contains equal quantities of purines and pyrimidines? There emerged — like Botticelli's Venus on the shell…the regularities that…are now known as base pairing." This was later called "Chargaff rules." He suggested "as far as chemical possibilities go, they [the nucleic acids] could very well serve as one of the agents, or possibly the agent, concerned with transmission of inherited properties" and realizing how unique his findings were he attempted to build models, but these were (in his own words) unsuccessful. And, in a characteristic mood of bitterness, Chargaff wrote, "Thus, I missed the opportunity of being enshrined in the various halls of fame of the science museums." Indeed in 1953, Maurice Wilkins (1916–2004), James Watson (1928–) and Francis Crick (1916–2004) using Chargaff rules, and the results of Rosalind Franklin's (1920–1958) X-ray crystallographic work[1] were able to build a structural model that solved how the bases and sugar and phosphate were arranged. It was a ladder-like molecule with two long chains of sugar-phosphate forming the uprights and the bases — a purine pairing with a pyrimidine — linking the two uprights, as would the rungs of a ladder. According to the pattern of the X-ray, the two strands of DNA were twisted around one another in the form of a helix (similar to a spiral staircase). They called it a double helix.

In his memoir [125] Chargaff recalled meeting Watson and Crick in 1952: "They impressed me with their ignorance. Crick didn't even know the chemical differences among the bases. I never met two men who knew so little and aspired so much. They didn't seem to know of my work but said they wanted to construct a helix, talking much about 'pitch'." Chargaff caustically wrote: "Two pitchmen in search of a helix." He added, "I told them all I knew. If they had heard about base pairing rules, they concealed it." When in 1953 Watson and Crick published their landmark paper on the double helix he claimed, "they did not acknowledge my help." He believed "that the double-stranded model of DNA came about as a consequence of our conversation." But when asked why he had not discovered the celebrated model he cynically replied he was "too dumb, but if I had collaborated with Rosalind Franklin, we might have come up with something of the

---

[1] In X-ray crystallography X-rays are shot at a crystal of material and the scattering recorded on a photographic plate reveals the three-dimensional form of that material.

sort in one or two years." Chargaff remained embittered for the rest of his life.

The Watson–Crick model of DNA provides a basis for the molecule to duplicate itself. One strand of the DNA is complementary to the other as when an A occurs in one strand a T occurs on the opposite stand, and when there is a G on one strand this is paired with a C in the other strand. The strands are able to separate by the paired nitrogenous bases moving apart; this "unzipping" of the strands allows a complementary strand to be formed from nucleotides and other molecules in the cell, under the direction of an enzyme, DNA polymerase. All the information necessary for arranging the bases in a linear sequence to complement the original strands of DNA is provided for by the mechanism of complementary base pairing — Chargaff's rule. For example, if one strand contains A then a T nucleotide will pair with it, and if the next base is a G then a C will pair with it. In this manner the old strands of DNA direct the sequence or order of the nucleotides in the new sister strands. The new strand is a complementary copy of the original strand, and the two new strands are identical to the original two; importantly, because of complementary base pairing the exact sequence in the original double helix will be faithfully reproduced. This process is called DNA duplication or replication. A particular sequence of bases in the DNA comprises a gene, and that particular sequence codes for the specific product, a protein, consisting of a string of amino acids. The genetic code is "written" as a three-letter code because each "word" is composed of a three (triplet) base sequence [602]. Each triplet codes for a single amino acid. For example, the three letter sequences CGA, CGG, CGT, and CGC code for the amino acid alanine. This is similar to the English language where several different words can be used to specify a similar object: house, abode, home. A code in which several different code words describe the same thing is called a degenerate code and thus DNA is a degenerate code. The differences between living things are not that they have different nucleotides in their DNA but are due to the differences in the sequential order of the bases. It is much like using the letters in the alphabet to make different words.

The letters used may be the same; however, the words formed and their meaning (message) can differ depending on the way the letters are arranged. A small DNA-containing virus contains as many letters as half a page of newsprint (8,000 words), the bacteriophage T_2, a virus, has the capacity of a 300-page paperback novel (say a half million letters), whereas we have in our genes 3 billion letters, roughly 500 times the word count in the Bible. A malaria parasite with its 23 million letters (bases) contains enough information for a hefty book with 4,000 pages!

The genetic code in the DNA molecule is like a set of blueprints housed in a library, but the blueprints cannot leave the library (nucleus). Consequently, in order to direct the manufacture of the gene product (protein) there is a blueprint transcriber as well as a translator of the code, the molecules messenger RNA (mRNA) and transfer RNA (tRNA), respectively. These RNAs also contain four nitrogenous bases but instead of T there is uracil (U), the sugar is ribose not deoxyribose, and the molecules are not in the form of a double helix but are single strands. When a gene is to be transcribed (copied) the DNA double helix unzips and a faithful copy (using complementary base pairing!) is made into mRNA by means of an enzyme, RNA polymerase. The mRNA leaves the library (nucleus) and moves to the cytoplasm where it attaches to an RNA-containing particle, the ribosome. At the ribosome the mRNA is translated into a protein. Each amino acid, corresponding to an mRNA triplet, is ferried by means of a tRNA, of which there are 20 different ones for each of 20 amino acids. The ribosome acts like a jig to hold the mRNA in place and guides the tRNA-amino acid into proper alignment for coupling. As the ribosome moves along the strand of mRNA "reading" the message a string of amino acids (= protein) is formed. This process is called translation. In short, genetic information flows: DNA→RNA→protein [602].

A change in the sequence of DNA, either by deletion of a base or by the substitution of a different base, may lead to a change in the mRNA sequence and in turn to a different sequence of amino acids and hence a different protein. Such a genetic change — a mutation — may result in either a meaningful message or a meaningless one.

The nucleus of each of our cells with 3,000 million bases is packaged into 46 chromosomes; during cell division, each chromosome becomes visible with a light microscope [602]. The malaria parasite chromosomes are different: they do not condense during division and hence they are not visible with a light microscope. Detailed information on the number and size of malaria parasite chromosomes became available only after the development of pulsed-field gel electrophoresis, which is a clever technique that permits separation of very large-sized molecules. Using this novel approach, it was possible to separate the 14 chromosomes of *P. falciparum*. All malaria species have 14 chromosomes although the size of individual chromosomes can vary amongst the various kinds of malarial parasites [599].

In the early 1970s, recombinant DNA technology made possible the isolation of genes from a variety of sources; later technology enabled mapping these same genes to a particular chromosome, and even to a region of that chromosome. The recombinant DNA method (as originally developed by Boyer and Cohen) used plasmids — small circles of naked DNA. Both strands of a circular plasmid double helix could be severed at a specific location with molecular scissors known as restriction enzymes. When mixed in the same test tube with a second DNA (such as *Plasmodium* DNA) that had been similarly cleaved, and in the presence of another enzyme (called ligase), the snipped ends could be "glued" together; the end result was the creation of a hybrid or recombinant plasmid. These mixtures of DNA could then be used to infect bacteria. After bacterial growth in a Petri dish containing antibiotics, only the recombinant DNA plasmids would enable the bacteria to survive. In some instances the bacteria containing the recombinant plasmid would express the protein encoded by the foreign DNA. Although plasmid-based vectors were used early on, later it was found that engineered viruses called bacteriophage (especially λgt11) were more efficient vectors and the number of clones generated was greater than that found for plasmids. In addition, handling of large numbers of phage and screening of recombinant clones was far easier.

Gene expression libraries can be constructed using genomic DNA, that is, the DNA in the chromosomes. These expression libraries are often constructed from sheared DNA or DNA partially digested with nuclease, including restriction enzymes, able to chop up the chromosomes into small pieces of DNA. To be able to handle these pieces of DNA it is necessary to copy and store them, just as a book needs to be printed and bound. In the copying process, each fragment of DNA is attached to the DNA of a bacteriophage. After the chromosomes of the malaria parasite have been chopped up into small pieces of DNA and attached to phage DNA, the phage can be used to infect the bacterium *Escherichia coli*. Each phage carries its own DNA as well as fragments of malaria parasite DNA. When the surface of a Petri dish is covered with a lawn of *E. coli* infected with such a phage clear spots appear on the bacterial lawn where the viruses have killed (lysed) the bacteria. These spots, called plaques, contain millions of virus particles with millions of copies of the original pieces of malarial DNA. Theoretically, genomic libraries have all the DNA sequences present at equal frequency [599].

Another kind of library, a complementary DNA (cDNA) expression library, is somewhat more difficult to prepare because it first requires the isolation of the transitory and unstable mRNA; however, in this library the sequences are present in proportion to their abundance as mRNA molecules and thus represent differentially expressed genes. In other words, the mRNA is a concise working copy of the DNA code. The mRNA can be faithfully copied into a stable and complementary form, cDNA, using the enzyme reverse transcriptase. As with the method for producing a genomic DNA library, when the cDNA is inserted into a plasmid or a phage and the *E. coli* infected, the DNA is copied.

Using filter paper it is possible to remove DNA from Petri dishes growing *E. coli* with either genomic or cDNA libraries. The filter paper is soaked overnight in a solution of the DNA probe for the DNA of interest. To determine whether it has bound to the cDNA in the library the probe is labeled — made radioactive. The filters are then washed to remove any of the radioactive probe not attached to DNA fragments, dried, and placed onto an X-ray film for a few days. Positive colonies or plaques — areas where the radioactive probe has bound onto the target DNA fragment — are revealed on the X-ray film as black spots. Once that colony has been identified and sufficient DNA is produced it is possible to "read" the message of the DNA by working out the order of the bases (i.e. CGAT) that form the "rungs" of the DNA double helix. The reading of the bases is called sequencing, and Frederick Sanger (1918–) at the Medical Research Council's Laboratory of Molecular Biology in Cambridge, England, described the basic method. In May 1975, Sanger reported on the complete gene sequence of a virus (ϕX174), consisting of 5,368 bases, and later he went on to sequence another virus. In 1980, Sanger received the Nobel Prize for his achievements in gene sequencing methods [599].

Powerful as the recombinant DNA methods are, there is a significant limitation: they require growing large volumes of bacteria to amplify the DNA of interest as well as a considerable investment in time and research funds to identify the recombinant clones of interest. By 1985 they were largely replaced by the polymerase chain reaction (PCR), a technique called "molecular photocopying." In brief, two short stretches of single-stranded DNA (called primers), corresponding in nitrogenous base sequence to the regions bracketing a DNA expanse of interest, such as a specific gene, are synthesized. The primers are added to the DNA template, i.e. total genomic DNA or a cDNA population of interest, and the DNA is "melted" by heating to 90–95°C to separate the helical strands. Upon cooling, the primer can bind to its complementary stretch of single-stranded template DNA. In addition, present in the test tube is an enzyme, DNA polymerase, and all four bases. The polymerase will only begin incorporating bases where the DNA is already double-stranded, and so it begins adding bases at the end of the primer and synthesizes the DNA region that follows. By using a thermal cycler, one that heats ("melts") and cools ("anneals"), the process can be repeated every five minutes and the stretch of DNA of interest will be copied again and again; in two hours the DNA of interest will be increased about 34 million-fold. This amplified DNA can then be sequenced (= read).

In the Sanger sequencing method, the strands of DNA are duplicated by means of DNA polymerase in the presence of a mixture of the normal nucleotides A, T, G, C plus some dideoxy (dd)A, ddT, ddG or ddC. If the polymerase incorporates the normal base the DNA chain grows, but when it encounters a dideoxy nucleotide it stops lengthening. The result is four different samples, each containing a series of DNA chains of varying length, depending where in the growing chain the different dideoxy nucleotides were incorporated opposite the complementary T, A, C, or G template bases during the replication process. Each sample is placed on a gel and the fragments separated in an electric field: short chains move faster and longer chains slower. The positions can be read off such that the shortest fragment will contain the first base, the next larger the second and so on. Later, this four-lane method of manual sequencing was replaced by one that could automatically read out the order of bases in a stretch of DNA using a single lane; a different colored dye for each type of chain-terminating dideoxy nucleotide was added to the polymerase mix and incubated. By subjecting the single-lane sample to an electric field all the DNA pieces can be sorted according to size. With ultraviolet light illumination each fragment fluoresces differently depending on its terminal dideoxy nucleotide; by scanning the fluorescent pattern and feeding this into a computer the base sequence of a gene or a piece of a gene can be printed out.

Using an automated system many thousands of bases could be sequenced in a day. Indeed, in 1996 an international consortium of scientists from more than a dozen institutions set out to determine the 23 million base pairs in the nuclear DNA of *P. falciparum*. The 14 chromosomes were physically separated by pulsed-field electrophoresis. The DNA was then mechanically sheared into random fragments; after the fragments were cloned in bacteria they were sequenced using high-throughput, automated dideoxy sequencing and the nucleotide order determined for individual chromosomes by assembling overlapping sequences using a computer. In 2002 the results of the *P. falciparum* Genome Project were published: 5,279 genes were postulated [249].

Chapter 3

The Nature of *Plasmodium falciparum* and its Genome

Cells, the stuff of which life is made, were first seen 400 years ago by Antony van Leeuwenhoek (1632–1723), a Dutch linen merchant who had a hobby of grinding magnifying lenses. Using a crude microscope (with a magnification of 300 times), Leeuwenhoek was the first person to see sperm, frog and fish red blood cells, bacteria, microscopic roundworms and rotifers, the green alga *Spirogyra* and a variety of "animalcules" (today these are called protozoa). It was Robert Hooke (1635–1703), however, not Leeuwenhoek, who coined the term "cell" after observing a very thin slice of cork under a microscope and finding "little boxes" resembling a monk's cell. All the processes of life occur within cells and, although the "cells" Hooke observed, described and illustrated were not alive, the name stuck [597].

All cells are basically similar to one another, having many structural features in common [602]. We can more easily understand the concept of a cell by making an analogy and comparing them to the rooms of a house. Rooms and cells have boundaries with exits and entrances — rooms have walls, floors, ceilings, doors and windows, whereas cells have walls and membranes with pores of various sizes. Rooms and cells come in a variety of shapes and sizes, and each kind of room may have its own particular kind of use, function or specialty. A house may be composed of one room or many rooms. If a plant or animal is composed of one cell (one room) it is considered one-celled or unicellular (uni meaning "one"); if made two forming a membrane complex. In addition to rhoptries and micronemes, the merozoite contains small rounded vesicles (dense granules), a single nucleus, mitochondrion, an apicoplast and some ribosomes — the minimal equipment needed for maintaining itself in a host cell [43, 44].

The invasion of red blood cells occurs in a series of steps beginning with loose attachment to the red blood cell surface and reorientation to bring the tip of the merozoite into closer contact with the red cell membrane, followed by secretion of a complex cocktail of chemicals from the micronemes and rhoptries causing a deep membrane-lined pit to appear in the red cell's surface. The parasite moves into this ever-enlarging depression, the fuzzy coat is removed, and the red cell membrane closes over the merozoite so that the now "naked" parasite is left in a membrane-lined space, the parasitophorous vacuole (PV) [43, 44].

After it has entered the cell, the merozoite loses its invasive organelles and becomes cuplike, with a thinner center, which under a light microscope and in some images obtained using an electron microscope looks like a hole, hence the name "ring" stage. The parasite begins eating the hemoglobin-rich cytoplasm within the red blood cell using a special organelle, the cytostome (cyto meaning "cell" and stoma meaning "mouth"). As feeding progresses, the parasite becomes plumper and, at this time, the trophozoite feeds more voraciously, depositing the digested hemoglobin in small vacuoles where the hemoglobin is broken down into the golden brown-black malaria pigment. Eventually these smaller food vacuoles fuse into a single large pigment-containing food vacuole [3].

The nucleus of the schizont divides repetitively, with the final number depending on the particular kind of *Plasmodium*. In *P. falciparum* the nucleus undergoes four rounds of multiplication yielding 16 nuclei, and each enters a merozoite bud, with the other organelles of invasion assembled near it. The mature merozoite detaches from the parent schizont leaving a vacuole (called the residual body) filled with malaria pigment. Enzymes released from specialized organelles trigger a complex series of chemical changes in the infected red blood cell allowing the merozoites to escape from the red blood cell. The remnant of the red cell, now depleted of its hemoglobin-rich cytoplasm, and with its pigment-containing residual body, is called a ghost.

The reason why malaria parasites switch from being asexual to becoming sexual forms is not yet understood but the "decision" seems to occur at or before the merozoite stage. Male and female gametocytes are long and slender, surrounded by a three-layered pellicle, and enclosed by the red cell membrane; both contain microtubules, as well as unique secretory organelles used by the parasite to leave the red cell. When gametocytes are ingested by a suitable mosquito or are subjected to low temperature they escape from the red cell. In the mosquito, chemicals in the stomach provoke microgamete release. Within 15 minutes the nucleus of the male gametocyte divides into several motile, sperm-like gametes each with a long tail — exflagellation — a process first witnessed by Laveran (see p. 4). Using the lashing tail, the male gamete swims to make contact with the female gamete to fertilize it. Fertilization results in a zygote with a double set of chromosomes (= diploid). At 18 hours after the blood meal, the zygote elongates into a motile ookinete (described by MacCallum, see p. 9), lacking rhoptries but with large numbers of micronemes and polar rings at the front end and an inner membrane complex containing the muscle-like protein, myosin. Leaving the mosquito stomach is essential for parasite survival, for were the ookinete to remain in the mosquito stomach it would be digested with the blood meal. The parasite leaves using its invasion organelles; the ookinete penetrates the stomach wall of the mosquito, rounds up, and secretes a cyst wall, the oocyst, on the outer surface of the stomach. Within the diploid oocyst, the number of chromosomes is halved (by a process called meiosis) to produce a single set or haploid number of chromosomes. The DNA of the haploid chromosomes is replicated many times by repeated asexual (mitotic) divisions, forming a massive nucleus containing hundreds of genomic centers. The groups of chromosomes separate into individual nuclei around the periphery of the parasite and move into fingerlike projections at the parasite surface, each of these generating new organelles, and eventually the fingerlike projections detach as sporozoites. As Ronald Ross discovered a century ago, the process of sporozoite formation takes about one week and a few thousand sporozoites are found in each oocyst. When mature, the sporozoites migrate through the cyst wall into the mosquito's fluid-filled body cavity, and eventually ~25% reach the salivary glands where they undergo further maturation to become infectious.

Although the sporozoite contains a similar array of organelles as a merozoite, it is much longer, slightly curved and tapered at both ends; the elongate sporozoite is shaped more like a hot dog, whereas the smaller and plumper merozoite is pear-shaped. Each sporozoite contains a single nucleus, a mitochondrion, an apicoplast, four rhoptries, numerous micronemes, and an elongate pellicle complex stretching back from the polar rings to form a spiral cage-like skeleton, the structural element for gliding movements [44].

There are many thousands of sporozoites in the salivary glands; however, fewer than 25 sporozoites are deposited into the skin of the vertebrate with the mosquito's anti-coagulant saliva, preliminary to its taking a blood meal and a requirement for mosquito egg maturation. (It has been shown that fewer than 10% of *Anopheles* regurgitate more than 1,000 sporozoites.) The deposited sporozoites move about, penetrate the skin, enter the bloodstream or lymphatic vessels and, within minutes to an hour, infect the liver (mammals) or the cells lining capillaries (endothelial cells) and macrophages (birds and lizards). The blood-borne sporozoites may penetrate several cells before entering their final destination, a liver cell. Within the liver cell the parasite feeds, grows and multiplies, generating thousands of merozoites. The time for merozoite formation in the liver varies according to the kind of malaria: in *P. falciparum* it takes five days to produce 30,000 merozoites and in *P. vivax* it takes eight days to produce 10,000 merozoites; in *P. ovale* and *P. malariae* the merozoites develop more slowly, taking nine and 15 days, respectively. Eventually clusters of these are shed into the bloodstream inside membrane-lined packets (called merosomes), and pass into the lung circulation, where the merosome membrane breaks down to release the merozoites into the general circulation. The final destination of the merozoites released from the liver is the red blood cell, not another liver cell.

Location, location, location

The haploid nucleus of *Plasmodium* with 23 million nucleic acid bases packaged into 14 chromosomes codes for ~5,300 proteins (as determined by gene-finding computer programs). About 60% of these proteins lack any similarity to proteins found in other living things. As a consequence, these have been referred to as hypothetical proteins. Perplexing and frustrating as the finding of a large number of proteins with unknown function may be, there could be a silver lining: these unique hypothetical proteins might represent the parasite's Achilles' heels for designing novel drugs. In addition to the nuclear genome, malaria parasites contain another genome in the apicoplast [355, 410]. The apicoplast, a relict plastid, is critical to the survival of the parasite. The 35,000 nucleic acid bases in the apicoplast genome code for about 30 proteins, none of which is involved in photosynthesis, and proteins encoded by the nuclear genome supplement the apicoplast proteins.

Approximately 77 (14%) of all the predicted malaria parasite proteins have been identified as enzymes. This is fewer than the number of genes found in yeast (with a genome of approximately similar size to *Plasmodium*) and suggests this genetic "slimming down" may be a result of the parasitic lifestyle. Genes for the enzymes involved in the breakdown of glucose to lactic acid — an energy-yielding process needing no oxygen — were identified in the genome. However, genes critical to the synthesis of carbohydrate storage products, such as glycogen, were not found. The third genome of *Plasmodium* is found in the mitochondrion [694] and, although all the genes necessary for the full oxidation of glucose into carbon dioxide and water were identified in the genome, it is clear these do not function for energy production when coupled to an electron transport system as do the mitochondria of other cells.

In studies undertaken by my laboratory in the 1960s we found malaria parasites (*P. lophurae* in this case) to be unable to synthesize purines, whereas pyrimidines could be made from much simpler molecules such as glutamic and aspartic acids and bicarbonate [719]; consistent with the findings from this bird malaria, the genes for each of the steps in the pyrimidine synthesis pathway were found in the *P. falciparum* genome, whereas those for purine synthesis were lacking [249].

The genome of *P. falciparum* contains a repertoire of transporters to import glucose, nucleosides, para-aminobenzoic acid, amino acids and other essential metabolites into the *Plasmodium* as well as those involved in the export of lactic acid and antimalarial drugs; these transporters are located on the parasite plasma membrane. Then there are transporters on the membrane of the mitochondrion and the apicoplast. Of special interest are the unique trafficking pathways that allow the parasite to extend itself beyond the boundaries of its own plasma membrane. These pathways may be a consequence of the parasite being enclosed within a membrane-lined PV. Two prominent membrane features of the export machinery are the tubulovesicular network (TVN), formed by extensions of the PV membrane, and compartments known as Maurer's clefts [59, 398]. The trafficking system is postulated to function in the transport of materials destined for the surface of the infected red blood cell. Highly variable gene families (called var) have been found clustered at the ends of chromosomes (called telomeres). These genes, of which there are approximately 208, or 3.9% of the genome, code for proteins that are exported via Maurer's clefts to the outer surface of the infected red blood cell where they are involved in both adhesion and immune evasion.

With the publication of the complete genome sequence of *P. falciparum* in 2002, it was predicted that: "The present genome sequence will stimulate vaccine development by the identification of hundreds of potential antigens that could be scanned for desired properties such as surface expression or limited antigenic diversity" [249]. This prediction is yet to be realized.

The malaria genome capers

Craig Venter (1946–) has been described by his critics as an egotistical, narcissistic and ruthless entrepreneur bent on making billions of dollars from sequencing genomes, and by his supporters as a passionate, brilliant visionary who tamed the power of genomics to radically transform healthcare by sequencing the human genome. Venter received his B.S. in Biochemistry from the University of California, San Diego, in 1972 and his Ph.D. in Physiology and Pharmacology from the same institution in 1975. He worked for a short time at the State University of New York (SUNY), Buffalo, and then in 1984 Venter joined the NIH, where he learned the technique of rapidly identifying all the mRNAs present in a cell using short cDNA sequence fragments called expressed sequence tags (ESTs).

An EST is produced by one-shot sequencing of several hundred base pairs from an end of a cDNA clone taken from a cDNA library. These clones consist of DNA that is complementary to mRNA, hence the ESTs represent portions of expressed genes. ESTs can be mapped to specific chromosome locations and, if the genome of the organism that originated the EST has been sequenced, one can align the EST sequence to that genome.

Originally, the use of ESTs was controversial as most geneticists felt it would not be accurate enough to sequence an entire organism. Unhappy with the support of NIH, Venter left in 1992 to establish TIGR, dedicated to the shotgun sequencing technique in which all of the DNA is chopped up, sequenced with redundancy, and then assembled for the sequencing of a variety of microbes. In a few short years the "whole genome shotgun sequencing approach" was successfully used to sequence the bacterial genomes of *Haemophilus influenzae* [225], *Mycoplasma genitalae* [231], and *Borrelia burgdorfi*, the causative agent of Lyme disease [230].

As early as 1994, frustrated with the slow progress and inefficient use of labor on the publicly funded Human Genome Project (begun in 1990), and unable to get his own funding, Venter decided the same approach could be used to sequence the human genome.

In 1998, Venter left TIGR to become the first President of Celera Genomics (a company established by Perkin Elmer for the purpose of generating and commercializing genomic information and to use their automated sequencing machines). One year later, his team at Celera applied the shotgun approach to the *Drosophila* genome as a proof of principle and then to the human genome using 300 DNA sequencers, a powerful computer, and a technique developed by Gene Meyers, a mathematician (from the University of Arizona and recruited to Celera), for reassembling the sequences from the shotgun approach. Meyer's algorithms, which allowed the fragments of DNA to be aligned and "stitched" together, was scorned by his scientific peers for being error-prone and unworkable, but at Celera he proved the technique could work. Two years later, the race to sequence the human genome ended in a tie. On 26 June 2000, Venter and Francis Collins (then head of the NIH-sponsored Human Genome Project) stood next to President Bill Clinton and heard him declare: "Today we are here to celebrate the first complete survey of the entire human genome...with this profound knowledge humankind is on the verge of gaining immense, new power to heal. Genome science...will revolutionize the diagnosis, prevention and treatment of most, if not all, human diseases" [147].

By 1992, in the UK, the Wellcome Trust was beginning to develop an interest in genome sequencing. In that same year, a chance meeting between David Kemp (then at the Walter and Eliza Hall Institute) and Alistair Craig (then at Oxford University) at a Keystone Conference in Utah resulted in the suggestion that there be a coordinated effort by several laboratories engaged in sequencing to sequence the malaria genome [476]. This dovetailed nicely with the subsequent establishment (1994) of the Wellcome Trust Sanger Institute, dedicated to genome sequencing, within the Wellcome Trust Genome Campus (at Hinxton). At a meeting on severe malaria sponsored by the Wellcome Trust, Craig, Kemp and Chris Newbold (Oxford University) approached the Director of the Trust, Bridget Ogilvie, as to whether the Trust would support sequencing the genome of the malaria parasite, *P. falciparum*.

The Trust agreed to fund a multi-center collaboration to map ESTs that included Alan Cowman and David Kemp from Australia, from the US Jeffrey Ravetch (then at Rockefeller University), and, from the UK, Newbold and Tony Holder (Naval Medical Research Institute [NMRI]). In 1995, a Trust-sponsored meeting at the new Sanger Institute discussed establishing a Pathogen Sequencing Unit, under Bart Barrell. At this meeting, priorities for sequencing the bacterium causing tuberculosis and *P. falciparum* were established and monies for a pilot project to sequence *P. falciparum*'s very smallest chromosome (number 1) were provided. In 1996, Barrell and colleagues at Sanger demonstrated that it was possible to assemble gene sequences from small clones of the *P. falciparum* chromosome 1. This was a proof of principle that was needed for the feasibility and initiation of large-scale sequencing of the entire *Plasmodium* genome.

Independently, "across the pond," Stephen L. Hoffman (1948–), a lean, green-eyed, exuberant optimist, entertained similar ideas about sequencing malaria genomes. Hoffman received a B.A. from the University of Pennsylvania (1970) and an M.D. from Cornell University Medical College (1975), followed by a residency at UC San Diego. Armed with a Diploma in Tropical Medicine from the London School of Hygiene & Tropical Medicine, he spent nearly five years in Jakarta, Indonesia (NAMRU-2), studying tropical diseases, and was then rotated back to the US where he was based at NMRI (later named the Naval Medical Research Center [NMRC]). There, as the Director of its Malaria Program from 1987 to 2001, he built a team of over 100 individuals in the US and overseas working on all aspects of malaria but especially vaccine development [599]. Hoffman read Venter's papers on the genomes of *Haemophilus* and *Mycoplasma*, musing, "Wouldn't it be interesting to do the same with *Plasmodium falciparum*?"

At the NMRC was Sarah L. French (1951–). In 1993 French, who had received her Ph.D. from the Scripps Institution of Oceanography in La Jolla, was doing post-doctoral training with Oscar L. Miller, Jr. at the University of Virginia, studying chromosome structure and function. Miller was then approaching retirement and, in order to continue the service contract on the electron microscope and her research, French submitted, on her own, a grant application to the NIH. This narrowly missed funding and, without support, her work couldn't continue. Unable to wait for another round of grant reviews, she applied for various advertised positions that sounded interesting. When she read about an open position in the malaria program at NMRC, she felt it dovetailed with her interests in adaptation and, in July, Hoffman hired her. In the spring of 1995, she went to TIGR for a seminar that the UC San Francisco biochemist Bruce Alberts was giving and to speak with him about shared interests in the interaction between DNA replication and transcription. She also had the chance to speak extensively with Craig Venter about a connected interest concerning genome organization and the orientation of genes relative to the direction of replication. It was during this chance conversation that issues regarding the possibility and importance of sequencing the malaria genome were discussed. Contact information was then shared with Hoffman. Hoffman called Venter to ask whether TIGR had an interest in sequencing the *P. falciparum* and *P. vivax* genomes. When Venter said yes, Hoffman approached one of the scientists on his staff, Malcolm Gardner, to give him a tutorial on ESTs and genomics, before he went to see Venter. Gardner received his Ph.D. (1985) studying murine retroviruses and oncogenes [598] and then moved to the NIMR in Mill Hill, UK, where he spent six years studying the *P. falciparum* apicoplast. Doing a great deal of manual sequencing and learning how to use computer programs to collect and analyze sequence data were skills that proved useful in the Malaria Genome Project. In 1995, he moved to Hoffman's group at the NMRC to work on *P. yoelii* DNA vaccines. In addition to Gardner, the sequencing efforts at the NMRC would be aided by Daniel Carucci. Carucci, who joined Hoffman's team in January 1996, received his Medical Degree from the University of Virginia School of Medicine supported by a scholarship from the U.S. Navy [22]. His military stint exposed him to tropical diseases and thereby began a lifelong interest in their cure. In 1991, he was awarded a Master's of Science in Clinical Tropical Medicine and in 1995 a Doctor of Philosophy from the London School of Hygiene & Tropical Medicine. At the NMRC, he was responsible for the creation of the Malaria Genomics and Applied Genomics Laboratory, and was the Director of the Malaria Vaccine Program leading the U.S. Navy's efforts in the development and testing of malaria vaccines.

The fond hope of the Malaria Genome Project was that the gene sequence would "provide a road map for malaria research in the 21st century, research that will lead to improved treatment and prevention" [314]; however, at the time there was no common fund to support the sequencing efforts in the US. Hoffman went over to TIGR, discussed the project with Venter, and they agreed a concerted effort would have to be made to obtain the $28 million in funding Venter estimated the project would cost.

In late December of 1995, Venter and Hoffman elected to hold a meeting at TIGR's headquarters to "set the ball in motion" and to delineate the necessary steps to obtain funds; however, it was a meeting that nearly did not come about. The meeting was scheduled for Saturday 6 January 1996. On that day, snow began falling on Washington, DC, ushering in the blizzard of 1996. "The entire region was shut down. High winds caused snow to blow back onto plowed roads and many streets were not plowed at all. The Metro system fared no better as frozen rails crippled the outside portion of that system. One train with 100 passengers on board, got stuck for five hours near Takoma Park, Maryland " [16]. A four-car Red Line Metrorail train over-ran the Shady Grove Station platform and plowed into a six-car train, killing the operator. Thousands of people were without electricity [475]. President Clinton shut down the government. Undaunted by the fierce weather, the government's Navy men (Hoffman, Gardner and Carucci), and William Bancroft, Head of the Military Infectious Disease Program at Ft Detrick, Maryland, trudged through drifts of snow (which in some places had accumulated to two feet) to get to TIGR's headquarters in Maryland. The justification for the involvement of the Department of Defense (Navy and Army) was that sequencing would accelerate the development of a protective vaccine for malaria (as well as drugs).

In March, after giving a 20-minute presentation at the NIH on the Project, Hoffman was able to garner $1 million in funding. TIGR wrote an addendum to its NIH grant on sequencing the genome of tuberculosis and received additional monies for the malaria project. Venter contacted the Burroughs Wellcome Fund and they too promised money. By May, with interest already expressed in the UK by the Sanger and the Wellcome Trust, an international consortium of scientists and funding agencies was established. The NMRI was to provide DNA and chromosomes, and TIGR, Stanford University and the Sanger would do the sequencing. Coordination of activities would be through e-mails, conference calls and two meetings a year — one in the US, at the meeting of the American Society for Tropical Medicine and Hygiene (ASTMH), and another in the UK. By September, in conjunction with the annual meeting of the ASTMH, it was announced that $28 million had been committed by the various donor agencies for the Malaria Genome Project.

The Sanger Institute and TIGR were not only competitors — some would describe them as enemies — the two had very different philosophies. In 1993, John Sulston was the Director of the Sanger Institute, and during a visit to the US he met Venter who was focused on sequencing more and faster, and intent on moving into the private sector [657]. Sulston was personally offended by Venter's competitive streak and he regarded the research being done at TIGR as a challenge to what the Sanger Institute was doing. There were other problems. TIGR was set up as a private, non-profit genome center funded by a venture capital group that established Human Genome Sciences Inc. to develop and market products from TIGR discoveries. Venter predicted finding a thousand genes daily. Sulston regarded Venter's approach to be a compromise of his academic integrity and felt Venter wanted it both ways: recognition and acclaim from his scientific peers, but also secrecy to help his business partners and ultimately to enjoy the profits. Sulston's view was that the results of a genome project were to provide as much information as possible that could be used by everyone, public and private, to advance understanding.

The real danger, as he saw it, was that if companies selling drugs or diagnostic kits were to gain exclusive rights to genomic sequences it would prevent others down the line from having any incentive to use the information in creative ways, thus impoverishing science and medicine. The Wellcome Trust, as a charity, shared this view. Indeed, its absolute rule was that the sequence data generated at the Sanger Institute would be immediately released into the public domain. The Trust went a step further and categorically stated they were opposed to patenting the basic genome information and would be prepared to contest such in the courts. Despite this, patenting of gene sequences (initially a practice at the NIH but later discarded) would be open to those who made the discovery of a sequence's function and for the development of commercial products. There were deep suspicions among the gene sequencers regarding Venter's motives. Sulston has written caustically: "Part of the reason why people find Craig hard to stomach, and why others admire him greatly, is his cavalier disregard for academic niceties...Craig was in a business and the priority for business is not scientific credibility but share price and market penetration" [657]. The Sanger Institute was on its guard for the challenge from TIGR, especially as Barrell's team had beaten them in completing the sequence of the tubercle bacillus. However, to some participants — in what has been described as a "mud-wrestling match" — it wasn't a matter of racing to be the first and claiming glory, there was the looming danger that control of a genome sequence might reside in private hands [657].

There followed many meetings (see below) in an attempt to reconcile the differences between the "gene hunters" and their respective institutions and, in time, several funding agencies came together to finance and help manage the groups that would undertake the sequencing efforts as well as to assist in establishing the policies on data release and access. The result was a policy that balanced the rights and responsibilities of the sequencers to analyze and publish their results with the opportunities that rapid release of the genomic sequence data would provide to other investigators. This policy on data release and access has since become the guideline for other sequencing projects.

Due to the large size of the *P. falciparum* genome and the large cost of the project, it was reasoned that the efforts of all the members of the consortium would be needed to sequence all the genes in all of its 14 chromosomes [119]. In May 1996, at a meeting sponsored by the NIH and the Burroughs Wellcome Fund, scientists and potential funders met to discuss the possibility of using a chromosome-by-chromosome shotgun sequencing approach. The result of the meeting was a plan to divide the work by chromosome with about half the work being done in the US and half in the UK. A pilot project at TIGR and the NMRC, funded by NIH and the U.S. Department of Defense, was dedicated to sequencing chromosome 2. Sequencing of chromosome 12 was undertaken at Stanford University with support from the Burroughs Wellcome Fund and, at the Sanger Institute (Cambridge, UK), Patricia Goodwin from Wellcome acted effectively and collegially with the National Institute for Allergy and Infectious Diseases (NIAID) on coordinating the effort, and a pilot project was launched at the Sanger Institute to sequence chromosome 3. Eventually, the distributed parasite genome sequencing projects were: the Sanger Institute with chromosomes 1, 3–9 and 13, TIGR with chromosomes 2, 10, 11 and 14, and Stanford University with chromosome 12. Early on, the sequencers encountered surprisingly difficult problems. Indeed, Malcolm Gardner said the project to sequence the nuclear genome of *P. falciparum* was a "real nightmare," because of the parasite's unusually high proportion (~89%) of adenine (A) and thymine (T) in the coding region of the DNA that made it difficult to clone the genetic material in bacteria and to sequence it [208]. Nevertheless, with the development of new techniques and software, the problem of sequencing an AT-rich genome was eventually overcome. Chromosome 2 was the first chromosome fully sequenced, assembled and annotated, and was published in 1998 by the TIGR–NMRC team [251]. The sequence of chromosome 3 came a year later by the Sanger–Oxford team [82] and the entire *P. falciparum* sequence in 2002 [249, 250].

On occasion, the meetings at which members of the sequencing centers, funders and collaborators met to discuss progress and exchange data could be quite tense affairs. A notable one was at the Fourth Malaria Genome Sequencing Meeting, Orlando, Florida, on 11–12 December 1997, where a heated discussion between the sequencing partners at Oxford University and the NMRC led to some interactions in which neither party came out well. The Eleventh Meeting was held at the Sanger Institute in 2001. After the day-long formal discussions, some of the attendees repaired to the local Red Lion pub. After the pub closed, informal discourse continued back at the bar in the main house, with the bartender manning the bar until midnight, when the door to the bar was locked. Not content with the amounts of alcohol that had been consumed during the evening, the remaining sequencing center revelers scaled the top of the bar and grabbed bottles of single malt whiskey to continue the "discussions" into the wee hours. At the next year's meeting, a metal security gate appeared over the bar!

In 1998, Venter left TIGR and the *P. falciparum* Genome Project to become the President of Celera Genomics, a for-profit company dedicated to sequencing the human genome. A year prior to Venter's departure it was decided that it would be optimal to have Gardner move from the NMRC to TIGR to work on the *P. falciparum* genome project full-time. A Brazilian scientist, Leda Cummings, joined Gardner and a very talented post-doctoral fellow, Herve Tettelin, in the sequencing project. Tettelin had received a degree in Industrial Engineering in Biotechnology from the Institut Supérieur Industriel de la Province de Liege, Belgium, and a B.S. and Ph.D. in Applied Sciences from the Université Catholique de Louvain, Belgium. During his Ph.D. thesis, he coordinated the work of 25 laboratories in Europe for the sequencing of yeast chromosome 7. At TIGR, Tettelin became the primary person responsible for the sequence analysis of the *P. falciparum* chromosome 2. Hoffman was also keen for TIGR to sequence the genome of the rodent malaria model *P. yoelii*, as a means to jump-start gene finding in a rodent model used for immunological studies at NMRC — and at TIGR Cummings took on the role as the principal investigator of this project. However, things did not quite go as planned. The working relationship between Gardner and Cummings was, to say the very least, strained. Indeed, Venter had, on more than one occasion, summoned both of them into his office and demanded they work their issues out, or else. This deteriorating relationship culminated in Cummings leaving TIGR in the spring of 2001 to join the Computational Biology Branch (CBB) of the National Center for Biotechnology Information (NCBI) at NIH. Ironically, Jane Carlton (1967–) was also at the CBB but desperate to find a way out and be a part of the malaria genome projects. The two female scientists overlapped at NCBI for three weeks before Carlton left to join TIGR. At the time, it was rumored that the ever-competitive Venter had said that TIGR won the Cummings–Carlton swap; in the process, Carlton inherited the project to sequence the genome of *P. yoelii* and also worked on the sequencing of the falciparum and vivax genomes. Carlton subsequently went on to publish the yoelii genome in the special *Nature* issue devoted to the genome projects [109] and Cummings was given a middle authorship. Thereafter, Tettelin became a TIGR faculty member, sequencing the genomes of several bacterial pathogens including meningococcus (*Neisseria meningitidis*), pneumococcus (*Streptococcus pneumoniae*), *Ehrlichia*, *Anaplasma* and *Neorickettsia*. Presently he is an Associate Professor at the University of Maryland School of Medicine using microarrays for functional genomics on these and other pathogens.

As was Tettelin, Carlton was admirably suited for the task of genome sequencing. Carlton graduated in 1990 with a B.S. in Genetics from Edinburgh University. She then joined David Walliker's malaria genetics group at Edinburgh University for her Ph.D., studying the inheritance of particular genes using the rodent malaria *P. chabaudi*. After receiving the Ph.D. (1995), she spent another two years in Edinburgh analyzing genetic crosses for chloroquine (CQ) and mefloquine resistance. In 1997, Carlton moved to John Dame's laboratory at the University of Florida where it was realized, for the first time, that ESTs (short ~500 nucleotide tags) would be useful for gene discovery; Carlton and Dame began using ESTs to start sequencing *P. vivax* and *P. berghei*, as well as *P. falciparum*. After a brief stay at the NCBI, she was appointed to the faculty of TIGR where she helped to complete the *P. falciparum* genome sequence.

In 2000, Hoffman was recruited to Celera Genomics to become Vice President of Biologics with the intent of turning the human genome sequences into immunotherapeutics for cancer. While there, and with Venter's support and a $9 million NIH grant, the sequencing of the *Anopheles gambiae* genome was begun. Despite the genome of *A. gambiae* being 10 times larger than that of *P. falciparum*, the sequencing was completed in 40 days. On 2 October 2002, just a day before publication of the *Plasmodium* and *Anopheles* genomes, a press meeting was called to announce this landmark event. This required an intense amount of coordination, as it involved several genome sequencing centers (TIGR, Stanford and Sanger), several funders (Wellcome Trust, Burroughs Wellcome Fund and U.S. Department of Defense and NIAID/NIH), both *Nature* (publishing the *Plasmodium* genome papers) [249] and *Science* (publishing the *Anopheles gambiae* genome) [326], as well as numerous collaborators. The press meetings were held in both the United Kingdom and United States, with representatives from each of the centers flying to both places to provide an equal presence. After the press meetings, the UK celebrations were predictably raucous. The TIGR/NMRC team, in contrast, had a rather short and staid lunch with the *Science* editors and Berriman of the Sanger Institute. The celebration was completed by mid-afternoon — most likely in time for all concerned to get back to their desks and continue hunting genes.

The press interest in the completion of the malaria and mosquito genome projects was intense; however, it was short lived. On 3 October 2002, a shooting spree in a suburb of Maryland by the Beltway Snipers made the headlines and bumped the malaria genomes off the front page. Gardner ruefully quoted the pop-artist Andy Warhol and said this ended his 15 minutes of fame.

Undoubtedly the success of the *Plasmodium* genome projects was accelerated by the injection of "new blood" into the project in

2000/2001. In early 2000, in addition to Barrell, the project at Sanger was taken on by Neil Hall, a young, motivated and highly agreeable scientist who made friends easily especially among those he respected. Hall's particular friendship with Director of Bioinformatics Owen White at TIGR (fueled by their mutual interest in a convivial night on the town and good science) helped smooth the interactions between TIGR and Sanger. (Hall left Sanger and joined TIGR in 2003, spent two years there, and then returned to the UK.) At TIGR, Carlton's camaraderie with her fellow compatriots at the Sanger facilitated the exchange of ideas and greater collaboration. Gardner too was considered an "honorary Brit," having British heritage and having spent a few years in London working at Mill Hill. Finally, Matt Berriman, number two at the Sanger, had spent several years as a parasitology postdoctoral fellow in the US. The camaraderie between the two groups was cemented even further during the week after the 11 September 2001 terrorist attack on New York and Washington, DC. Hall and Berriman were stranded in Boston after the meeting they were to attend was canceled. As air-space was shut down, the TIGR team invited them down to Washington, DC, so that their trip would not be in vain. Driving through the back roads between Boston and Washington, DC, they made it to TIGR and spent the next few days accepting TIGR hospitality, before flying back to the UK. They were enormously pleased to have been invited and the relationship between the two sequencing centers flourished.

Authorship positions on the falciparum and yoelii genome papers remained a sticky subject. So that each genome sequencing center would be able to get full credit and also because *Nature* articles are notoriously short, it was decided that each center would have its own paper describing the chromosomes assigned to that center, in addition to an overall whole genome paper. The yoelii paper was to be a comparative analysis between it and falciparum, the first comparative genomics paper between two *Plasmodium* species. Hall graciously accepted Gardner as being the first author on the whole *P. falciparum* genome paper, followed by alternating Sanger, then Stanford authors. Carlton as lead investigator on the yoelii project was primary author on this paper. The choice of final author positions, however, was a little more fraught. Despite requests that Gardner be the last-named author on the yoelii paper due to his work on the project, this was not to be.

Chapter 4

Chipping Away at the Genome

In 1998, a time before the genomic sequence of *P. falciparum* had been completed and published, Su and Wellems stated: "Progress has been made in the genetics and genomics of malaria parasites. Numerous genes have been cloned and millions of base pairs of the parasite deoxyribonucleic acid (DNA) have been deposited in public databases; genetic maps of the *P. falciparum* 14 chromosomes and hundreds of linkage markers have been reported; yet despite these advances, we still know very little about the function of most genes!" [655].

Indeed, by the early 1980s, only partial [206] and complete sequences of the sporozoite surface protein (CSP) genes from a variety of malarias were available [499] and a *P. falciparum* surface antigen gene had been cloned and sequenced [364]. Each gene could, by a variety of laboratory techniques, be used to synthesize a particular protein and that protein could be made in quantities sufficient for use in immunization. This was, according to some, the breakthrough needed for developing protective vaccines. For example, Araxie Kilejian, working with *P. lophurae* in Trager's laboratory, isolated a unique protein, the histidine-rich protein (HRP). Believing the HRP could serve as a vaccine, Kilejian used its published gene sequence [540] to identify the HRP gene by base pairing using a synthetic nucleotide probe, and the gene was cloned and expressed in a cell-free system [367]. Although initial tests showed some protection [366], subsequent work proved HRP to be useless as a vaccine [600]. Moreover, even if it had been successful, critics of her research would certainly have argued: who needs a vaccine to protect white Pekin ducks against malaria? Instead, they would state the real objective of gene sequencing was to provide a basis for a vaccine against human malaria, in particular the lethal *P. falciparum*. The challenge for the gene cloners was to find vaccine candidates lurking in the *P. falciparum* genome, and to find how these might be discovered amongst the more than 5,000 malaria parasite genes.

The early methods of cloning individual genes and their expression were useful in identifying and characterizing a limited number of genes expressing a particular protein antigen; however, with the availability of the complete genome sequence of *P. falciparum*, it was believed it would open up the possibility of finding many other putative vaccine candidates. Sequencing of entire genomes (rather than individual genes) began with the work of Frederick Sanger in England. In Sanger's method (see p. 38), the time- and labor-intensive nature of preparing large amounts of DNA, cutting this into smaller pieces, preparing and running gels, as well as the absence of suitable computer programs, drastically reduced the efficiency of the process and increased the costs. As a consequence, these factors limited the complexity of the genome that could be sequenced. However, using his method in 1977, Sanger was able to manually sequence the bases in a virus's genome. Sanger then used the method of "shotgun" sequencing, a strategy based on the isolation of random pieces of DNA from the genome to be used as primers for the PCR amplification of the entire genome. The amplified portions of DNA could then be assembled by their overlapping regions to form contiguous transcripts (known as contigs). The final step involved the utilization of custom primers to elucidate the gaps between the contigs, thus giving a completely sequenced genome. In 1982, Sanger used "shotgun" sequencing to complete the genomic sequence of the 3,569 nucleic acid bases in the RNA of the bacteriophage MS2. Since then, other and more complex viral genomes have been sequenced using similar techniques, such as the 230,000 bases in the genome of cytomegalovirus (CMV), and the 186,102 bases in the genome of the smallpox virus.

In 1989, a consortium to sequence the genome of the yeast *Saccharomyces cerevisiae* — a eukaryote — was set up. The *S. cerevisiae* genome was approximately 60 times larger than any genome sequence previously attempted; however, it was reasoned this would be the next logical step towards the eventual characterization of the human genome, a task beyond the scope of the then current technology. Despite this, the following year saw a bandwagon approach among gene hunters with the initiation of a plethora of ambitious genome sequencing projects.

By 1995, the 1,830,000 nucleic acid bases in the free-living bacterium *Haemophilus influenzae* had been sequenced at TIGR [225]. "The image of the entire genome, a kaleidoscopic wheel marking all the genes that it takes to be *Haemophilus influenzae*...had a hypnotizing effect" [757]. The TIGR researchers were able to list every gene and to sort them according to likely function — a "guilt by association" approach. There were not a lot of surprises. It had genes for metabolism, for attaching to cells and for sensing the environment. All this was already known; however, what was remarkable was the fact that now large chunks of DNA could be sequenced so quickly. Previous sequencing projects had been limited by the lack of adequate computational approaches to assemble the large amount of random sequences produced by "shotgun" sequencing. In conventional sequencing, the genome is broken down laboriously into overlapping segments, each containing up to 4,000 bases of DNA. These segments are "shotgunned" into smaller pieces and then each is sequenced and aligned to reconstruct the genome. Venter's team utilized a more comprehensive approach by "shotgunning" the entire *H. influenzae* genome [231]. Previously, such an approach would have failed because the computer software did not exist to assemble such a massive amount of information accurately. Software, developed by TIGR, called the TIGR Assembler, was up to the task, "stitching" together the approximately 24,000 DNA fragments into a whole genome. After the *H. influenzae* genome was "shotgunned" and the clones purified, the TIGR Assembler software required approximately 30 hours of central processing unit time on a computer containing half a gigabyte of RAM, testifying to the enormous complexity of the computation.

Earlier, Venter's *H. influenzae* project had failed to win funding from the NIH because of the serious doubts by the members of the NIH Study Section (composed of molecular biologists) surrounding his ambitious proposal and the use of ESTs. It simply was not believed that such an approach could accurately sequence the large genome of this bacterium. Venter proved everyone wrong and succeeded in sequencing the genome in record time and at a cost of 50 cents per base, which was half the cost of and dramatically faster than conventional sequencing. This method of sequencing led to a multitude of completed sequences over the ensuing years by TIGR. *Mycoplasma genitalium*, a bacterium that is associated with reproductive-tract infections, was sequenced by TIGR in a period of eight months between January and August 1995. TIGR subsequently published the first genome sequence of a sulfur-metabolizing organism, *Archaeoglobus fulgidus*, as well as the genome sequences of the pathogens involved in peptic ulcer disease, Lyme disease [230], and then the genomes of the bacteria causing plague, cholera, syphilis, anthrax, and whooping cough. By 2005, the genomes of 185 prokaryotic as well as eukaryotic organisms — including the roundworm *Caenorhabditis elegans*, the malaria mosquito vector *Anopheles gambiae* and two malaria parasites, the deadly human malaria *P. falciparum* and the rodent malaria parasite *P. yoelii* — were sequenced (http://www.genomenewsnetwork.org/resources/sequenced_genomes/genome_guide_p1.shtml). Today, the number of sequenced prokaryotic genomes is ~1,000 (http://www.ncbi.nlm.nih.gov/sites/genome) and that of eukaryotes is ~800 (http://www.ncbi.nlm.nih.gov/genomes/leuks.cgi).

Sequencing of a genome is only the beginning for understanding the functional nature of a cell or a virus. "What remains truly exciting is the kind of research that starts after the genome is sequenced: discovering what the genes do, mapping out the networks in which the genes cooperate and reconstructing the deep history of life" [757]. Further, to understand gene function, it is necessary to know what the signal is that turns a gene "on" and what is the "off" switch. Functional genomics is the study of the processes involved in turning "on" genes (= gene expression) and has as its goal not simply to provide a catalogue of all the genes and information about their functions but also to understand how the components work together. Knowing which genes are associated with particular biochemical pathways is important for understanding the function of a cancer or normal cell and, in the case of malaria parasites, could provide a wealth of novel drug and vaccine targets, as well as the biochemical basis for the disease process. It was hoped that, using systematic genome-dependent approaches, it would be possible to study evolution, characterize protein complexes, understand drug action and resistance, develop vaccines and better understand the immune response.

To examine the functional role of individual genes and their relationship to other genes, high-density DNA microarrays were developed in the Stanford University (Palo Alto, California) laboratories of Ronald W. Davis and Patrick O. Brown. In 1995, an engineering student, Dari Shalon, developed a simple, low-cost way of printing thousands of long gene fragments onto a glass slide, using a robotic microarraying device. Following this, Mark Schena, a post-doctoral fellow in the laboratory of Davis, and Shalon from the Brown laboratory, used PCR-amplified cDNAs from the weed *Arabidopsis thaliana* and this cDNA library was placed onto the glass slide to study gene expression. The mRNA obtained from the *Arabidopsis* was labeled with a fluorescent dye and hybridized to the array, and then the fluorescent signal was measured. Although this microarray technology was efficient for studying gene expression in a large number of clones, its full potential was reached when the complete genomic sequence of the yeast *Saccharomyces* was released (1996), largely due to an international collaboration of more than 600 scientists from over 100 laboratories under the aegis of Davis. Therefore, not surprisingly, Davis chose to devote a considerable amount of attention and effort by himself and his graduate and post-doctoral students to undertake a functional analysis of the *Saccharomyces* genome. Two of these students were Elizabeth Winzeler and Joseph DeRisi. Elizabeth

Winzeler (1962–) grew up in Reno (Nevada), attended Lewis and Clark College in Portland, Oregon, and, upon completing a B.A. (1984), she was still a bit confused about what she wanted to do after graduation so she went to work as a computer programmer for the U.S. Bureau of Labor Statistics in Washington, DC, for several years. At 25 years of age, she decided to seriously try science. First, she obtained an M.S. in Biophysics (1990) from Oregon State University (Corvallis, Oregon), and then transferred to the Developmental Biology Department at Stanford University where she worked on the stalked bacterium *Caulobacter*, receiving her Ph.D. in 1996. When the time came to do a post-doctorate, Winzeler's interests had turned to parasitology; her thesis advisor told her that parasitology was fine, but she really thought the laboratory of Davis (in the Department of Biochemistry) would best suit her interests and temperament. This turned out to be excellent advice. Winzeler told Davis that she was interested in his work on *Arabidopsis* but he said that, as the yeast genome had just been sequenced, there might be some opportunities using this organism. Being an opportunist, Winzeler decided to work with yeast. Having a background in both computer science and molecular biology turned out to be fortuitous. During her time at Stanford she helped with some of the first yeast microarray gene expression experiments and developed methods for studying DNA replication.

Joseph DeRisi (1969–), after receiving his B.S. from UC Santa Cruz (1992) in Biochemistry and Molecular Biology, joined the Brown laboratory intending to study how retroviruses integrate into the host genome. DeRisi and a fellow graduate student, Vishy Iyer, developed methods to miniaturize the microarrays and were able to print the first DNA microarrays for the yeast genome on a glass slide no bigger than a computer chip. The first result was a success and enabled whole gene expression studies for yeast by looking at the pattern of mRNA expression. In the generation of DNA microarrays, called "chips," short stretches of nucleotides are either directly synthesized and spotted on a glass slide or, using photolithographic chemistry, are synthesized *in situ* on a glass slide.

After the isolation of mRNA, it is labeled (usually with a fluorescent tag) and then allowed (by Chargaff rules) to hybridize to the array. Using a fluorescence scanner, it is possible to quantitatively estimate the degree of fluorescence for each spot and to calculate transcript, i.e. mRNA abundance. By this approach, a functional analysis of the yeast genome was carried out [174, 741].

When it came time to find a real job, Winzeler considered microbiology departments and was asked about future directions. She said that she really wanted to work on malaria (even though she had no experience in this). She thought that, although malaria parasites had been relatively difficult to work on, the availability of the genome sequence would change this and that soon one would be able to use systematic genome-dependent approaches to predict transcriptional regulatory networks, study evolution, characterize protein complexes and understand drug resistance. Finally, in 1999, she joined the fledging Genomics Institute of the Novartis Foundation (La Jolla, California). One year later, she became a faculty member at the Scripps Research Institute (also in La Jolla, California) where she is currently an Associate Professor.

To help develop the malaria program, Winzeler recruited Karine Le Roch (1971–), a very talented post-doctoral student, to her laboratory. Le Roch received her B.A. in Biochemistry (1995) from the University of Paris VI (Paris, France). This was followed by an internship (1995) in a European exchange program (Erasmus) at North Wales University (Bangor, UK), where she worked on the development of a vaccine against the blood fluke, *Schistosoma mansoni*. Le Roch obtained a Master's degree in Parasitology at the University of Lille II (Lille, France) and Oxford University (Oxford, UK) (1997), working at the Institute of Molecular Medicine at the John Radcliffe Hospital (in Oxford) on the genetic diversity of *Pneumocystis carinii*. She completed her Ph.D. (2001) at the University of Paris VI. Her Ph.D. research concerned the cell-cycle regulation of *P. falciparum* and, more specifically, cyclin-dependent kinases. In 2001, she joined the Scripps Research Institute and began work with Winzeler to set up the functional analysis of the *P. falciparum* genome using microarray technology.

Zbynek Bozdech (1967–) grew up in Czechoslovakia. In 1990, he received his Master's degree in Biochemistry from Charles University in Prague (Czech Republic). From here he moved to McGill University in Montreal, Canada where he received his Ph.D. (Studies on the "Molecular characterization of ABC [adenosine triphosphate (ATP)-binding cassette] proteins in *Plasmodium falciparum*"). After completing his Ph.D. in 1999, he moved to the University of California at San Francisco and, in 2000, he settled in the laboratory of DeRisi to explore (as were Winzeler and Le Roch) gene transcription of *P. falciparum* using the fast-emerging microarray technology.

Le Roch et al. [401] and Bozdech et al. [83] were the first to use DNA chips to analyze gene expression in *P. falciparum*. The first study analyzed total RNA from sporozoites, merozoites, gametocytes and six stages in the asexual cycle of parasites growing and reproducing within the red blood cell. Using gene expression clustering and "guilt by association" inferences followed by temporal patterns of expression with genes of known function or cellular processes, it was possible to provide functional clues for more than 60% of the genes that the *P. falciparum* genome project had identified as lacking in sequence similarity to other organisms and called "hypothetical." In the second study, a cascade of gene expression was found and 75% of the genes were turned on only once during an asexual cycle with activation correlated to time-specific processes. The wave of transcription, i.e. synthesis of mRNA, began as rings grew into early trophozoites. Some 950 genes appeared to be involved in general or housekeeping functions and were induced with broad expression profiles. In the transition from trophozoite to schizont, 1,050 genes were maximally transcribed and, in the mid-to-late schizonts, a further 550 genes were induced. Bozdech et al. [83] concluded, "The lack of continuous chromosomal domains with common expression characteristics suggests the genes are regulated individually…[and in a] fundamentally different mode of regulation from other eukaryotes. This…implies disruption of a key transcriptional regulator…may have profound inhibitory properties."

Turning "on," turning "off"

By the 1960s, with the failure of CQ to cure malaria, there was a renewed interest in developing other drugs for treatment and to understand the mechanisms that were behind the parasite's resistance to antimalarials. One of the drugs to replace CQ was pyrimethamine (Daraprim) developed by George Hitchings at the Burroughs Wellcome laboratories in Tuckahoe, New York.

In 1988, George Hitchings shared the Nobel Prize with Gertrude Elion for "discoveries of important principles for drug treatment." Hitchings had an early interest in nucleic acids and, for his doctoral research (1933, Ph. D), developed microanalytical methods for detecting the purines in the high-energy compound ATP. In 1942, he joined the Burroughs Wellcome Company as the sole member of the Biochemistry Department housed in a converted rubber factory in Tuckahoe, New York. There, in collaboration with Elion, he began to explore synthetic mimics (commonly called analogs) of purines and pyrimidines to prevent the multiplication of cancer cells by inhibiting nucleic acid synthesis. In the late 1940s — the time of their work — the knowledge of nucleic acid synthesis was limited save for the fact that purines and pyrimidines were constituents of DNA and RNA. Instead of using cancer cells to test the analogs, they used a simpler model system: the folic acid-requiring bacterium *Lactobacillus casei* that also needed purines and pyrimidines for its growth. By 1947, Hitchings and Elion found a diaminopurine, an analog of adenine, to inhibit the growth of *L. casei*. It had a promising effect on experimental leukemia in mice as well as in early human clinical trials; however, when toxic effects appeared, further clinical work was abandoned. (Subsequent studies, however, did lead to successful chemotherapeutic drugs such as 6-mercaptopurine and thioguanine for leukemia.) Hitchings and Elion prepared a series of 300 substituted 2,4-diaminopyrimidines, some of which inhibited *L. casei* growth. To their surprise, reversal of the effect of the analog was achieved by folinic acid. Indeed, folinic acid was 500 times more effective than folic acid in reversing growth inhibition. This suggested that the 2,4-diaminopyrimidines were not acting as pyrimidine analogs but instead were interfering with the synthesis of folinic acid from folic acid. Hitchings sent some of his substituted compounds to Len Goodwin at the Burroughs Wellcome Laboratories in the UK for testing against the chicken malaria, *P. gallinaceum*, a screening program developed to find a replacement for quinine, which was unobtainable during World War II due to the Japanese occupation of Java, where Cinchona trees — the world's source of quinine — were cultivated. When malaria-infected chickens were treated with the analogs, the most effective was the 2,4-diamino-5-p-chlorophenyl-6-ethyl pyrimidine. It was, for obvious reasons, named pyrimethamine in 1950. If infected blood from a pyrimethamine-treated chicken was examined with a microscope, the effects were seen on nuclear division — it was abnormal and merozoite formation was blocked.

Using radioactive tracers, it was shown that, with pyrimethamine treatment, the synthesis of parasite DNA was inhibited, but RNA was unaffected [581]. The precise mode of action of the drug, however, remained a mystery. At the time of this report, my laboratory was focused on the characterization of malaria parasite enzymes and so we elected to study the enzymes involved in DNA synthesis to better understand why pyrimethamine was an effective antimalarial. Using *P. lophurae*, growing in ducklings (see p. xiii), pyrimethamine blocked the formation of parasite DNA by preventing the synthesis of the pyrimidine, thymidine, from much smaller molecules. In effect, this key pyrimidine was made from scratch utilizing the key pyrimidine pathway enzymes orotidine 5'-monophosphate phosphorylase and thymidylate synthetase. The lack of effect of pyrimethamine on RNA was due to the parasite's inability to make purines from scratch because it lacked the appropriate enzymes; instead it salvaged the intact nucleoside. Meanwhile, other laboratories (including those of Hitchings) had shown that pyrimethamine exerted its effect by binding to a folic acid enzyme, dihydrofolate reductase (DHFR), thereby depleting the cell of a critical vitamin cofactor tetrahydrofolate or folinic acid necessary for the synthesis of thymidine. Subsequent studies found that what applied to bird malarias also applied to other kinds of

Plasmodium spp., including rodent and monkey malarias. Still later, through recombinant DNA technology and the availability of laboratory-grown *P. falciparum*, cloning of the genes and characterization of all the enzymes in pyrimidine synthesis were achieved.

It wasn't long after the introduction of pyrimethamine as an antimalarial that parasite resistance was observed, and this was found, in some cases, to be due to decreased binding of the drug to DHFR as a result of mutations (changes in the DNA base sequence) in the parasite DNA, whereas in other cases resistance involved an increased synthesis of DHFR. That was the situation in the late 1990s. As might have been anticipated, the problem of drug resistance and drug mode of action was revisited after 2003 when DNA chips (microarrays) became available. It was expected by the gene hunters that microarrays would "permit a quick, unbiased look at the mode of action of antimalarial drugs...the underlying premise was that the parasites would sense metabolic perturbations from a drug and make compensatory changes in the synthesis of mRNA to adjust for the perturbations" [248]. This was anticipated based on experience with the yeast *Saccharomyces*, the fungus *Candida*, and the tuberculosis bacterium, *Mycobacterium*, as well as mammalian cells and even plants where specific drugs were found to elicit as much as a tenfold increase in mRNA by dozens of genes. The microarray results with *P. falciparum*, however, were not only surprising, they were frustrating. Indeed, gene expression by the malaria parasite was unresponsive to specific, lethal perturbation by a drug (a more potent anti-folic compound, WR99210) directed at DHFR. When *P. falciparum* was grown in glass vessels and samples taken periodically during the 48-hour period of growth and reproduction, and mRNA isolated and used for binding to a parasite DNA microarray chip, the levels of mRNA remained virtually unchanged even as the parasites died from exposure to the drug. In contrast, the individual enzymes for pyrimidine and folic acid synthesis increased tenfold during drug treatment. This suggested that the overall mRNA levels (called the transcriptome because it encompasses all the mRNA transcripts) for metabolism (and in this particular case for DHFR and pyrimidine enzymes) were largely pre-determined. It was speculated that, because the parasite lives within the stable milieu of a red blood cell, the hard wiring of the *Plasmodium* genome may be sufficient for managing transcription without employing a mechanism to sense alterations in the environment. Based on these experiments (which, of course, disappointed the gene hunters), it was clear that microarrays were unable to assign a mechanism for the action of anti-folic acid drugs by simply measuring changes in mRNA levels.

Some gene hunters, however, were not convinced that malaria parasites were hard wired and believed these studies of DNA synthesis were the exception rather than the rule. Here too they would be thwarted.

In the 1970s, in collaboration with the laboratory of George G. Holz at the Upstate Medical Center in Syracuse, New York, lipid metabolism by *P. lophurae* was studied [52]. Later, when *P. falciparum* could be grown in the laboratory, a post-doctoral fellow in my laboratory, Patricia Maguire, did similar work [419]. The reason for studying malaria lipids was that the parasite synthesizes vast amounts of membranes during its growth and reproduction, and for this it needs large amounts of lipid precursors. Again, the underlying reason for studying malaria lipids was to discover whether there might be an Achilles' heel for parasite lipid synthesis (different from the red cell or other tissues in the body) that could be exploited for drug development. In 2008, Karine Le Roch and co-workers studied the action of the bizthiazolium drug, T4, a choline analog, on parasite lipid metabolism [400]. T4 was active against both drug-sensitive and drug-resistant lines of *P. falciparum*, and was presumed to target lipid metabolism; however, the mode of action of T4 was unknown. Using microarray chips and techniques similar to those used for studying anti-folic acid drugs, the finding was that T4 inhibited the synthetic pathway of the dominant parasite lipid, phosphatidylcholine from choline. Despite this, "no significant transcriptional changes were detected after addition of T4 despite the drug's effects on metabolism." However, when protein abundance was measured, a decrease in the enzymes necessary for the synthesis of phosphatidylcholine was found.

The studies using chip technology had shown that most *P. falciparum* genes are turned on only once during the 48-hour developmental cycle. Presumably the turning on of the genes occurs at a time when the particular gene product is required. Further, unlike most other cells, malaria parasites have few transcriptional changes when exposed to external stimuli such as antimalarial drugs. This rigidity in transcription, as well as the lack of putative specific transcription factors in the genome sequence of *P. falciparum* [249], has suggested there might be other mechanisms beyond the genome itself that regulate the parasite's life cycle. Using evidence from other eukaryotes, the gene hunters have focused attention on whether gene expression could be altered by changes beyond the genomic sequence itself. The study of changes in gene expression that do not involve alterations in a DNA sequence is called epigenetics (from the Greek "epi" meaning "on" or "in addition to"). Epigenetic changes are not passed down from one generation to the next, but instead occur within a single lifetime. The most commonly identified mechanisms for epigenetic change involve changes in the structure of chromatin.

Recall that in Miescher's work with sperm he found an arginine-rich protein, named protamine, associated with nuclein (DNA). Today it is believed that protamine is critical to DNA stabilization and allows for a denser packing of the DNA in the head of the sperm cell. Unlike sperm, however, most cells lack protamine; instead they contain another kind of protein called histone, first described by Albrecht Kossell in 1884. Histone proteins are rich in the amino acid lysine and they are the principal protein found in chromatin. The histone acts like a spool around which the DNA winds. Without such "spooling" the double helical DNA in chromatin (for example in our own cells) would be a very long attenuated thread some 1.8 m in length; however, by being wrapped around histone, the DNA thread is condensed to 90 mm and, in some organisms (but not malaria parasites), the chromatin may be further coiled and compacted to form the colored staining bodies called chromosomes and described a century ago. Simply put, chromatin is the combination of histone with DNA. The association of DNA with histone protein is not random; instead there are 50 bases of DNA wrapped around a core of eight histone molecules, numbered H2A, H2B, H3 and H4; this repeating unit, as seen with an electron microscope, is a small spherical body, called a nucleosome; a terminal tail of lysines protrudes out of the nucleosome and these are subject to post-translational modifications.

Using sophisticated biochemical methods, the architecture of the nucleosome of *P. falciparum* has been studied during its asexual life cycle. The histones in a nucleosome go through rounds of binding and unbinding to the DNA rather than sliding along the DNA strand [522]. At the ring stage, nucleosomes are removed locally and specific gene transcription occurs; this is followed at the trophozoite stage by disassembly of histones allowing DNA replication and massive transcription of many genes, and finally at the schizont stage the histones reassemble and the DNA is once again maximally packed. The pattern will repeat in the next asexual developmental cycle after merozoites invade new red blood cells. The described changes in the nucleosome structure in *P. falciparum* are much greater than is usually observed in cancer and other cells.

Clearly, histones are not simply packing material for DNA (as was the belief prior to the 1990s) but are critical to the way genes are turned "on" and "off," i.e. regulation of gene activity. In general, the genes that are active (turned "on") have less bound histone and conversely the inactive genes (turned "off") have more bound histone. Post-translational remodeling of chromatin involves principally histones H3 and H4 by enzymes (present in the falciparum genome and identified through sequencing of its genome). Although histone-modifying enzymes occur in the cells of our body, the human counterparts are not identical in their amino acid composition to those of the parasite and, because of this, it has been suggested they could be likely drug targets. Indeed, some years ago, through a natural product-screening program, Merck scientists isolated a cyclic tetrapeptide from a fungus, *Fusarium pallidoroseum*, found on twigs collected in Costa Rica. The peptide, named apicidin, is an antibiotic similar to penicillin and streptomycin; however, in this case, its target was cancer cells, parasites related to malaria and malaria itself. Injected into the body cavity of mice infected with *P. berghei*, it attenuated the numbers of parasites in the blood [20] and, at exceedingly low (nanomolar) concentrations, showed activity against *P. falciparum* cultures. Apicidin is far from the perfect antimalarial: it has poor selectivity for *P. falciparum*-infected red cells over mammalian cells, poor bioavailability, general toxicity and there was significant hemorrhagic toxicity in rats after oral dosing; however, analogs of apicidin have increased potency against *P. falciparum*, suggesting that structural modifications of apicidin could represent potential drugs for malaria therapy [19].

How does apicidin work? The enzymes involved in histone modification are named histone methyl transferase and histone deacetylase, which either adds a methyl (–CH3) group or removes an acetyl (-OCH3) group, respectively, from the lysine "tail" of histone H3. These modifications influence how tightly or loosely the chromatin is compacted and consequently gene expression is affected. Acetylation of lysine residues weakens the electrostatic interaction between the histone and the negatively charged phosphate backbone of the DNA. Acetylation of histones H3 and H4 is correlated with chromatin or active chromatin, allowing various transcription factors access to promoters of target genes. By contrast, when the acetyl group is removed, the chromatin condenses, gene inactivation occurs and transcription is turned off. Apicidin inhibits the deacetylase enzyme. When this occurs, there is a concomitant increase in histone acetylation, particularly of histones H3 and H4, and a demethylation of histone H4. This disrupts expression of 50% of the parasite's genes and leads to a collapse of the cascade of gene expression. (This effect differs from cancer cells where apicidin inhibits only a limited number of genes, usually 10%.)

Evasion

On 29 September 1890, a farmer, feverish and near death, was brought to Santo Spirito Hospital in Rome where he was examined by the attending physicians Amico Bignami (1862–1929) and Giuseppe Bastianelli (1862–1959). At the beginning of the fever paroxysm they took a drop of blood from the tip of the farmer's finger, smeared it out on a microscope slide and stained it. When the smear was examined, there were crescents and small ring-stage parasites. Two days or so after the ring forms had been seen, they began to grow, and then these developing forms disappeared from the circulation. Where did they go? Looking at samples of tissues, adult forms of the parasite were found, forms that radically altered the red blood cell. These altered cells, trapped in the small blood vessels deep in the body, were clumped together, sticking to the walls of the vessels, blocking and slowing the movement of the blood. All this was occurring while, inside the red blood cell, the malarial parasite was doing what parasites do so well — reproducing. They called the disappearance of the more mature parasites from the bloodstream (and thus not being present in a drop of blood taken from the patient's finger) sequestration, and noted that the phenomenon was a signal characteristic of *P. falciparum* infections [62].

What these Italian clinicians discovered more than a century ago was that red blood cells containing the ever-enlarging parasite could accumulate in large numbers to clog the blood vessels, depriving that part of the body of oxygen, and, as was the case with their patient, when the congestion of the vessels was massive enough in the brain, i.e. cerebral malaria, there was coma and death. It is this special propensity of *P. falciparum*-infected red cells to develop and deprive the inner organs of the body that makes it the world's greatest killer.

Although Bignami and Bastianelli's observation explained why the larger, growing asexual stages of the parasite were absent from a drop of blood taken from the fingertip, whereas the young, non-pigmented ring-stage parasites circulated freely in the bloodstream, it did not reveal the mechanisms for such preferential binding. Almost a century later, when more powerful (electron) microscopes — those able to magnify objects 100,000 times or more — became available, the differences on the surface of the

P. falciparum-infected red blood cell would be seen, providing clues to the reason for their stickiness. Examination with the electron microscope showed that red blood cells containing mature asexual stage parasites lose their normal biconcave disc shape, become distorted and covered with minute pimples called "knobs" [9, 274], whereas red cells infected with ring-stage parasites retain a smooth, pimple-free surface, i.e. they are knobless.

The knobs, first found in Trager's Rockefeller University laboratory using blood samples from patients from a malaria-endemic area [685], were later seen in laboratory-bred *Aotus* monkeys infected with *P. falciparum* [454]. In the monkey, it could be seen that there was attachment to the cells lining the small blood vessels via the knobs. Additionally, when *P. falciparum*-infected red blood cells were exposed to "immune serum" and only the knobs were recognized [397], it became clear that the knobs were "non-self," that is, they were antigenically distinct. These early "picture studies" led to a search for the "non-self" antigens that were the "glue" of the knob. Moreover, there was a stated practical goal: when the "glue" was identified and characterized it might be the basis for a vaccine to prevent sequestration and cerebral malaria.

Identification of the surface antigens of *P. falciparum* can be traced to studies carried out by Monroe Eaton in the 1930s. Eaton showed that serum from monkeys infected with *P. knowlesi* could clump (agglutinate) schizont-infected red cells. This schizont-infected cell agglutination reaction (SICA), was a clear indication that the surface of the malaria-infected red cell had been altered. As the SICA reaction was species- and stage-specific, it implied (but did not prove) that the reaction was due to the presence of a parasite antigen on the surface of the red blood cell. Eaton went on to show that, over time, the expression of SICA changed and, with each relapse, the newly appearing parasitized red blood cells were new variants, each having a different antigen. Thirty years later when Neil Brown and Ivor Brown at NIMR explored this phenomenon of antigenic variation, they were able to show that relapse parasites from a single isolate differed in their SICA antigens [91]. Using a pure-bred line of *P. knowlesi* (derived from a single infected red cell and hence a clone) proved that the parasite had the genetic capacity to express all the antigenic variants. The SICA antigens were polymorphic (literally "many forms") parasite proteins that differed in size and shape.

It was only natural after the identification of the *P. knowlesi* SICA antigens and the availability of laboratory-grown *P. falciparum* to see whether similar polymorphic antigens were present on the surface of *P. falciparum*-infected red blood cells, and whether these played a role in sequestration and immune evasion. By tagging the red blood cell surface with radioactive iodine and precipitating with isolate-specific immune sera, a consistent protein named *P. falciparum* erythrocyte surface protein 1 (PfEMP1) was identified [404]. PfEMP1, like SICA, could be removed from the surface of *P. falciparum*-infected red cells by exposure to low concentrations of the protein-digesting enzyme trypsin and, after "stripping," the infected red cells lost their stickiness. That is, trypsin removed the surface "glue." Sera taken from people immune to one malaria strain precipitated PfEMP1 from the same strain but not from others. These sera also blocked binding to cells from blood vessels in a strain-specific way. The conclusion was inescapable: the parasite made a molecule, PfEMP1, responsible for both blood vessel attachment and antigenic variation, i.e. it was the strain-specific "glue."

It has been speculated that, by living within a red blood cell, the parasite had evolved to find a perfect hiding place as red cells, unlike most other cells, lack a recognition signal (the histocompatibility complex used for typing before tissue transplantation) critical to the immune system for removal of foreign materials, including parasites. This in turn, the hypothesis goes on, may have led to an "impasse…because unchecked parasite proliferation may have killed the host before it could transmit itself to a mosquito" [583]. To counteract unbridled parasite multiplication, the infected red cells express a surface molecule (or molecules), recognized by the immune system; however, to prevent its complete elimination from the body, the parasite sequentially expresses a variable protein on the outer surface of the red blood cell. By switching from one variant protein to a different variant protein the parasite is able to keep one step ahead of the host's immune system with its potential for killing the invading malaria parasite. This survival mechanism allows *P. falciparum* to establish a prolonged chronic infection (without sterilizing immunity) with waves of parasites appearing in the blood that may recur sometimes over several months or years. Each wave of infection represents the rise and fall of a distinct population of parasites expressing a particular antigen. A consequence of antigenic variation is that *P. falciparum* is able to avoid clearance by the human immune system and can be transmitted over a longer period of time by susceptible *Anopheles*. Further, the observed need in humans for many years of exposure to falciparum malaria infections before adults develop clinical immunity may be the result of having to develop an immune response to a wide array of antigenically distinct parasites.

For almost a decade, the PfEMP1 protein itself could not be isolated by standard biochemical methods (presumably because of its low abundance). Indeed, because of this, I boldly (and unwisely) questioned the very existence of PfEMP1 [608]; however, in 1995 the gene for PfEMP1, named var (for variant), was cloned and sequenced independently in three laboratories [49, 634, 656]. Today, thanks to genome sequence analysis, we know that the PfEMP1 proteins are coded for by ~60 var (variant) genes [391]. The var genes are localized primarily to the ends (telomeres) of all 14 chromosomes of *P. falciparum*, are extremely divergent in their DNA sequence and theoretically could code for very large proteins (~200,000–500,000 molecular mass [Mr]). Each var gene encodes a large variable surface-exposed region and a smaller conserved region that encodes the portion of the protein anchored inside the red cell membrane. Each gene also includes two transcriptionally active promoters, the first giving rise to mRNA and the second expressing non-coding RNAs of unknown function.

Since its discovery, PfEMP1 has been considered an attractive vaccine candidate. Indeed, PfEMP1 has been an especially tantalizing target because it is exposed on the surface of the infected red blood cell for ~20 hours and correlations have been found between antibody recognition of it and clinical protection against severe malaria. However, it remains a mystery exactly what role PfEMP1 plays in limiting an infection. Despite this, some workers have hypothesized that vaccination with PfEMP1 will promote an early build-up of immunity in children to the most virulent parasites and significantly reduce malaria-related deaths. However, based on vaccination studies using recombinant proteins and DNAs, and their inability to protect, the persistent argument for inclusion of PfEMP1 as a component of a protective vaccine rings hollow. Regrettably, the selection of the "correct" antigen that mediates sequestration in cerebral malaria has not been identified despite decades of intensive effort.

So why the continued focus on gene expression? The gene hunters believe it is critical to discover how gene expression is altered during infection in order to better understand how this produces the pathology of the disease. In consequence, most recent research on PfEMP1 has focused on the mechanisms of switching of antigens [583]. This, it is contended, would allow for a better appreciation of the way immune evasion is achieved, the manner by which PfEMP1 is trafficked from its place of synthesis within the parasite to the outer surface of the infected red cell [421] and perhaps, they argue, when the molecular events in gene expression are fully understood, it may be possible to develop truly novel therapeutic approaches that will turn off specific genes that allow for immune evasion.

By the parasite switching expression between different var genes, *P. falciparum* is able to avoid antibody recognition. Only one var gene is actively expressed at a particular time and the other var genes remain silent. Silencing of the var genes occurs when the ends of chromosomes, the telomeres, localize to the periphery of the nucleus. Here, in what is called a repressive center, the var genes are silenced by the addition of three methyl residues to a H3 histone lysine; in contrast, activation requires location to another area within the nucleus (and outside the repressive center where the silent var genes reside) where the chromatin is open for transcription, and there is replacement of the methylated H3 histone by acetyl residues [198, 226, 415]. Despite this, for var expression to occur, it takes more than localization within the nucleus. Specific transcription factors, including a TATA-box-binding protein and DNA elements that display the proper timing of expression, have been found, along with specific nuclear factor or factors that could be involved in transcriptional activation, as well as the ways in which the mutually exclusive expression of var genes is achieved. "However, the nature or identity of the ...factor...remains to be determined" [333].

Despite years of study of antigenic variation and var gene expression, there still remain questions. Neither the mechanisms for localization within the nucleus periphery for silencing the var genes, nor how the activation site is chosen is known. It is not clear whether there is an intrinsic switch rate and order of var gene expression adjusted to the time it takes for an immune response to be mounted or if the parasite actually senses changes in the environment and then alters its switch rate. Is there a predictable order of gene switching or is the process entirely random? It appears that the parasite in a single infected red cell is able to count var gene expression so that there is a coordinated turning "on" of different variants so repetition does not occur. This capacity to remember the expression of a particular var gene for successive parasite generations seems to result from a particular "marking" of the histone by trimethylation — a phenomenon called epigenetic memory — but the mechanism for this remains undefined. What proteins or protein complexes act to "tether" var genes to the periphery of the nucleus? Is localization at a particular site a requirement for regulation? The list of questions goes on.

Var genes, however, are not the only genes that contribute to antigenic variation in *P. falciparum*. The *P. falciparum* genome project identified repetitive interspersed family (rif) and sub-telomeric open reading frame (stevor) genes, adjacent to the var genes, that also appear to have a role in immune evasion. The products of these genes may act to camouflage the merozoite proteins needed for invasion and, as such, "could protect the Achilles' heel of merozoites against immune attack by masking critical invasion molecules" [583].

■ Chapter 5 ■

The Importance of Import

A malaria-infected red blood cell consists of several compartments much like those seen with nesting wooden Russian dolls where each doll, smaller than the previous, fits inside another. In the malaria-infected red cell, the larger compartments are separated from each other by a series of membranes. The outermost is the red cell membrane, beneath this is the parasite's parasitophorous vacuolar membrane (PVM), inner to this is the parasite plasma membrane, and finally there are the smaller membrane-lined compartments of the various organelles including the large food (digestive) vacuole, the single mitochondrion bounded by a double membrane, the vestigial plastid, the apicoplast, bounded by four membranes, and a complex of sheets of membranes studded with ribosomes, the endoplasmic reticulum. Nutrients as well as vitamins and inorganic ions necessary for growth and reproduction of the malaria parasite must be obtained from the host and, for this to occur, these critical substances have to pass across some of these membranes; in addition, for effective treatment, antimalarial drugs must also pass into the parasite across these membrane-lined compartments to hit their intended target (see Chapter 14). Conversely, the waste products of parasite metabolism must move from their place of origin outward across membranes and into the bloodstream, lest these accumulate to toxic levels.

"Molecular movers" — specific integral membrane proteins — called transporters — allow for the flux of materials into and out of the parasite, as well as between the organelles. Although transporters may play a critical role in the life of cells, only 24 of the

FDA-approved oral drugs for humans have transporters as their target. Despite this, it has been argued, a fuller understanding of the transport mechanisms could serve as targets for the design of novel drugs that would block import and/or export, and either starve the *Plasmodium* or let it perish immersed in its own metabolic wastes.

The trafficking of materials into malaria parasites was first recognized in 1948 with a report [495] that provided "inferential evidence of altered permeability...in malaria;" however, it would take another 20 years before the first study on the mechanism of import of substances into malaria-infected red blood cells would be conducted [607]. The impetus for my work on nutrient import was to test a hypothesis formulated by Clark P. Read of Rice University. Familiarity with the work of Read stemmed from several summers (beginning in 1964) when I was an instructor in the Invertebrate Zoology course at the Marine Biological Laboratory (MBL), Woods Hole, and he was the course director. At the time, Read was one of the "giants" of parasitology and he and his students were actively testing his hypothesis that tapeworms were unable to control their import of amino acids and so a proper balance of amino acids, provided in the gut of the host, enabled the parasite to synthesize its proteins. Indeed, he postulated, if the array of amino acids provided were not balanced the parasite would be unable to grow. In 1967, stimulated by Read's ideas, and being young and naïve, my laboratory set out to examine the applicability of Read's hypothesis for malaria parasites.

As a malaria-infected red cell contains compartments separated by membrane interfaces, the plasma membrane of the red cell, the PVM and the plasma membrane of the parasite, studying the import of materials into the parasite cytoplasm is not straightforward as would be the case with the surface of a tapeworm. In blood-stage malaria parasites, materials for entry have to traverse these membranes, each of which may contain multiple "molecular movers" for a particular substance. To complicate matters further, when we (and others) embarked on such studies of membrane transport we were naïve and unaware of the complexities of the molecules that might be encountered. Indeed, at the time, we were unappreciative of the technical difficulties as well as possible artifacts that might result from using parasites removed from the red blood cell, so-called "free" parasites, as well as changes from the natural state of growth induced by the use of heavy infections obtained from unnatural hosts or by using parasites grown outside the body in glass vessels, i.e. Petri dishes.

To those uninitiated with transport across membranes, a word about it may be in order. Transport is mediated by proteins that allow for the translocation of molecules and ions across the membrane. There are two classes of transporters — channels that are essentially gated water-filled pores and carriers (also known as permeases) that bind solutes and undergo a conformational change to move the substance across the membrane. Channels and carriers differ from one another in their functional characteristics as well as in their mechanism of action. A channel, when open, typically shows high throughput by means of a diffusion pathway and the substrate moves down the electrochemical gradient. Carriers, like enzymes, have specific substrate-binding sites, do not provide an open diffusion pathway and, for the substrate to cross the membrane, the carrier must undergo a conformational change; as a result, there is usually a much lower throughput rate than with channels. Carrier proteins are of three types: facilitative, which allow transport down an electrochemical gradient; secondary active carriers, which use energy derived from the electrochemical gradients of H^+ and Na^+ to drive other solutes against their own electrochemical gradient; and primary active carriers, which use chemical energy (predominantly ATP) to drive transport. Channels and carriers may provide delivery routes for drugs that target essential parasite mechanisms and, because the carrier can be susceptible to inhibition by pharmacologic agents, these drugs may assist in discrimination between the different types of transporters.

As noted above [596], our earliest studies with *P. lophurae* were designed to examine the validity of Read's hypothesis as well as to answer the following question: by what mechanisms do malaria-infected red cells and malaria parasites accumulate amino acids? Five of the 10 amino acids (alanine, leucine, histidine, methionine, and cysteine) that we tested entered the uninfected duck red cell by a saturable, non-uphill transport system; however, these entered infected cells *via* a mechanism that showed diffusion kinetics. When presented with a balanced or an unbalanced array of amino acids, there was no difference in parasite protein synthesis. Therefore, we found no evidence to support Read's hypothesis; however, what we did discover was that the *Plasmodium* was able to change the permeability properties of its host red blood cell. Moreover, the parasite appeared to have induced a differential permeability in the red blood cell for some but not all amino acids. This phenomenon, which we described as "leakiness" of malaria-infected red cells, would later be attributed to a transport pathway termed the new permeability pathway (NPP) (see below).

A decade after the bird malaria experiments were conducted, and with the availability of *P. falciparum* in continuous *in vitro* culture, it became possible to study the transport properties of a human malaria. Beginning in 1982, a group at the Hebrew University in Israel, led by Hagai Ginsburg and Ioav Cabantchik, reported that the permeability of *P. falciparum*-infected red cells was increased to a range of amino acids including tryptophan and isoleucine [389]. (Isoleucine and tryptophan are essential amino acids both for us and for malaria parasites.) Elford *et al.* [205] found no effect on isoleucine transport when human red cells were infected with *P. falciparum*; however, a recent re-evaluation [433] has shown that isoleucine uptake by uninfected human red cells was rapid and, upon infection with *P. falciparum*, there was a five-fold increase in the rate of influx; the parasite-induced transport component was wholly blocked by the inhibitor furosemide, consistent with isoleucine uptake being *via* an NPP (see below). Indeed, it was claimed that the NPP served as the major route for the uptake of isoleucine into *P. falciparum*-infected red cells and presumably for other species of malaria as well, although for many of these no studies have been carried out.

It takes energy in the form of the high-energy compound ATP, produced (from adenosine monophosphate [AMP]) as a result of metabolizing carbohydrates or fats, for proteins to be fabricated by a cell. Malaria parasites do not have an energy reserve such as glycogen (as do we in our liver) and therefore they are dependent on a continuing supply of glucose to meet their energy needs. Blood serves as a readily available and abundant source of glucose and it is rapidly consumed by malaria parasites; in some instances, the rate of consumption has been calculated to be as much as 100 times greater than that of the uninfected red cell. Herman *et al.* [308], using the chicken malaria *P. gallinaceum*, postulated that the rate of uptake of glucose into an uninfected red blood cell could not support the rate of glucose consumption by infected red cells and, to permit increased entry, malaria parasites would either have to change the permeability of the red blood cell, alter the metabolic pathway by removing an inhibitor or enhance the movement of glucose across the red cell membrane. Using the biochemical techniques available at the time, it was not possible for Herman *et al.* to discriminate between glucose transport and glucose metabolism. However, when we embarked on our studies, we were fortunate in that radioactive tracers were readily available. Using radioactive glucose as well as an analog, carbon-14-labeled methyl glucose that the parasites could not metabolize, we found increased entry to be due to the infected red cell becoming "leaky" (as had been the case with some amino acids) with the rate of entry being directly related to parasite size [604]. The larger the parasite the faster the glucose entered the infected red cell. Although the mechanism for the induced change was unclear to us, it was obvious that the increased rate of glucose uptake was separate from its metabolism. Our work was confirmed [327, 474] for *P. berghei*-parasitized cells and glucose uptake was described to be "diffusion-like." Glucose entry into parasites removed from the red blood cell ("free" *P. lophurae*) differed from that of the uninfected and infected duckling red cell and the "free" parasites differed from the host red cell in the number and nature of specific glucose transporters, i.e. there were plasmodial transport proteins.

Later, with the availability of *in vitro*-grown *P. falciparum*, glucose uptake was shown to be a saturable, equilibrative process attributed primarily to a membrane transporter for six-carbon (hexose) sugars, specifically fructose and glucose. Studies of the *P. falciparum* genome have identified well over 100 known and putative transporter sequences termed the *Plasmodium* "permeome" and some of these may have therapeutic potential. One of these, the hexose transporter named PfHT, was localized (using antibody) to the surface of the parasite plasma membrane [351], not the red cell membrane. The *P. falciparum* genome sequence revealed that PfHT is a single-copy gene located on chromosome 2; the gene for PfHT has been cloned and sequenced. To prove its function, the gene was injected into the eggs of the clawed toad, *Xenopus*, where it was expressed and able to function in the transport of both glucose and fructose, unlike the mammalian glucose transporters GLUT1 and GLUT5, which transport glucose and fructose, respectively. The observation that 3-O-methyl glucose strongly inhibited the uptake of glucose by PfHT1 suggested that other 3-O-derivatives might act as inhibitors [630]. Indeed, the compound 3361 [3-O-(undec-10-en)-yl)-D-glucose] was found to be a selective inhibitor of PfHT and a very poor inhibitor of human GLUT1. The addition of this compound to *in vitro*-grown *P. falciparum*-infected red cells resulted in a rapid decline in levels of ATP. As a consequence, the *in vitro* growth of *P. falciparum* was blocked and parasites were killed with an IC_{50} of 16 μM [351, 571]. Twice daily treatment of mice infected with *P. berghei* with 25 mg/kg body weight of compound 3361 for four days suppressed the increase in blood parasites. Genetic disruption of the hexose transporter gene in this rodent malaria rendered asexual parasites non-viable [631]. In addition, the glucose transporter gene from *P. vivax* has also been cloned and expressed in *Xenopus* eggs, and it too was inhibited by compound 3361 [350]. The mode of action of quinine, used for the treatment of malaria for centuries, remains a mystery. However, based on the quinine-induced inhibition of glucose uptake of the yeast, *Saccharomyces*, it has been postulated that quinine may exert its antimalarial effect by its inhibition of PfHT1 [183].

It is noteworthy that the genome sequence of *P. falciparum* encodes two other potential glucose transporters in addition to PfHT1, although these have not been characterized. The hexose transporter of *P. berghei* can transport galactose and mannose as well as glucose and fructose. Compound 3361 attenuated liver-stage parasite growth and ookinete development. Daily intravenous injections of compound 3361 (2 mg/kg body weight) reduced blood infections by ~45%; however, survival enhancement was modest. These findings suggest that compound 3361 may have low *in vivo* efficacy [74].

Work undertaken in our laboratory in the 1960s established that the parasite required pre-formed purines for the synthesis of DNA and RNA, and that these salvaged purines were supplied by the host. This led quite naturally to our investigation of purine import. Purine acquisition occurs in a series of steps: first, they are transported from the blood plasma across the red cell membrane into the red cell cytoplasm; then, after crossing the PVM, they are imported into the parasite by specific transporters. A graduate student, Susan Tracy, and I found that infected red cells had a high uptake (= transport plus metabolism) of adenosine, inosine and hypoxanthine, and the degree of accumulation was directly related to the size of the parasite [679]. Adenosine diphosphate (ADP), AMP, and ATP were taken up to a very limited degree. Adenosine transport was inhibited by the presence of hypoxanthine and inosine but such mutual inhibition was not seen in uninfected red cells. We suggested that adenosine, which is present in the blood plasma, was converted into inosine shortly before or during uptake, and concluded that the infected red blood cell membrane had a common transport site for adenosine, inosine, and hypoxanthine.

Human red cells express high-affinity nucleoside transporters that mediate the sodium-independent uptake of a broad spectrum of purine and pyrimidine nucleosides. Searches of the *Plasmodium* genome sequences with the sequences of known mammalian nucleoside transporters led to the identification of the gene for a nucleoside (purine) transporter of *P. falciparum*, PfNT1. The gene, on chromosome 14, was cloned and sequenced and, using an antibody to the protein encoded by PfNT1, the transporter was localized specifically to the surface of the parasite plasma membrane. It is expressed during the entire asexual cycle with the highest levels at the late trophozoite and schizont stages. When the gene was expressed in *Xenopus* eggs [111, 536], PfNT1 had a broad specificity for transport of the nucleosides adenosine and inosine, but no affinity for nucleobases [111]. Another study identified a gene identical to PfNT1 (called PfENT1) that was localized to chromosome 13; the gene was cloned, sequenced and expressed in *Xenopus* eggs [504] and had a low affinity for adenosine, adenine, and hypoxanthine [118, 186]. In yet another study that directly measured transport into the *Plasmodium*, it was suggested that the uptake of hypoxanthine and adenine was comprised of two components: rapid equilibration *via* a low-affinity transport mechanism and accumulation *via* a higher affinity process [185]. Quashie *et al.* found additional transporters: a distinct low-affinity adenosine transporter (PfLAAT) and a high-affinity (PfADET1) and low-affinity (PfADET2) adenine transporter. PfNT1 served as a high-affinity transporter for hypoxanthine [535].

The importance of PfNT1 in the economy of the parasite was clearly established using genetically disrupted, i.e. "knockouts" of, *P. falciparum* [204, 535]. If the gene for PfNT1 was "knocked out," it affected only the high-affinity hypoxanthine/nucleoside transporter, whereas the low-affinity transport of adenosine and adenine was unchanged. Further, parasites lacking PfNT1 were unable to use hypoxanthine, inosine or adenosine and *in vitro* growth of *P. falciparum* was impaired but could be restored by provision of high concentrations (50 μM or higher) of these nucleosides in the medium. Additionally, because *P. falciparum*-infected red cells were able to convert adenosine into hypoxanthine and the hypoxanthine was found in the parasite cytoplasm, it suggests that the parasite exploits the enzymes of the host red cell for conversion of adenosine and inosine into hypoxanthine before these are imported across the parasite plasma membrane. Hypoxanthine is the principal purine salvaged by the human malaria *P. falciparum* [204] and this confirms work performed with the bird malaria *P. lophurae* [679, 748] a decade earlier.

The *Plasmodium* sequencing project revealed three additional transporters, PfNT2, PfNT3, and PfNT4. Of these, PfNT2 has the highest homology to PfNT1. Unexpectedly, it localized not to the parasite plasma membrane but to the parasite's endoplasmic reticulum [184]. It is suggested that the endoplasmic reticulum may serve as a purine store. The identification of PfNT1, located on the parasite plasma membrane, which mediates the highly efficient uptake of hypoxanthine (and adenosine) and guanine as well as adenine, suggests that there may be potential to exploit nucleoside transport for chemotherapy [33]. However, as of the time of writing, no suitable inhibitor has been found.

For more than 50 years it has been recognized that erythrocytes infected with the asexual stages of malaria parasites have an altered permeability to a wide variety of low molecular substances [205, 369, 375, 596]. This increase, attributable in the main to the NPP, provides a route for the influx of nutrients and also allows for the efflux of metabolic wastes such as lactic acid (from glycolysis) and excess amino acids (from digestion of hemoglobin) [261]. The NPP also changes the salt environment of the red cell from one that is low Na^+/high K^+ to one that is high Na^+/low K^+ [403, 603, 643]. In *P. falciparum*-infected erythrocytes, glucose permeates through the NPP but the rate is much lower than the rate at which it is transported *via* the red cell glucose transporter, GLUT1. As with glucose, the endogenous nucleoside transporter works rapidly; however, despite this, the flux of nucleosides via the NPP seems not to be crucial for serving the parasite's nucleoside needs.

As the continuous *in vitro* growth of the blood stages of *P. falciparum* requires the presence of several critical exogenous substances (i.e. pantothenate, isoleucine, choline, methionine), it has been contended that these must be acquired in increased amounts from host sources such as the blood plasma or the red cell cytosol. However, it should be noted that for many, perhaps most, of the substances required by the parasite, the endogenous transport activity of the red blood cell is (more than) sufficient to meet the parasite's requirements [373]. Indeed, for falciparum, there are very few nutrients for which this is not the case [60, 433, 570], but for other species of malaria parasites and/or their host cells the situation may be different [370].

The asexual stages of malaria living within the red blood cell ingest and digest up to 75% of the red cell's principal protein, hemoglobin. However, less than 20% of the amino acids derived from hemoglobin are utilized by *P. falciparum* for the synthesis of its proteins. The majority of free amino acids liberated from hemoglobin are exported from the red cell into the blood plasma, whereas the rest (~20–30 mM) remain within the cell.

Not all of the amino acid requirements of *P. falciparum* can be obtained from hemoglobin. Particularly lacking in hemoglobin are methionine and isoleucine, and both of these were shown to be required by some strains of *P. falciparum*. Normal human red blood cells take up isoleucine and methionine *via* a saturable transporter; however, upon infection and maturation of *P. falciparum*, there is a dramatic increase in the rate of influx of these amino acids and this is attributable to the NPP. On entry, isoleucine is taken up by the parasite *via* a high-capacity saturable transporter [433]. There is also a dramatic increase (~15-fold) in the uptake of methionine by the infected red cell, which is again attributable to the furosamide-sensitive NPP [138]. Within the red cell, the parasite takes up methionine *via* a transporter yet to be characterized, but clearly slower than the isoleucine transporter. The isoleucine transporter is able to operate with maximal efficiency when exchanging one neutral amino acid for another because the substrate-loaded transporter undergoes the necessary conformational change more rapidly than the unloaded transporter. This provides the parasite with a mechanism for taking up neutral amino acids that are under-represented in hemoglobin, such as methionine and isoleucine, in exchange for the surplus of amino acids liberated from hemoglobin, i.e. leucine and alanine, which the parasite has to release in order to avoid osmotic overload [138].

If the assumption is correct, that the survival of the parasite requires the upregulation of existing transporters and/or creates new permeation pathways, then compounds that would prevent such acquisition might be effective antimalarials [370, 373, 681]. Regrettably, despite intensive research over the past three decades with in vitro-grown *P. falciparum*, questions remain regarding the number and kind of channels that make up the NPP and whether they originate from parasite-encoded proteins or from modifications of host proteins.

During the growth of *P. falciparum* within the red blood cell there is a fourfold increase in the phospholipid content of the infected red cell and ~45% of this is in the form of phosphatidylcholine synthesized by the parasite. In the uninfected red cell, choline is taken up *via* a saturable carrier, whereas in *P. falciparum*-infected red cells there is an additional (non-saturable) component, i.e. the NPP [60, 405, 644]. (Parasitized red cells from rodents infected with *P. vinckei* [645] and monkeys infected with *P. knowlesi* [17, 715] also show an increase in the rate of influx of choline *via* the saturable component, but precisely how this occurs is still undetermined.) Once choline is inside the infected red cell, it crosses the PVM and is transported across the parasite plasma membrane. Using "free" *P. falciparum*, the choline carrier was localized to the parasite plasma membrane. The transporter has a much higher transport capacity than the NPP [60], is sodium independent and appears to be energized by a proton pump. The effects of pharmacologic agents support the notion of a parasite membrane proton pump, i.e. furosemide was without effect on this in "free" parasites and bafilomycin A1, an inhibitor of V-ATPase proton pumps and the protonophore carbonyl cyanide m-chlorophenylhydrazone (CCCP) reduced choline transport in "free" parasites [60, 405]. Although phospholipid synthesis is sensitive to quaternary ammonium compounds such as T16 and G25, this appears not to be due to a direct inhibition of choline transport. Significantly, choline influx was inhibited competitively by quinine [405].

Pantothenate, a precursor of coenzyme A, is an essential nutrient for the growth of several species of *Plasmodium*. A dietary deficiency of pantothenate in chickens as well as inhibitors of pantothenate reduced the number of blood parasites in chickens infected with *P. gallinaceum* [85] and, when pantothenate was added to *in vitro* cultures of *P. lophurae,* infectivity was demonstrated after eight days *in vitro* and the presence of male gametocytes capable of exflagellation after 16 days. In the absence of pantothenate, both infectivity and exflagellation persisted for only five days (summarized in [680]). In short-term *in vitro* experiments with *P. falciparum,* the pantothenate analogs and pantothenol inhibited the *in vitro* growth of *P. falciparum* [569, 639]. Pantothenate does not enter normal red blood cells; however, it does enter *P. falciparum*-infected red cells *via* the NPP as attested to by its import being sensitive to furosemide and non-saturability [570].

The earliest biochemical studies of malaria parasites were conducted just prior to and during World War II. James Fulton and co-workers at the NIMR in the UK indicated that glycerol could serve as an energy source for malaria parasites, but it would take almost 50 years to show that the pathway for entry of glycerol into the *Plasmodium*-infected red cell was *via* the *P. falciparum* aquaglyceroporin (AQP) [530]. In *P. berghei* there is only a single gene encoding the AQP and the protein shows 62% amino acid homology to that of the falciparum transporter; when the gene was expressed in *Xenopus* eggs, they became permeable to water and glycerol. Using antibody, the AQP was localized to the parasite plasma membrane. In a screen of more than 60 isolates of *P. falciparum,* with six having been in culture for many years and 59 isolated from patients, only two mutations were found. These rare mutants had no effect on AQP function [388]. AQP "knockouts" were viable but grew more slowly, presumably because glycerol, which is used by the parasite for its synthesis of lipids, can also be obtained *via* glycolyis. The proliferation of *P. falciparum* in culture was inhibited by methylglyoxal (IC_{50} 200 µM) and dihydroxyacetone, which enter the infected red cell *via* the AQP transporter [511]. In *P. berghei,* the knockout of PbAQP was viable; however, a total deficiency did not lead to death of the parasite although glycerol uptake was inhibited. Survival of the knockout was enhanced, suggesting that AQP may serve for the uptake of toxic materials rather than it being the basis for the design of novel antimalarials. PfAQP-blocked parasites may survive as do genetically attenuated parasites (GAPs) and in this regard might serve as the basis of a vaccine.

As noted earlier, the observation on the movement of Na^+ into the infected red blood cell was made in the late 1940s [192, 495–497] yet, until recently, little was known about whether this was of any value to the *Plasmodium*. In 2006, it was discovered that the parasite exploits not an H^+-coupled transporter but an Na^+ electrochemical gradient to energize the transporter for inorganic phosphate, an essential nutrient for the malaria parasite. The gene for the transporter was cloned, and the transporter protein localized to the parasite plasma membrane. When the gene was expressed in *Xenopus* eggs it resulted in uptake properties similar to those observed in the parasite [572].

In 1982, Kutner and Ginsburg, working at the Hebrew University in Jeruslaem, began using an amino reactive reagent [101], 4,4'di-isothiocyano-2,2'-dinitrostilbene sulfonate (DIDS). This does not enter normal red cells and it specifically labels surface lysine residues of the anion transporter, band 3 protein; however, Kutner *et al.* and Ginsburg *et al.* found that *P. falciparum*-infected red cells (at the time when ring stages were becoming trophozoites) were highly permeable to this reagent and a variety of otherwise impermeable substances [262, 263, 389, 390]. The authors claimed the increased permeability to be the result of a new pathway. (In fact, the "new pathway" was already known and had been referred to by us as "leakiness" [604].) When the Israeli group discovered that efflux of NBD-taurine (a substrate of the anion transporter, band 3 protein) was very rapid in infected red cells and that this efflux was not blocked by DIDS, they concluded that the increased permeability had a pore-like quality and that band 3 protein was probably not a constituent of the pore [262]. The positively charged pore was estimated to have a radius of 0.7 nm and, as such, facilitated the movement of small negatively charged molecules (anions) and excluded those molecules that were positively charged (cations) [263]. Selectivity by the pore was unchanged with parasite growth [262] and pore development was asynchronous [262, 383]. Other peculiarities were found with the transport of L-tryptophan: in fresh human red cells uptake could be resolved into linear and saturable components but, upon infection or storage of red cells, the linear component was substantially increased. Further, the changes in the permeability of the *P. falciparum*-infected red cells were unaffected by p-chloromercuribenzoate (PCMB) and cytochalasin B, known inhibitors for glucose transport, as well as the anion transport inhibitors 4,4-dinitrostilbene-2,2-disulfonic acid (DNDS) and DIDS, but were inhibited by phloretin, a modifier of the membrane dipole potential shown to block a variety of mediated and non-mediated transport mechanisms. When added to cultures, phloretin also inhibited the *in vitro* growth of *P. falciparum* [261].

It became apparent that, as the parasite began to grow in size within the red blood cell, its metabolic activity was also increased and there was a concomitant increase in the permeability of the red cell. This increase in permeability has been ascribed to the NPP, and its characterization is based on the inhibitory effects of a range of pharmacologic agents, including broad-specificity inhibitors of anion transport as well as quaternary ammonium compounds and a biotin derivative. The NPP has also been studied using the import and export of radioactive compounds, and more recently by electrophysiology ("patch clamp"). It has generally been assumed that these techniques, though quite different from other methods, would provide a similar determination of the character of the NPP and its origin. Unfortunately, this has not been the case.

The NPP, a poorly selective channel for negatively charged substances (anions) with many characteristics of volume-sensitive anion channels, has been postulated to be a single channel type [371]. However, a recent analysis using radioactive tracers suggests that the NPP may consist of two kinds of channel: one that is present in low copy number (~4/cell) and selective for charge and size; the other more abundant (~400 copies/cell), allowing the movement of anions (Cl^- and lactate) and nucleosides (adenosine) [264].

The evidence from electrophysiology for the NPP is contradictory and some of the discordant conclusions certainly could stem from differences in techniques and methods used by the different groups involved in such studies, i.e. degree and type of cell washing, presence or absence of serum, cell–cell contact, age of cells, ionic composition of the medium, nature of the substance being studied, numbers of parasites and cell lysis [172, 642, 667]. Desai and co-workers at the NIH contend that the NPP consists of ~1,000 parasite-encoded anion-selective channels that are of a single type [11, 175]. They have named the channel the *Plasmodium* surface anion channel (PSAC) and claim that it is specifically inhibited by dantrolene [359, 412]. Despite this, until recently the *P. falciparum* genome project found no suitable candidates. Hill *et al.* [311] suggested that, because resistance to blasticidin S was correlated with a reduction in permeability to multiple solutes, was tightly linked to defects in PSAC activity and underwent rapid reversion, there was evidence for "changes in the expression of genes, one or more mutation hotspots or transcriptional switches." Following on from this, Desai and co-workers used a high-throughput screen to identify PSAC inhibitors. Screening of two strains of *P. falciparum* (Dd2 and HB3) with a 50,000 compound chemical library yielded compounds that inhibited PSAC. One compound, ISPA-28, inhibited PSAC activity with an 800-fold higher affinity in the Dd2 strain when compared with the HB3 strain. HB3 and Dd2, used previously in a genetic cross, yielded 35 recombinant progeny that were used to evaluate the effect of IPSA-28. Genetic mapping of these, using quantitative trait loci, identified a region on chromosome 3 with 42 candidate genes. When 14 of these were assayed for ISPA-28 inhibition, two genes were found to be inhibited: clag 3.1 and clag 3.2. This was surprising as these clag (cytoadherence-linked antigen) genes had previously been implicated in red cell adhesion and not transport. Protease treatment of intact infected red cells cleaved the protein product of clag3, suggesting that PSAC was exposed on the surface of the parasitized red cell. Antibody to a fragment of the clag product localized to the cytoplasm, the red cell membrane and the rhoptries [477]. Although this work provides evidence for PSAC being an anion transporter, questions remain [201]. How is it that clag is not homologous to known anion channel gene families? Does the product of clag act as an auxiliary channel complex with either a parasite-encoded protein or an endogenous red cell channel? How does the PSAC traffic to the red cell membrane? How does the PSAC relate to the NPP?

However, experiments conducted by several groups support a red blood cell membrane channel rather than a parasite one. The Thomas group in Roscoff, France, using the patch-clamp technique has shown three channels with at least two being intrinsic to the red cell membrane [79]. The channels could be activated in uninfected red cells by a combination of protein kinase A and ATP [172, 200] or by membrane stretch [200]. The Huber-Lang group in Tübingen, Germany, using similar patch-clamp techniques as Desai and co-workers, as well as inhibitors, claimed that there are two channels and these could be induced by oxidation of the red blood cell [194, 334]. Another study that did not support the hypothesis of Desai and co-workers was that of Verloo et al. [712], where it was shown that P. falciparum developed normally in red blood cells from patients with cystic fibrosis in which an inwardly rectifying conductance was lacking and yet such a conductance (as measured by patch clamping of ring-infected red blood cells) is activated long before the appearance of the NPP.

In a further attempt to determine whether the NPP is of red cell or parasite origin, Baumeister et al. [50] studied the effects of proteases. Pre-treatment of P. falciparum-infected red cells with chymotrypsin (but not trypsin) abolished NPP activity as well as the increased anion conductance. Although it would appear that band 3 protein (which is cleaved by chymotrypsin) might be implicated in the NPP, these investigators discounted its involvement for the following reasons: (1) pre-treatment of intact infected red cells with chymotrypsin cleaved band 3 within two minutes yet there was no effect on NPP activity until >10 minutes and (2) when chymotrypsin treatment was carried out at 22°C, band 3 was cleaved as it was at 37°C; however, at the lower temperature, the NPP- and the parasite-induced conductance were unaffected. Further, the protein synthesis inhibitor cycloheximide and secretory blocker brefeldin inhibited the appearance of NPP in the hours following chymotrypsin treatment, indicating that parasite viability and secretion were involved in NPP formation. Possibly, they suggest, the NPP channels are intrinsic to the membrane of the red blood cell and are either activated and/or maintained by parasite-derived, chymotrypsin-sensitive auxiliary proteins (PSAC?) exposed at the erythrocyte surface, or they become chymotrypsin-sensitive only after activation by parasite proteins. These experiments tend to provide tantalizing evidence for the involvement of a parasite-encoded protein in the formation of the NPP; however, it has also been shown that the induced transport pathways that were chymotrypsin-sensitive could be restored in uninfected red cells when the enzyme was removed and when activation was repeated [51]. The suggestion from this work is that chymotrypsin treatment inactivates only a sub-set of the transporter proteins and these silent channels can be activated possibly by kinases secreted by the parasite [449]; if true, this would explain the effects of inhibitors of protein secretion and parasiticidal drugs on the reappearance of the NPP in infected red blood cells [51]. It is noteworthy that, on exposure to oxidative stress, uninfected red cells show properties similar to the NPP [334] and, were the NPP to be formed in part by modified band 3 protein, only a small number of molecules would need be involved; however, their recruitment would still require a growing parasite to incur the necessary oxidant stress to produce it [57] and for there to be export of kinases.

The report of the complete genome sequence of *P. falciparum* noted that the parasite had a limited repertoire of membrane transporters, particularly for the uptake of organic nutrients [249] and no sodium, potassium or chloride ion channels. However, Kirk and co-workers have attempted to describe the full range of channels and transporters encoded by the *P. falciparum* genome sequence [372, 432]. In this approach, more than 100 putative transport proteins were identified. For the majority, the import specificity and membrane location remain unknown. Despite this, Kirk and Saliba [373] note that this base can be valuable for "as we learn more about the pathways involved in the uptake and metabolism of key nutrients by the parasite it may be possible to identify/develop combinations of inhibitors that attack these pathways at multiple sites. The rational design of such combinations will be dependent on our having a detailed knowledge of the import mechanisms."

Chapter 6

Remodeling the Genome's Home

Unlike all the other cells in our body, a red blood cell lacks a nucleus and any other organelles. It has been described as a "simple sack" containing a syrup-like solution of hemoglobin, specialized for carrying oxygen to and carbon dioxide from the tissues. Thanks to the durability of its membrane skeleton, it is able to make a 1,000-mile journey a million or more times through our bloodstream without repair. The red cell's flexible skeleton, composed of a two-dimensional meshwork of the protein spectrin, is connected to the overlying plasma membrane by a series of vertical "struts" that includes the proteins ankyrin, band 3 protein, rhesus factor antigen and glycophorin but, as *P. falciparum* grows and enlarges, the red cell's ability to deform itself, thereby enabling it to squeeze through the smallest confines of the capillaries without tearing itself apart, is so altered that it becomes both rigid and adhesive. Adhesiveness of the *P. falciparum*-infected red cells results in sequestration of these red cells in the smaller blood vessels leading to life-threatening complications; if the sequestered infected cells block the vessels in the brain it can lead to coma and death (cerebral malaria) and when blockage occurs in the placenta low birth weight and death of the offspring may occur (placental malaria). The increased "stickiness" of the *P. falciparum*-infected red cell coincides with its loss of biconcave shape and the appearance of pimple-like distortions, called knobs, on its outer surface. With an electron microscope, a knob can be seen to consist of an elevated plasma membrane beneath which there is an electron-dense plaque. Knobs contain the sticky "glue" responsible for sequestration, which in turn prevents clearance of the *P. falciparum*-infected cells by the spleen.

As mentioned earlier (see p. 44), merozoite invasion of a red blood cell occurs with the invagination of its plasma membrane. As a consequence, the parasite is enclosed in a membrane-lined PV. The PVM, derived from the red blood cell membrane, is devoid of red cell membrane proteins, and initially its lipid composition is similar to that of the host cell during invasion [724]. Over time, and as the PVM enlarges to accommodate the growing parasite, it is modified though the insertion of parasite proteins, and the PVM puts out membranous branches. The red blood cell — the home for the parasite's genome — is itself remodeled by the activities of the growing parasite. This remodeling depends on the parasite's ability to export hundreds of proteins [166]. By understanding the molecular mechanisms involved in protein export into the red blood cell, the malaria gene hunters argue, it may be possible to develop novel therapeutic approaches to prevent the unfavorable consequences of a malaria infection.

Because red blood cells lack a nucleus and are transcriptionally and translationally inactive, they lack the protein-secreting machinery present in other cells. As a consequence, the malaria parasite cannot rely on the red cell to solve the protein export problem and instead it must establish its own secretion machinery. In 1997, Kasturi Haldar's laboratory at Stanford University studied the movement of fluorescently labeled molecules into the *P. falciparum*-infected red cell and a unique membrane system, the TVN, in the red cell cytoplasm was discovered. Using an electron microscope, the TVN is seen to consist of interconnected tubules and vesicles that grow out from the PVM during asexual development. The TVN is involved in the recruitment of lipids and proteins that make up the ever-enlarging PVM. The import of red cell raft proteins was found to be restricted when tubule development was arrested by blocking the enzyme sphingomyelin synthase. Indeed, using a threo-phospholipid analog, 1-phenyl-2-palmitoylamino-3-morpholino-1-propanol (PPMP), tubule formation was retarded, raft accumulation decreased and nutrient uptake blocked [399]. When the enlargement of the TVN was blocked by exposure to PPMP, the parasite was not killed outright but instead development was impaired. The block was, however, reversible by washing out the drug. Fifteen years after its discovery it was possible, using microarray chips after perturbation of the parasite by PPMP, to identify the role of a hypothetical gene necessary for the function of the TVN. The gene codes for an exported protein found in a novel compartment — a vesicle — derived from the red cell membrane and critical to the assembly and maintenance of the TVN [661]. The protein, named erythrocyte vesicle protein 1, is required for growth of the *Plasmodium* and, as, it has been suggested that it could be a suitable therapeutic target.

In addition to the TVN, there is another membrane system in the *P. falciparum*-infected red cell, named Maurer's clefts after its discoverer Georg Maurer. In 1902, decades before electron microscopes were invented, Maurer used light microscopy and Giemsa staining of *P. falciparum*-infected red cells to describe dots in the red cell cytoplasm. These structures are a feature of red cells infected with mature stages of *P. falciparum* [284] and, half a century later, when these "dots" were magnified 100,000 times with the transmission electron microscope, they were seen to be membranous sacs with a translucent lumen and an electron-dense coat [28, 381, 396, 685]. Under the electron microscope, Maurer's clefts appear twisted and branched and, near the PVM, they are tethered to it by a stalk. With parasite growth the clefts approach the inner surface of the red cell membrane where they are also tethered [292, 670, 671]. Recent experiments make it clear that "Maurer's clefts bud from the PVM and move within the cytoplasm of the red cell before taking up residence at the cell periphery" [640]. There appears to be no direct connection between the TVN and Maurer's clefts. However, the rarity of finding connections between the TVN and the PVM and Maurer's clefts may be due, as Haldar *et al.* [288] claim, to technical difficulties associated with electron microscopy as well as the undulating nature of the TVN tubules, or more likely that such connections are only transitory.

What role do Maurer's clefts play in the life of *P. falciparum*? As the parasite grows within a red cell it is at a distance from the red cell membrane. Despite this, the parasite is able to export "remodeling proteins." Among the proteins involved in remodeling are PfEMP1 (see p. 80), the knob-associated histidine-rich protein (KAHRP) (a constituent of the knob's electron-dense plaque), rifins and stevor — all implicated in the parasite's ability to evade the immune system. Maurer's clefts function as a highway for the sorting and export of these (and other) proteins from their place of origin in the *Plasmodium*, across the parasite plasma membrane, across the PVM that envelops the parasite, into the red cell cytoplasm, and then to the red cell membrane where they specifically associate with components of the red cell skeleton [291, 292]. The reason for studying the processes involved in remodeling falciparum's home is that, were it possible to block export of these proteins, parasite growth might be limited and the chronic nature of the disease thereby prevented.

The passage of parasite proteins into the red cell cytoplasm is a multi-step process involving protein synthesis in the parasite endoplasmic reticulum, followed by trafficking to the parasite plasma membrane, translocation across the PVM into the red cell cytoplasm, assembly into the Maurer's clefts and finally insertion into the red cell membrane [280]. For parasite proteins to be exported beyond the PVM (and into the red cell cytoplasm) requires selective translocators in the PVM. Recent studies with KAHRP and PfEMP1 (and other proteins) have shown that a short parasite-encoded peptide (consisting of five amino acids) is necessary for export of these proteins across the PVM into the red cell cytoplasm. The motif, termed *Plasmodium* export element (PEXEL) by Marti *et al.* [430] or host targeting by Hiller *et al.* [312] has the consensus amino acid sequence R/xLxE/Q/D with positions 1, 3 and 5 being critical to the export function.

Identification of the PEXEL motif has allowed a search for its presence in the *P. falciparum* genome sequence. The findings predict ~400 proteins (i.e. ~8% of all the parasite's proteins) that theoretically could be exported to the red cell surface where they may be "involved in antigenic and structural modifications of the erythrocyte membrane and cytoplasm...and provide the machinery for protein export to the erythrocyte" [430]. Identification of this signal motif has allowed for the prediction of exported proteins for other malarias including *P. yoelii, P. berghei, P. chabaudi, P. vivax* and *P. knowlesi* as well as *P. falciparum*, but the role of these proteins in malarias other than falciparum remains a mystery.

For the successful transfer of proteins across the PVM, it is necessary for the PEXEL motif to be processed (cleaved) and it was hypothesized that a protein-digesting enzyme, a protease, might be responsible. Cowman and colleagues noted that the human immunodeficiency virus (HIV) has a protease that cleaves proteins with a motif similar to PEXEL. To test whether this was the case with malaria, *P. falciparum*-infected red cells were incubated with known inhibitors of the HIV protease — lopinavir, nelfinavir, ritonavir, sequenavir — currently the mainstay for therapy for patients with AIDS. Cells so treated had the *P. falciparum* PEXEL reduced in amount. Although these drugs did not inhibit parasite growth, the results did suggest that processing might also occur in *P. falciparum* by a similar protease. Indeed, a search of the *P. falciparum* genome sequence found a nucleotide sequence for a unique protease, named plasmepsin V; using antibody to this protease it was localized to the *Plasmodium*'s endoplasmic reticulum. What plasmepsin V does is that, when the PEXEL is cleaved, it exposes an export signal (xE/Q/D) from the protein destined to be exported, this frees the protein from the endoplasmic reticulum and allows the unfolded protein to be deposited in the vacuolar space and to be threaded through a pore (a protein complex named a translocon) in the PVM. Movement of the protein through the pore is powered by ATP. Once across the PVM, the protein, now present as a soluble complex in the red cell cytoplasm [376], diffuses away and is then inserted into the Maurer's cleft membrane, either within or attached to the cytoplasmic face of the cleft; the protein then moves to its final destination, the red cell skeleton, where it becomes anchored at the knob structure. The final step in the delivery to the red cell surface may involve fusion of the protein with the cholesterol-rich microdomains in the red cell membrane [229].

For PfEMP1, its final translocation to the surface of the red cell involves a Maurer's cleft protein, skeleton-binding protein 1 (SBP1). SBP1 is a 48 kDa integral membrane protein that spans the cleft membrane with its N-terminus within the cleft and the C-terminus exposed to the cytoplasm [71]. In certain parasite lines where SBP1 was not expressed (i.e. SBP1 "knockouts"), PfEMP1 was not found on the outer surface of the red cell (and the cells were not adhesive) yet the number of knobs appeared to be the same in the knockout as in the wildtype; the number of Maurer's clefts and their structure were also unaffected. Other proteins associated with Maurer's clefts were trafficked normally to the underside of the red cell membrane. It was hypothesized [150] that the effect of SBP1 on PfEMP1 translocation is indirect and subtle, affecting the morphology of the Maurer's cleft and its movement to the infected red cell surface such that the final translocation step is inefficient. Further, they reason that, as other proteins were unaffected, it appears these proteins have a larger margin of error for binding to the red blood cell skeleton.

Several exported proteins lack a PEXEL motif. The best characterized of these so-called PEXEL-negative proteins are SBP1 [577], Maurer's cleft associated histidine rich protein, MAHRP [640], and the ring-exported proteins, REX 1 and REX 2. It has been suggested that, in the PEXEL-negative exported proteins, there is an N-terminus that is comparable to that found in the processed PEXEL proteins and, because of this, there may be a common trafficking factor-binding site (i.e. a chaperone) that allows the PEXEL-negative proteins to have a convergent pathway with the PEXEL proteins [638].

PfEMP3 and KAHRP, deposited on the cytoplasmic face of the red cell's skeleton, are components of the knob's electron-dense plaque and act as a platform for the presentation of the membrane-embedded adhesion protein PfEMP1. KAHRP binds to the skeletal proteins spectrin and ankyrin, and with its deposition the deformability as well as the adhesiveness of the red cell is reduced. Neither PfEMP3 nor KAHRP, however, were found to be required for PfEMP1 trafficking [265, 291].

A variation on the trafficking of PfEMP1, KAHRP and PfEMP3 is seen with STEVOR. STEVOR proteins are expressed in gametocytes and sporozoites as well as on the outer surface of red cells bearing asexual stages [447]. STEVOR proteins contain two transmembrane domains and a predicted signal sequence. The transmembrane domains are crucial for the targeting of STEVOR to Maurer's clefts and, when this is deleted, it accumulates in the red cell cytoplasm in a soluble form [532]. The function of STEVOR variants, it is claimed, is to contribute to antigenic variation. Consistent with STEVOR playing a role in immune evasion is the observation that the protein is not required for *P. falciparum* survival in culture, and more than 90% of infected red cells isolated from patient samples express STEVOR, whereas fewer than 10% of samples from those maintained in the laboratory express this protein [478].

PfCG2, a non-integral high-molecular weight protein (320–330 kDa) located on the cytosolic face of the PVM, is trafficked to the digestive food vacuole [151]. With the transmission electron microscope, PfCG2 is seen as electron-dense patches along the PVM, the cytoplasm and the digestive food vacuole, and it increases in amount through the trophozoite stage and then declines rapidly in the schizont. As PfCG2 does not have a PEXEL motif, the mechanism for its trafficking to the digestive food vacuole remains an open question. Once inside the food vacuole PfCG2 is subjected to proteolysis by a plasmepsin, and then it becomes associated with the membranous material between the hemozoin crystals; it could serve as a nucleation site for hemozoin pigment formation. Although a precise function for PfCG2 is unknown, its importance is suggested by the fact that attempts to knock out the gene have been unsuccessful.

Defining the molecular mechanisms involved in the regulation of trafficking of malaria parasite proteins had been a black box before the *P. falciparum* genome sequence; however, now it is clear that this is a complex and unique system that could be exploited for the development of antimalarials. For example, because plasmepsin V is responsible for the cleavage of PEXEL and is unique and critical for the export of parasite proteins, its identification possibly provides an important target for novel antimalarials.

■ Chapter 7 ■

Getting on the Inside

Malaria parasites need to live within a cell. This being so, getting inside is a must. As such, the molecular mechanisms of invasion may be the Achilles' heel to be exploited for the development of new therapies. Despite half a century of "invasion research," however, a practical and effective means for interrupting the entry process into red blood cells has not been achieved. Nevertheless, the hope for novel interventions to prevent the parasites from getting inside remains and the *Plasmodium* genome sequences may aid in that regard.

In 1969, the invasion of erythrocytes by merozoites was described using the electron microscope [8, 40, 41]. At the apical end of the *Plasmodium* merozoite and within its cytoplasm is a pair of pear-shaped and ducted structures called rhoptries; the cytoplasm contains more numerous (up to 40) fusiform bodies that converge around the tips of the rhoptry ducts, called micronemes, and spheroidal membranous vesicles, the dense granules. The surface of the merozoite is covered with an electron-dense fibrillar coat. Thirty years ago, in an elegant series of studies using both video [196] and electron microscopy, the dynamics of the complex multi-step process of invasion were described in considerable detail. Following this, there were additional findings critical to understanding the invasion process [39, 529, 674]. In 2011, there was a "super-resolution" of merozoite invasion of the red blood cell [554]. Initial attachment of the merozoite occurs at any point on the merozoite surface. Presumably this involves merozoite surface proteins (MSPs) with the best characterized being MSP-1 (see later).

The merozoite then reorients itself so that the apical end (where the rhoptries and micronemes are located) is juxtaposed to the erythrocyte surface. At the point of contact, the merozoite is irreversibly attached to the surface of the red blood cell and is now committed to "getting inside." An electron-dense area is seen beneath the erythrocyte membrane bilayer [8]. The electron-dense area, termed a "tight junction," forms a circumferential ring around the merozoite. (The components and constituents of the junction are not known.) Active invasion proceeds through a depression in the red blood cell surface — an invasion pit — and the parasite's actomyosin motor drives the merozoite deeper and deeper into the pit. As the tight junction progresses rearwards, the fibrillar coat surroundiung the merozoite is shed, suggesting that protein-digesting enzymes (proteases) are at work. One of these is the micronemal subtilisin-like protein SUB2. During the process of invasion, the rhoptries discharge their contents, the tips of the ducts fuse with each other and the merozoite plasma membrane, and then they collapse.

Crucial to entry into a red blood cell is the timely and sequential secretion of microneme and rhoptry proteins onto the surface of the merozoite. The trigger for this process is when, prior to their release, merozoites are exposed to the low concentrations of potassium ions in the blood plasma. This low-potassium environment triggers a rise in calcium levels within the merozoite from internal stores, i.e. endoplasmic reticulum, and involves the enzyme phospholipase C. Indeed, the *P. falciparum* genome encodes a putative enzyme and an inhibitor of phospholipase C — U73122 — is able to block blood-stage development by inhibiting the rise in calcium and preventing translocation of specific rhoptry proteins (see below). U73122 also inhibits invasion by *P. falciparum* merozoites *in vitro*. (In anticipation of invasion, the rhoptry proteins may already be on the surface of the merozoite at the time of schizont rupture and prior to merozoite release from the infected red cell.) The secretion of the microneme protein apical membrane antigen 1 (AMA-1) onto the merozoite surface (and by its binding to a red cell receptor) restores the basal level of calcium in the merozoite and triggers the release of rhoptry proteins (such as cytoadherence-linked asexual gene [CLAG3.1] and PfRH2b). AMA-1 is a partner to a complex containing rhoptry proteins that are now located at the tight junction and this association has been shown to be critical for merozoite invasion [554]. It has been hypothesized that "identification of the key microneme and rhoptry proteins that play crucial roles in invasion could enable the development of efficacious multi-component blood stage malaria vaccines that target distinct steps in the invasion" [626].

Concomitant with rhoptry–microneme discharge, the tight junction moves from the apical end to the posterior end of the merozoite [363]. In the process, the merozoite's fuzzy coat is shed *via* the action of a micronemal serine protease, SUB2 [293]. As the tight junction moves rearward, it is thought that the molecular linkers (= ligands) mediating the invasion process (see below) are removed by one or more serine proteases of the rhomboid family [156, 481, 482]. With the discharge of the contents of the rhoptries, the red cell membrane pit begins to expand into an ever-enlarging pocket and, when this is completely sealed, it forms an enclosing parasitophorous vacuole membrane (PVM) that separates the merozoite (and subsequent developmental stages) from the red cell cytoplasm. Initially, the PVM phospholipids are derived from the host [724] but, after the dense granules release their contents into the PV and these materials are inserted into the PVM, the vacuole begins enlarging. It should be noted that the protein and lipid contents of the PVM remain largely uncharacterized, but host-derived proteins appear to be absent. However, detergent-resistant raft proteins containing 10 raft proteins, including flotillins, B2 adrenergic receptor, aquaporin, G-proteins and tyrosine kinases, are present [287, 471, 472]. The entire process of a merozoite getting on the inside is quick, taking less than 30 seconds [196]; however, it is a highly organized cascade of molecular events set in motion by the formation of the tight junction [554].

Investigations using the light microscope as well as a visualization of the steps in red blood cell invasion, i.e. merozoite attachment, apical reorientation, junction formation and merozoite enclosure, as seen with the electron microscope, suggested the existence of parasite ligands and red cell receptors. However, to unravel the molecular mechanisms that allow entry to proceed in a stepwise fashion required functional and genomic studies.

In the late 1970s and early 1980s — a time when *P. falciparum* was already being grown continuously in the laboratory — there was mounting evidence that the red blood cell receptor for *P. falciparum* differed from that of *P. vivax* and *P. knowlesi*. Several laboratories reported that a receptor for *P. falciparum* merozoites was sialic acid (N-acetylneuraminic acid [Neu5Ac]) as human erythrocytes treated with neuraminidase or trypsin resisted invasion by *P. falciparum* merozoites and this was unlike the situation with *P. vivax* and *P. knowlesi* where "stripping" by chymotrypsin but not trypsin blocked merozoite invasion. Further, Tn erythrocytes, which lack a glycosyltransferase that results in cells deficient in sialic acid, were also refractory to merozoite invasion, as were En (a–) cells; when S-s-U (–) cells, which were relatively resistant to invasion, were treated with trypsin they became totally resistant to invasion [505, 507]. These studies suggested that a sialic-acid-bearing glycoprotein, glycophorin, was the red cell receptor for the *P. falciparum* merozoite [506]. Glycophorins occur in three main types: A, B and C, with glycophorins B and C representing ~10% and glycophorin A representing ~90% of the total. Indeed, it is estimated that there are two million molecules of glycophorin A on the surface of a human red cell. Because En (a–) cells are deficient in glycophorin A, and S-s-U (–) cells lack glycophorin B, the former were almost totally resistant and the latter partially resistant to merozoite invasion. S-s-U (–) cells are found at a frequency of >1% in West Africa, suggesting a degree of protection for populations having this trait [507, 508].

Shortly after these studies appeared in print there were disturbing reports from different laboratories showing that merozoite invasion could occur with enzyme-stripped or naturally occurring and poorly sialylated red cells. The discrepancies were partly resolved when a young post-doc in the laboratory of Sydney Cohen at Guy's Hospital (London), Graham Mitchell, was invited to visit Miller's NIH laboratory (Bethesda) and given the task of examining the process of invasion into both enzyme-stripped and Tn cells. Starting with *P. falciparum* parasite populations that were uncloned, and so retained some natural diversity, cultures were maintained with only Tn cells added. Selection occurred for parasites able to use a non-sialic-acid-dependent invasion pathway, as could be shown by subsequent comparison with sibling cultures that had been maintained with "normal" red cells, and when both populations were isolated and allowed to invade into stripped and other experimental cell preparations. Clearly, Mitchell had discovered that there was receptor diversity and heterogeneity in at least one step in the merozoite entry pathway [462].

Contemporaneously with the findings of red blood cell receptor heterogeneity, attempts were made to identify the merozoite ligand for glycophorin using classical biochemical methods (reviewed in [282, 283]). When radioactively-labeled merozoites were extracted and allowed to react with immobilized glycophorin A, several proteins, including those of 155, 140, 70 and 35 kDa, were found to bind [513]. As the latter two proteins could be eluted by *N*-acetylgalactosamine (GlcNAc) (previously reported by Jungery et al. [354] to block merozoite invasion), it provided suggestive evidence that they were merozoite ligands. However, when it was found that inhibition by GlcNAc was due to its toxicity rather than receptor blockade and that GlcNAc did not block invasion by free merozoites [282, 331], it was concluded that GlcNAc was not involved. Indeed, the role of the 140, 70 and 35 kDa antigens has never been determined. Although the merozoite soluble proteins of 175, 120, 65 and 46 kDa bound to normal human red cells and these required the presence of sialic acid, except for the 175 kDa protein, the role of these other proteins as putative glycophorin-binding ligands remains suspect [282].

The 175 kDa protein [named erythrocyte-binding antigen (EBA)-175], however, is different. It is a merozoite ligand for glycophorin, first described by Camus and Hadley working at the Walter Reed Army Institute of Research (WRAIR) in 1985 [103]. To identify it, schizont-infected red cells were radioactively tagged, culture supernatants collected and incubated with intact erythrocytes (to preserve the natural environment of the receptor and to avoid the use of detergents), and then the red cells were washed, lysed and precipitated with a serum from an immune Nigerian donor. Of the four antigens recovered, only EBA-175 bound to intact normal human red cells but not to neuraminidase-treated cells, Ena (a–) or Tn cells and it specifically bound to glycophorin A. Later, the assay was modified to avoid washing and immunoprecipitation: erythrocytes with bound proteins were centrifuged through a silicone gel cushion to remove unbound proteins and then the bound proteins were eluted with 2 M NaCl [300]. To add further support for EBA-175 being the falciparum ligand, when antibodies were made against EBA-175, they were able to block the *in vitro* invasion of *P. falciparum* merozoites by 90% [503]. Although glycophorin B contains the same O-linked oligosaccharides as glycophorin A, EBA-175 did not bind to it. Later, the reason for this would be shown to be that the binding of EBA-175 to glycophorin A is not determined solely by sialic acid residues but specific amino acid sequences are also needed [620].

EBA-175 is a micronemal protein encoded by a single-copy gene located on chromosome 7. It is the signaling agent for the secretion of the rhoptry proteins CLAG3 and RH2b [576]. EBA-175 has extracellular *N*- and *C*-terminal cysteine-rich regions in addition to transmembrane and cytoplasmic domains. The two cysteine-rich regions (referred to as II and VI) bear numerous conserved cysteine residues with region II having the red-cell-binding function. Targeted disruption of the gene for EBA-175 in the clone W2mef, whose invasion is sialic acid-dependent, was associated with a switch to a sialic acid-independent pathway of invasion [193, 260, 542]. However, when the EBA-175 gene was disrupted in another cloned line (Dd2/Nm) selected on neuraminidase-treated red cells, the knockouts invaded the treated cells, as did the wildtype. Such findings suggested that *P. falciparum* could invade red cells without EBA-175 except in one parasite clone [255, 257]. Indeed, the *P. falciparum* reticulocyte homology 4 (PfRH4) protein is uniquely upregulated in the Dd2/Nm line when compared with the Dd2 line, and both native PfRH4 and a recombinant 30 kDa protein to a conserved region of PfRH4 (rRH4(30)) bound to neuraminidase-treated erythrocytes. rRH4(30) blocked both the erythrocyte binding of the native PfRH4 and invasion of neuraminidase-treated erythrocytes in the Dd2/Nm line [256]. This finding indicates that PfRH4 is a parasite ligand involved in sialic acid-independent invasion of red cells. As antibodies to rRH4(30) blocked binding of this protein to erythrocytes, but failed to block invasion, it suggests that, although PfRH4 was required for invasion of neuraminidase-treated erythrocytes by the Dd2/Nm line, it is inaccessible for antibody-mediated inhibition of the invasion process.

In field isolates from the Gambia and Brazil, the sialic acid-dependent pathway seems to predominate, whereas in India and Kenya, neuraminidase- and trypsin-sensitive chymotrypsin-resistant receptors (characteristic of EBA-175- and EBA-140-mediated invasion) were rarely used [168].

The *P. falciparum* genome sequencing project identified four analogs (= genes related by duplication within a genome that, during the course of evolution, evolve new functions) of EBA-175 (named EBA-140 [BAEBL], EBA-181 [JESEBL], EBA-165 [PEBL] and EBL-1). These, on the basis of sequence similarity to EBA-175, are encoded by a single-copy gene on chromosome 13, and the proteins have been localized to the micronemes [686]. Using enzyme-treated red cells as well as mutant erythrocytes deficient in surface molecules, the red cell receptors for BAEBL and several of the other analogs were identified. A glycophorin C variant known as Gerbich negative (Ge-) results from a deletion of exon 3 in the glycophorin C gene and occurs at a frequency of 47% in Papua New Guinea; certain falciparum lines invade Ge cells at ~60% of the rate of normal red cells [505]. In general, BAEBL binds to glycophorin C and the binding is sialic acid-dependent as BAEBL failed to interact with neuraminidase-treated red cells. The receptors for BAEBL are an N-linked oligosaccharide or a cluster of *N*- and *O*-linked oligosaccharides; addition of soluble glycophorin C blocked merozoite invasion *in vitro* [438]. It is the region II domain of BAEBL that binds to red cells and mutations in only four amino acids define the five BAEBL variants: valine serine threonine lysine (VSTK), valine serine lysine lysine (VSKK), isoleucine serine lysine lysine (ISKK), isoleucine asparagine lysine lysine (INKK), and isoleucine asparagine arginine glutamic acid (INRE). Three of the BAEBL variants (INRE, ISKK and INKK) lack specificity for glycophorin C, demonstrating that glycophorin C is the receptor for only the VSTK variant [438]. This finding contradicts that of Lobo *et al.* [413], who claimed that INKK was the ligand for glycophorin C. The BAEBL ligand from clone Dd2/Nm binds to glycophorin C between residues 14 and 22. However, in another strain (E12), the binding of BAEBL to red cells was sialic acid-dependent but resistant to trypsin, suggesting that its receptor is not glycophorin A, B or C, and knockouts of BAEBL in the 3D7 and W2mef clones were as invasive as wildtype parasites [420]. Clearly, there are differences between parasite lines, and the binding property between the BAEBL variants is due to a small number of amino acid changes in the region II of BAEBL and "suggests that a single...ligand can function in multiple invasion pathways" [255].

The gene for JESEBL was identified by a search of the *P. falciparum* genome. It is found as a single copy on chromosome 1 [6]. JESEBL is a 190 kDa micronemal protein whose binding was found to be dependent on a trypsin-resistant, chymotrypsin- and neuramindase-sensitive molecule that differs from glycophorin B. In the W2mef line, targeted disruption of the JESEBL gene did not affect invasion, suggesting that it is redundant and not critical to invasion. However, in the 3D7 line, JESEBL appears to be critical for invasion as disruption was lethal. Based on work with EBA-175 and other red cell-binding proteins, there is a strong likelihood that all these merozoite ligands need to be shed during invasion by the action of a rhomboid-like protease [32, 482], perhaps acting to disengage the tight-binding interactions required for binding to the red blood cell.

The fuzzy merozoite surface coat consists largely of glycosylphosphatidylinositol (GPI) anchored proteins. Currently, there are 10 GPI anchored proteins known or predicted to be located on the surface: MSP-1, MSP-2, MSP-4, MSP-5, MSP-8, MSP-10, Pf12,

Pf38, Pf92 and Pf113. Besides the GPI anchor, these proteins have other similarities: they contain cysteine-rich domains that are of potential significance in adherence, including epidermal growth factor-like modules. MSP-1 is the most abundant MSP [317, 318, 322]. In *P. falciparum*, MSP-1 is encoded by a single-copy gene on chromosome 9. Genomic sequence analysis shows that MSP-1 contains conserved regions interspersed with polymorphic regions. MSP-1 undergoes extensive proteolytic processing during schizogony. Indeed, just before merozoite egress, the MSPs are processed by the action of the enzyme SUB1, which is secreted into the PV space [378]. Soon after merozoite release, there is further protease processing of MSP-1 resulting in four fragments, 83, 30, 38 and 42 kDa, which remain as a complex on the merozoite surface with the 42 kDa fragment being attached via the GPI anchor. A final processing step mediated during invasion by the subtilisin-like protease SUB2 cleaves the 42 kDa fragment into 33 and 19 kDa fragments (= MSP-1_{19}) that remain on the merozoite surface during invasion. Full-length MSP-1 has been suggested to bind to red cells in a sialic acid-dependent manner and antibodies directed against MSP-1_{19} block invasion and also prevent proteolytic processing of MSP-1 [66]. It has not been possible to knock out the gene for MSP-1, suggesting that it is essential for invasion and parasite survival. MSP-1 is also found on the merozoite surface of *P. vivax* and its solution structure has been determined using nuclear magnetic resonance (NMR) [30]. As immunization with recombinant MSP-1_{19} protects mice and monkeys against challenge, it has been considered as an attractive blood-stage vaccine candidate.

Although the epidermal growth factor domain of MSP-1_{19} has been suggested to mediate the initial contact of the merozoite to the red cell, it can be altered without affecting invasion [188, 483], suggesting that perhaps the N-terminal regions may be involved in early recognition. It was proposed that MSP-1 mediates initial interaction with the red cell *via* specific regions of band 3 protein (reviewed in [486]). Indeed, peptides containing amino acids 720–761 and 807–826 from band 3 protein blocked *P. falciparum* invasion. Further, these peptides also interacted with two regions of MSP-9 and, when recombinant proteins of the two regions were added to cultures, merozoite invasion was blocked. Based on these findings, the authors hypothesized that, in the sialic acid-dependent pathway, band 3 complements glycophorin, whereas in the sialic acid-independent pathway, band 3 acts alone.

Peripheral proteins belonging to the MSP-3 group, the MSP-7 family, and the serine-rich antigen (SERA) protease family as well as acidic base repeat antigen (ABRA) and Pf41 have also been suggested to serve as merozoite ligands. Although there is an abundance of putative candidates as merozoite ligands, the MSP-7 family members as well as Pf41 are favored in effecting primary contact with the red cell largely because they are strongly associated with the merozoite surface. However, the possibility remains that MSP-3 and MSP-6 also play roles in merozoite invasion but what these are remains undetermined [156].

The apical merozoite antigen (AMA-1), referred to earlier, is a merozoite protein presently under consideration as a candidate for inclusion in a blood-stage vaccine [155]. First identified at the Walter and Eliza Hall Institute using human antibodies purified on a recombinant protein [518], its complete amino acid sequence was deduced from the gene sequence; initially incorrectly localized to the rhoptries [159], more recently it was shown to be in the micronemes [42]. AMA-1 is encoded by a gene located on chromosome 11 [676]. Early studies with the *P. knowlesi* homolog, Pk66, showed that merozoite invasion could be inhibited by a monoclonal antibody to AMA-1 [170, 666]. The gene sequence of AMA-1 has been determined for a number of different isolates of *P. falciparum* and several other *Plasmodium spp.* AMA-1 is an integral membrane protein with an N-terminal signal sequence, and a presumed transmembrane domain towards the C-terminus. AMA-1 contains a highly conserved cytoplasmic domain with an N-terminal ectodomain having 16 cysteine residues forming eight disulfide bonds; the entire ectodomain forms three closely packed structural domains, which include Plasminogen-apple (PAN) domains, known to be involved in protein–protein or protein–carbohydrate interactions [31, 520].

AMA-1 is expressed in the late schizont stage as well as in sporozoites; it is reported to be essential for the invasion of hepatocytes [619] as well as blood cells. PfAMA-1 is synthesized as an 83 kDa precursor from which the N-terminal "prosequence" is cleaved, by an as yet unidentified protease, to produce a 66 kDa micronemal mature form that is translocated to the merozoite surface where it is uniformly distributed. The AMA-1 ectodomain is then proteolytically shed during invasion in a 44–48 kDa form, by the action of the same SUB2 enzyme responsible for shedding MSP-1 and its associated proteins [293, 332]. AMA-1 has been suggested to be responsible for apical reorientation [464]; however, a more likely function is in partnership with the rhoptry protein that is associated with the moving junction that enables the parasite to gain traction to pull itself (via the actomyosin motor) into the cell. Cleavage of the tail of AMA-1 is believed "to be required for the parasite to release the host cell plasma membrane and complete invasion" [157]. This is achieved via proteases called rhomboids. It may also serve as a signal for the parasite to begin asexual division. It has not been possible to knock out the gene for PfAMA-1, suggesting that it is essential for invasion.

A large number of rhoptry proteins/peptides have been identified and some partially purified; however, the precise functions of the majority remain unknown. One of the major molecules within the rhoptries of merozoites is a complex of high molecular weight proteins, the RhopH complex [357]. In *P. falciparum*, the RhopH complex associates with the red cell surface on contact with the merozoite and then distributes to the PVM. Antibodies to the PfRhopH complex were found to inhibit falciparum growth *in vitro* and *in vivo*. RhopH1 proteins may serve to confer some degree of specificity to the roles of the individual complexes [358]. RhopH2 and RhopH3 are associated with Maurer's clefts. Attempts to disrupt (= knock out) the gene pfrhop3 were unsuccessful, suggesting that the complex is essential for parasite growth. The rhoptry-associated membrane antigen (RAMA) is synthesized as a 170 kDa protein in early trophozoites, several hours before rhoptry formation and is transiently localized within the endoplasmic reticulum and Golgi within lipid-rich microdomains. Regions of the Golgi membrane containing RAMA bud to form vesicles that later mature into rhoptries in a process that is inhibited by brefeldin A. RhopH3 and rhoptry associated protein 1(RAP1) are found in close apposition with RAMA. It has been suggested that RAMA is involved in trafficking of these proteins into the rhoptries. In rhoptries, RAMA is processed by a protein-digesting enzyme to give a 60 kDa form that is anchored in the inner face of the rhoptry membrane by means of the GPI anchor. The p60 RAMA form is discharged from the rhoptries of free merozoites and binds to the red blood cell membrane by its C-terminal region. In early ring stages, RAMA is found in association with the PV [673–676].

Proteases, such as the subtilisin-like protease SUB1 and the papain-like enzyme SUB2, and the serine proteases (rhomboids) may be promising targets for drug intervention as proteolytic events have been shown to be critical for successful invasion of the red cell by the merozoite [65, 481]. For example, *in vitro* invasion by falciparum merozoites can be blocked by the serine and cysteine protease inhibitors, chymostatin and leupeptin; moreover, these inhibitors act at different steps in the invasion process. Chymostatin inhibited both invasion and rupture of the schizont-infected red cell, whereas leupeptin inhibited only rupture [281]. Proteases may also affect merozoite release by intensive degradation of the red blood cell skeleton leading to abrupt vesiculation of the red cell membrane to allow merozoite egress [268, 557, 573]. Protein phosphatase 1 has also been implicated in the release of merozoites by modulating the phosphorylation levels of *P. falciparum* skeleton-binding protein 1 (PfSBP1), a transmembrane protein of Maurer's clefts [72]. Proteolytic modification of the red cell has also been shown to be necessary for invasion [86, 87]. Chymotrypsin treatment of red cells resulted in localized degradation of band 3 protein and, with disruption of the red cell membrane, may allow for the insertion of parasite-derived proteins/lipids into the PVM [556]. Significantly, peptides corresponding to the predicted cleavage site inhibited merozoite invasion [556].

Clearly, over the past 30 years, a staggering array of molecules involved in invasion have been identified [156, 243, 676]. In addition, some species such as *P. falciparum* and *P. knowlesi* (unlike *P. vivax*) use multiple pathways to gain entry into the red blood cell. The use of multiple pathways may provide malaria parasites with a "survival advantage when faced with host immune responses or receptor heterogeneity in host populations" [130] as well as a headache for those investigators who are attempting to produce vaccines or drugs to block merozoites from getting on the inside. Indeed, because of the heterogeneity of merozoite ligands, exploiting a specific invasion pathway as a vaccine target may be an impossible dream; however, other mechanisms involved in invasion (i.e. proteolytic processing) could be a suitable target for drug and other therapeutic interventions.

■ Chapter 8 ■

The Great Escape

One spring day in 1956, William Trager removed his jacket, rolled up the sleeves of his white shirt, removed his tie, sat down in front of the microscope, and turned on the microscope lamp. From a duckling previously infected with *P. lophurae*, he took a sample of blood by pricking the leg vein, and carefully placed the drop of blood on a microscope slide that was held in place on a warm stage of the microscope. He rotated the objective lenses over the red liquid, removed his wire-rimmed glasses, and rotated the fine adjustment wheel to bring the specimen into focus. What he saw was thrilling. It reminded him of what it must have been like for Alphonse Laveran 75 years earlier when he discovered that malaria parasites were animalcules and not bacteria. Trager described the infected red cell becoming rounded; the pigment clustered into the residual body and within a few minutes or less the merozoites were ejected with a kind of seething motion, which seemed to scatter them. The red cell did not explode; instead merozoite release was rapid and they exited not as clusters, but spread out [682].

Trager had been at the Rockefeller Institute for his entire career (since 1933), studying various aspects of nutrition of malaria parasites with the hope of defining the parasite's minimal daily requirements (provided by the red blood cell) to understand the *raison d'etre* for *Plasmodium* being parasitic. He was meticulous and self-effacing, a dedicated bench scientist, and an excellent microscopist. At the time, 30 years before he had been able to grow *P. falciparum* in continuous culture, he worked with *P. lophurae*, trying to grow the parasites removed from the red cell or developing within red cells kept in suspension by agitating the flasks containing a rich nutrient broth and gassed with air. Successful growth of *P. lophurae* within cells and outside was achieved, although Trager was never able to equal the growth of the parasite found in the duckling, and he was unable to completely replace the chemical constituents found within the living red cell. Further, no one, not even William Trager himself, who had witnessed merozoite release with his own eyes, understood the means of their escape.

Thirty years after Trager recorded his observations in a bound notebook in his own unmistakable and illegible handwriting, Lyon and Haynes, working at the Walter Reed Army Institute of Research in Washington, DC, provided a clue to the mechanism of merozoite release. At the time, they were not interested in release per se but rather their interest was in the characterization of the merozoite's surface coat and the protein that "may be required for attachment to erythrocytes and appears to be shed during...invasion" [417]. The proteins on the surface of the merozoite could be altered by protein-digesting enzymes (proteases) and "in view of the apparent proteolytic sensitivity of the antigens, we decided to determine if an increased yield of these antigens could be achieved by partially inhibiting the proteases..." [417]. To inhibit the proteases, *P. falciparum*-infected red cells bearing schizonts were incubated with a mixture of inhibitors including leupeptin, chymostatin, pepstatin, and antipain. None of these affected parasite protein synthesis but some did prevent merozoite release. The cysteine proteases, leupeptin and chymostatin, were the most effective in producing merozoite clusters surrounded by the red blood cell membrane.

Later, it was found that leupeptin on its own prevented merozoite release. When the "stalled" schizonts were examined with an electron microscope, it revealed that the rupture of the red cell plasma membrane had been prevented by leupeptin; however, disintegration of the PVM had already occurred, releasing merozoites into the hemoglobin-rich red cell cytoplasm. These experiments provided the first evidence that, in the schizont-infected cell, the PV and the red cell membranes behaved differently in the presence of protease inhibitors, and that the process of merozoite release was a multi-step process [267, 737].

Merozoite escape begins with the swelling of the PV as the erythrocyte shrinks, suggesting that there is redistribution of water and ions between these two compartments [266]. Vacuole swelling, it is believed, provides space for the separation of individual merozoites and ruptures the PVM. As, at this time, the red blood cell membrane does not lose its integrity, merozoite egress requires that pores form in the red cell membrane before it can rupture. Pore formation is independent of cysteine protease activity; however, examination of the *P. falciparum* genome shows it to have five genes coding for perforin-like proteins. Indeed, earlier studies with the electron microscope had noted that, in infected red cells bearing *P. falciparum* schizonts, an intact PVM was lacking, but the red cell membrane was intact. This suggested that the PVM was degraded first followed almost immediately by the red cell membrane [396]. The role for pore formation is not clear. It may be that it minimizes the viscosity in the host cell cytoplasm, leads to the influx of ions involved in the process of merozoite escape, or serves to weaken the barrier that the merozoites must breach to leave the confines of the red cell. It is possible to effectively inhibit the breakdown of the PVM by treating infected red cells with the protease inhibitors pepstatin or E-64; however, neither of these affects the red cell membrane. Clearly, proteases are responsible for the destruction of some membranes, including the skeletal proteins of the red cell, but their activity does not affect pore formation.

The genome of *P. falciparum* encodes more than 92 putative proteases falling into a half dozen classes [746]; of these, 88 were transcribed and 67 were translated at some stage of parasite development. Most of the proteases are involved in the digestion of hemoglobin in the parasite food vacuole; however, others play different roles [64]. A search of the *P. falciparum* genome identified three subtilisin-like serine proteases (subtilases), one of which, SUB2, named "sheddase," is responsible for the release of the MSPs, whereas another, SUB3, was found to be non-essential as its gene could be "knocked out" without effect. SUB1, however, is different. It is maximally expressed during the final stages of merozoite maturation, and attempts to knock out the gene for it were unsuccessful, suggesting it to be essential for the blood stages. Recent studies have localized SUB1 to previously unrecognized parasite vesicles or dense granules (called exonemes) that discharge SUB1 into the vacuolar space [750]. There, SUB1 cleaves two essential members of the SERA family of papain-like proteins and, interestingly, SERA is also found in the PVM. By screening 170,000 compounds for their action on SUB1, it was possible to identify a specific inhibitor MRT 12113 and, as might have been anticipated, when this inhibitor was added to falciparum cultures, rupture of the schizont-infected red cell was blocked. In other tests, another serine protease inhibitor, JCP104, was found to also prevent merozoite release and it too had SUB1 as its target. These investigators then used another inhibitor, staphylokinase (SAK 1), which inhibited yet another family of proteases, the dipeptidyl peptidases (DPAPs), all of which are related to the class of proteases called cathepsins. There are three DPAPs found in the falciparum genome sequence; however, only one, DPAP1, is involved in hemoglobin digestion. As SAK prevented merozoite release, the question was how do the DPAPs fit into the process of merozoite release? It was found that SUB1 is synthesized by the parasite as an inactive precursor (as are the enzymes from our pancreas, e.g. trypsinogen is the inactive form of trypsin) and this is trafficked to or through the exoneme, which contains DPAP3; by its action, SUB1 becomes mature and active [25]. (SAK has been shown to specifically inhibit infected red cell DPAP3.) Just before rupture, the mature SUB1 is released into the PV where it acts on SERA, causing it to become active and, as a result of its action, the PVM is degraded. Clearly SERA and DPAP3 are targets specific to *Plasmodium* and the fact that small molecules, such as the inhibitors of proteases, are able to perturb the process of merozoite release provides encouragement that it may be possible to use these as the basis for the development of novel therapeutics.

Insofar as the breakdown of the red blood cell membrane is concerned, there are other proteases, some of which are produced by the parasite that could possibly act on the red blood cell skeleton. For example, plasmepsin II cleaves spectrin and actin, and falcipain-2 cleaves ankyrin. A recent study, however, implicates not parasite proteases but one already present in the red blood cell. Again, a very specific inhibitor, DCG604 (related to E-64, an inhibitor of aspartic proteases or cathepsins), was found to affect human calpain-1 — found only in the erythrocyte. Calpain-1 is inactive in the red cell cytoplasm but, upon binding of calcium, becomes an active protein-digesting enzyme specifically associating with the red cell membrane [124]. Calpain-1 is able to digest the red cell skeletal proteins such as spectrin and integrin, and its activation occurs only during the late schizont stage. When cells were depleted of calpain or loaded with calpstatin, a specific inhibitor of calpain-1, parasite development was arrested and no merozoites were released.

Calcium ions play a role in many cellular functions; however, the concentration of calcium in the red blood cell is so low that it is incompatible with the survival, development, and multiplication of *Plasmodium*. To overcome this limitation, malaria parasites alter the permeability of the malaria-infected red cell and utilize the red cell plasma membrane calcium-ATPase (present in the PVM as a result of merozoite invasion, see Chapter 7) so that a high level of calcium ions is accumulated within the PV [258]. The source of calcium to activate calpain-1 is supplied from the PV and probably involves a pore-forming parasite protein, a porin. Searching the falciparum genome should enable the identification of the gene that encodes such a porin. However, even before this cascade of proteases is unleashed, there is evidence of a plant-like calcium-dependent enzyme (a kinase), identified as present in the falciparum genome and expressed in late schizonts, that triggers the ensuing steps of membrane degradation. In parasites that are deficient in this kinase, the red cells do not release merozoites although all of the other proteases involved in the process are present [197]. Further, in such kinase-deficient parasites, the number of merozoites formed and their ability to infect fresh red cells were normal. If it were possible to specifically inhibit this kinase, then there might be a possibility of developing an effective antimalarial.

The manner of merozoite release from the pre-blood stages in the liver is different from that described for the red blood cell. A sporozoite that enters a liver cell forms a PV and, within the vacuole, the sporozoite develops into a schizont with thousands of nuclei, each surrounded by the parasite plasma membrane. After several days or weeks there are thousands of infectious merozoites still enclosed by the PVM. The PVM disintegrates and the merozoites are liberated into the liver cell cytoplasm and, with this, the liver cell dies. The dying cell detaches from the neighboring liver cells, and forms host-derived blebs (called merosomes) filled with merozoites. The merosomes then bud off, leaving the liver to enter the bloodstream. Why this different strategy to escape from the liver? As the merozoites are covered by an intact liver cell membrane, they are not recognized by the immune system. By the transport of merosomes *via* the blood stream to the heart and then to the lungs where the merosomes are released in the lung capillaries, the merozoites are in a location where they can invade red blood cells. In some manner, which is not completely understood, the infected liver cell is able to prevent activation of its store of calpain by having a low concentration of calcium ions and this, in turn, blocks the externalization of membrane phosphatidylserine, the signal for phagocytes to "eat" dying cells. "This Trojan horse strategy allows...merozoites a safe journey from the liver tissue through the endothelium into the blood vessel without being recognized by the numerous macrophages patrolling the liver sinusoids" [309].

A previously identified hypothetical protein LISP 1 (liver-specific protein 1), localized to the PVM, is specifically involved in the egress of the parasite from the liver; however, its precise role is unknown [341]. (Liver-specific protein 1 is dispensable for parasite egress from the red blood cell.)

Transmission-blocking vaccines (TBVs) (see p. 227) have as their goal the disruption of parasite reproduction and/or impairment of development in the mosquito. In this regard, the genome sequencing project has identified new sexual-stage antigens as well as the molecular mechanisms that provoke blood-stage parasites to become sexual stages (gametocytes), how these become activated to form gametes and how the latter are able to egress [385].

Shortly after blood is ingested by the female *Anopheles* mosquito, gametocytes in the blood meal receive environmental signals (a 5 °C drop in temperature, mosquito-derived xanthurenic acid, a shift in pH from 7.2 to 8.0) and ~10 minutes after receiving these signals both male and female gametocytes round up and escape from the enveloping red cell. The egress of female gametocytes is linked to the presence of osmiophilic bodies that contain a hydrophilic protein, Pfg377. When the gene for Pfg377 is "knocked out," female gametocytes have fewer osmiophilic bodies and egress is blocked. Another protein, the male development gene/protein of early gametocyte 3 (MDV/Peg3), is present in both male and female gametocytes; however, it is more abundant in the females where it is associated with the osmiophilic bodies. This protein has been shown to play a role in the breakdown of the PVM and, when the gene for it is disrupted, there is a reduced capacity for the parasite to produce zygotes because male and female gametes were unable to disupt the PVM and therefore unable to egress from the red blood cell [523]. Indeed, as noted earlier, PVM destabilization must occur for release of the parasite as lysis of the red blood cell membrane does not occur if the PVM remains intact.

Sequences obtained from the Genome Project identified at least 86 hypothetical kinases in *P. falciparum*. One of these, the calcium-dependent kinase PfCDPK4, is activated by phospholipase C following xanthurenic acid release and elevation of calcium. PfCDPK4 is gametocyte-specific and is involved in genome multiplication and exflagellation of microgametes; a related kinase CDPK3 is required for ookinete motility and penetration of the mosquito stomach wall. Finally, because SERA 8, a cysteine protease of the serine repeat antigen family with a papain-like domain and encoded by one gene in a nine-member gene family, has been shown to be essential for the egress of sporozoites from the oocyst it has been dubbed the "egress cysteine protease" [15].

The goal for the future is to identify those parasite molecules (proteases and kinases and others?) that are involved in the rupture of the PVM to allow for parasite egress and/or those involved in gamete formation and to develop inhibitors for them.

Chapter 9

The Neglected Malaria, *Plasmodium vivax*

Of all the species of human malaria, the deadly *P. falciparum* is often regarded as the most important: it is the focus of most laboratory and field research, receives the greatest amount of funding, and its control is the centerpiece of public health measures. Yet, the most geographically widespread of the human malarias is not *P. falciparum*, but *P. vivax*. Outside Africa, vivax is the dominant species with 80–250 million cases out of the annual ~515 million malaria cases. *Plasmodium vivax* is usually found outside tropical areas and was present throughout temperate North America and Europe until the introduction of dichlorodiphenyltrichloroethane (DDT). Despite this burden of disease, vivax malaria is oftentimes left in the shadows of research [242]. The reason for this neglect is the widely held misconception that *P. vivax* is relatively infrequent, benign, and easily treatable. Further, studies of the biology of *P. vivax* have been hampered by a lack of a continuous cultivation system for any of its life stages and the absence of a readily available and economic primate model. However, with the advances in genome sequencing, it is hoped that some of these impediments will be overcome.

There is little doubt that, in 1880, when Alphonse Laveran examined the blood of soldiers in North Africa who were suffering with benign tertian malaria he saw *P. vivax*. He did not, however, regard it as distinct from all of the other kinds of malaria parasites he had previously observed with his crude microscope. The differentiation of vivax from other malarias was left for Camillo Golgi who, in 1885, examined both stained blood films and fresh blood droplets taken from patients in Pavia, Italy. Because Golgi was able to see the parasites within the red blood cell in active motion and with a tenuous cytoplasm — he termed the behavior "vivace mobilita" — this would later (in 1890) be used by Golgi's fellow Italians Battista Grassi and Raimondo Feletti to give this malaria parasite its specific name: vivax [253].

Gertrude Stein's famous observation "rose is a rose is a rose" certainly does not apply to malaria. Malaria parasites and the disease they cause are clearly unlike one another. The most obvious way in which *P. vivax* differs from *P. falciparum* is that it develops "sleeping stages," called hypnozoites, in the liver; these may lead to a reappearance of parasites in the blood, a relapse, many months or up to two years after an *Anopheles*, in taking a blood meal, has introduced sporozoites into the body. The blood of a *P. vivax*-infected individual may show gametocytes — and these are not banana shaped as in falciparum — even before there are any clinical symptoms such as chills and fever. There is a strong preference by vivax merozoites to invade the very youngest of the red blood cells, reticulocytes, and all developmental stages circulate, i.e. sequestration does not occur, the sticky knobs are absent, instead the membrane surface of the infected red cell is remodeled to produce numerous pits, called caveolae–vesicle complexes (seen with an electron microscope) and, in Giemsa-stained smears, these appear as a speckling, the result of the dye being deposited in the pits. The surface speckling of the *P. vivax*-infected red blood cell, called Schüffner's dots, is named after their discoverer Wilhelm Schüffner who described them in 1904. The *P. vivax*-infected red blood cell containing an active ameba-like parasite is greatly enlarged and more deformable than its uninfected or *P. falciparum*-infected counterpart [110].

Although vivax malaria is less likely to develop into a life-threatening disease (as does falciparum malaria), it has a substantial impact on an infected individual's health and economic well-being, with socioeconomic ramifications for the family, community, and nation. The chronic, long-lasting nature of the vivax infection — due to the persistence of hypnozoites — contributes to patient morbidity. Vivax infections are characterized by severe anemia with the destruction of both infected and uninfected red cells accompanied by a defective formation of red cells by the bone marrow despite there being low numbers of blood parasites in the circulation. Patients with vivax malaria and fevers of 40–41.6°C describe it as "feeling they are going to die" [242].

Resistance

Today, most *P. vivax* infections are found in Asia and this malaria is virtually absent from most of West Africa in the belt between Congo and Mauritania [253]. In P.C.C. Garnham's 1966 treatise, *Malaria Parasites and other Haemosporidia*, the doyen of malaria wrote: "the incidence in the continent of Africa presents a puzzle" [253].

Louis H. Miller (1935–) solved the puzzle. Miller received a B.S. from Haverford College (Pennsylvania) in 1956 and an M.D. from Washington University (St. Louis) in 1960. After an internship and residency in internal medicine at the Mount Sinai Medical Center in New York (1960–1964), he spent a year at Cedars–Sinai Medical Center in Los Angeles as a renal metabolism fellow, training that would stand him in good stead for later studies on renal pathology in human and experimental malarias. Returning to New York City, he received a Master's degree in Parasitology in 1965 from the Columbia University School of Public Health. Drafted into the U.S. Army as a Captain in the Medical Corps he was sent to WRAIR's SEATO (South East Asia Treaty Organization) Laboratory in Bangkok, Thailand, where he intended to study tropical sprue; however, after encountering only a single case in two years he felt it wise to change career. At that time, CQ-resistant malaria was a significant problem for the troops fighting in Vietnam and so Miller, the clinician, turned his attention from tropical sprue to malaria. Knowing little about malaria, he spent a month in the library in Bangkok learning all he could. Comparative studies on the pathology and renal physiology using monkeys and rodents, as well as deep vascular schizogony and sequestration, were carried out with Robert Desowitz (who had been a student of the great malariologist Colonel Henry E. Shortt at the London School). After completion of military service, Miller joined the faculty at the Columbia University School of Public Health as an Assistant Professor and he began to move further away from the clinical aspects of the disease and to focus on more basic questions that, he felt, would have an impact on malaria control. To follow up his work on sequestration, Miller contacted Craig Canfield, a friend from Bangkok and then Director of Experimental Therapeutics at WRAIR, to obtain two *Aotus* monkeys to be infected with *P. falciparum*. The result was the discovery of the presence of knobs on the surface of red cells in organs where sequestration occurs. In 1971, Miller left Columbia University to establish a malaria laboratory at the NIH. Shortly after joining the NIH, Miller attended a conference organized by Elvio Sadun at WRAIR and there he met Sydney Cohen (from Guy's Hospital in London) who was carrying out short-term cultures of *P. knowlesi* to study invasion. In 1973, Butcher, Mitchell, and Cohen (with support from the Medical Research Council [MRC] and World Health Organisation [WHO]) used an *in vitro* short-term culture system to study invasion and found that red cells from Old World monkeys (kra and rhesus) were susceptible to *P. knowlesi* merozoites, whereas New World monkey (*Aotus*) and human red cells were less susceptible and non-primates were completely resistant [98]. Although these workers were convinced that these results were due to differences in erythrocyte susceptibility and the basis of host-cell specificity, they did not identify the red cell receptor. However, in a series of what are now considered to be classic studies, Miller and co-workers [459, 460] used enzyme "stripping" of red cells to define the chemical nature of a red cell membrane receptor. When the surface of human red cells was treated with the protein-digesting enzymes, chymotrypsin or pronase, invasion of *P. knowlesi* merozoites was blocked, whereas other enzymes, neuraminidase and trypsin, were ineffective. However, rhesus red cells treated with the chymotrypsin remained susceptible. It was suggested that the differences in invasion between the two kinds of red cells were due to the "high affinity between *P. knowlesi* merozoites and [rhesus] monkey erythrocytes [which] may require greater alteration of the receptor to inhibit invasion" [457]. This interpretation would subsequently turn out to be incorrect. However, the study did allow for the formulation of a hypothesis for invasion [455] and eventually led to the identification of a human red cell receptor for the merozoite [459].

Miller and co-workers took advantage of the availability of human red cells that differed genetically in their surface molecules and were able to show that *P. knowlesi* merozoites could invade human red cells that have the Duffy factor antigen (called Duffy positive) on their surface but, if human red cells lacked Duffy antigen, they were refractory [460]. Then, using enzyme "stripping" [457], it was found that, when Duffy antigen was removed from human red cells by chymotrypsin (but not trypsin or neuraminidase) treatment, the erythrocytes became Duffy negative and these cells resisted invasion by *P. knowlesi* merozoites. These laboratory studies were extrapolated by Miller to explain the absence of *P. vivax* in West Africa (where most of the population is Duffy negative). Indirect evidence in support of Miller's proposal came from the following: (1) West Africans and black Africans, who are 95% and 70% Duffy negative respectively, are resistant to *P. vivax*; (2) Duffy-negative volunteers failed to become infected when bitten by mosquitoes carrying *P. vivax*; and (3) of 13 black Americans who had vivax malaria in Vietnam, all were Duffy positive [459]. Final and direct proof of Miller's postulate came when a short-term *in vitro* culture system for *P. vivax* showed that these parasites could not invade Duffy-negative red cells [47].

Human Duffy antigen is a 36 kDa acidic glycoprotein predicted to have seven transmembrane domains with an extracellular N-terminal domain of 60 amino acids having two N-glycosylation sites at residues 17 and 28. Duffy negativity has been determined to be due to a point mutation (cytosine to thymine) on the Duffy gene promoter 33 bases from the initiation codon that does away with the binding site for the guanine adenine thymine adenine (GATA-1) erythroid transcription factor and, as a result, transcription of Duffy mRNA in red blood cells, but not other cell types, is abolished [677]. Although the Duffy determinant has been shown to be crucial for the entry of vivax merozoites [678], recent observations indicate that this pathway is not absolute, i.e. *P. vivax* infections were found in two Duffy-negative individuals from South America [121] and in four Duffy-negative individuals from East Africa [568]. This suggests that *P. vivax* merozoites are able (albeit infrequently) to infect Duffy-negative red cells and, in this species as in *P. falciparum*, alternate pathways for erythrocyte invasion exist.

Recent studies using the electron microscope and recombinant DNA technology have shown that there are two Duffy-binding proteins (DBPs), PvDBP and PkDBP, found in *P. vivax* and *P. knowlesi* respectively, and these are present in the micronemes [130, 216, 458, 730]. The PvDBP, a 140 kDa protein, specifically binds to the human Duffy blood group antigen *via* a conserved N-terminal cysteine-rich region called region II [132]. Region II contains 330 amino acids and binds to a 35 amino acid region of the N-terminus of the Duffy antigen [131] with recognition by PvDBL/PkDBLα achieved by a single contiguous and fully exposed site on sub-domain II with residues Tyr 94, Asn 95, Lys 96, Arg 103, Leu 168, and Ile 175 being required for recognition [628]. Mutation of each of these residues abrogates or disrupts the Pv/PkDBLα binding [290, 702]. The sulfated tyrosine 41 of the Duffy antigen has been shown to be essential for binding of PvDBP and PkDBP [133]. "Knockouts" of PkDBPα invade rhesus erythrocytes as well as wildtype parasites; however, when the PkDBPα gene was deleted, *P. knowlesi* merozoites were unable to invade Duffy-positive human red cells. As both wildtype and "knockout" lines invade rhesus erythrocytes treated with neuraminidase, trypsin, proteinase K, and neuraminidase plus trypsin, it appears that deletion of the α-gene does not affect invasion of rhesus red cells by pathways other than the Duffy receptor pathway [623].

PkDBP, a 135 kDa protein, binds to both the human and rhesus Duffy blood group antigens. Antibodies to the PkDBP were used to screen a cDNA expression library from mRNA and the gene encoding PkDBP was cloned and sequenced [7]. PkDBP is encoded by an α-gene. Region II of the β and γ proteins, encoded by two other homologous genes, binds to rhesus red cells but not to Duffy-positive or Duffy-negative red cells. Initially, objections were raised to Miller's identification of the Duffy antigen as a determinant for *P. knowlesi* invasion as, paradoxically, Duffy-negative red cells treated with trypsin were susceptible to invasion yet they remained Duffy negative. Also, chymotrypsin-treated rhesus red cells that had lost the Duffy antigen could still be invaded by *P. knowlesi* merozoites. This paradox has now been explained by *P. knowlesi* having alternate pathways for invasion of rhesus red cells that do not depend on the Duffy antigen, which is mediated by the β and γ proteins [130]. The DBP-β binds to sialic acid residues in a stretch of 53 amino acids between the fourth and the fifth cysteines; the receptor for DBP-γ is still unknown.

Earlier, electron microscopic studies showed that, when *P. knowlesi* merozoites were treated with cytochalasin B (which blocks polymerization of actin to paralyze the invasion motor), there was an arrest of entry at the step of junction formation [456]. When cytochalasin B treatment was carried out with merozoites from the PkDBPα knockouts, no junction was visible with Duffy-positive human red cells (however, a junction formed with rhesus red cells), suggesting that the interaction of PkDBP with the human Duffy antigen is necessary for junction formation and, when this critical step is blocked, invasion does not occur [623].

As it is possible to produce large amounts of the recombinant receptor-binding domain of vivax PvDBP [749], and rabbit antibodies raised against a recombinant PkDBP inhibited the invasion of human and rhesus red cells by *P. knowlesi* merozoites, it has been suggested that PvDBP may serve as a possible vaccine candidate. Indeed, this potential for a vaccine has been supported by the finding of antibodies to PvDBP in children in Papua New Guinea where the levels of these antibodies remain stable for at least a year. The antibodies prevent invasion of reticulocytes and, although they do not completely abrogate infection, the infections were less severe and, when present, the children were less likely to develop clinical disease [368].

Plasmodium vivax preferentially invades reticulocytes. Two vivax reticulocyte-binding proteins, PvRBP-1 and PvRBP-2, have been identified and the genes for these have been cloned and sequenced [243, 244]. Both are large proteins — 325 and 330 kDa respectively — and both have a two-exon structure: the first exon encodes the signal peptide followed by a short intron and a large 6–9 kb exon. The RBPs have been localized to the apical region of the merozoite but it has not been precisely determined whether they are located in the micronemes or rhoptries, although the pattern of fluorescence with antibody is consistent with a microneme location. PvRBP-1 and PvRBP-2 have been postulated to form a complex that mediates adhesion and recognition of the reticulocyte independent of the Duffy phenotype (reviewed in [344]). As there is no evidence for the Duffy-like ligands being exposed at the surface prior to initiation and formation of a tight junction, it has been suggested that binding may be necessary to signal the timely release of the sequestered microneme proteins so that a tight junction forms and merozoite entry can proceed [243]. In this way, by acting prior to RBP binding and junction formation, i.e. during apical end orientation, the *P. vivax* merozoite is able to target reticulocytes, which usually represent less than 1% of the total red cell population. The receptor molecule on the reticulocyte to which the PvRBP binds remains unknown [243].

Genes sharing homology with the PvRBP were found in the genomes of *P. falciparum* and *P. knowlesi* although these species do not invade reticulocytes; to distinguish these from the RBPs they are called normocyte-binding proteins (NBPs). One ortholog of the PvRBP, *P. falciparum* normocyte-binding protein 1 (PfNBP1) has been shown to bind to a sialic acid-dependent trypsin-resistant receptor; antibodies to PfNBP1 blocked invasion of trypsin-treated red cells and two truncated versions were unable to invade trypsin-treated red blood cells [541]. Although the receptor recognized by PfNBP1 is a sialic acid-containing protein, its identity remains unknown, as does its role in invasion.

Secrets hidden in the genome

The nuclear genome of *P. vivax* (Salvador I) consists of 27 million nucleotides, packaged into 14 chromosomes, and codes for ~5,500 proteins, of which fully half — called hypothetical proteins — are of unknown function, having no counterpart in any other organism. Significant differences in the nucleic acid base composition between vivax and falciparum were found with 42.3% G + C in the former and ~19.4% in the latter. The key metabolic pathways, housekeeping functions and a repertoire of membrane transporters were conserved between vivax and falciparum, suggesting that the two species have similar metabolic functions [110]. Examination of the *P. vivax* genome suggests a candidate list of dormancy-associated genes that could be linked to the genetic switch for the formation of the hypnozoite upon sporozoite invasion of the liver. Paradoxically, *P. vivax* sporozoites showed the same stage-specific expression of the same genes needed for liver invasion and liver stage development as found in other species of *Plasmodium* [731]. *Plasmodium vivax* genes appear to be activated primarily at the schizont/ring stage, which suggests that the differences between vivax and falciparum derive mainly from events that occur at merozoite invasion and with early development within the red blood cell [84].

The largest gene family in the vivax genome is composed of members of several known and novel gene families, the best known of these being the vir genes. Variations in the length as well as motif composition of the vir genes suggested that they function in immune evasion and in the establishment of chronic infections [160]. However, others claim that, "whereas some features of the vir genes are consistent with a role in immune evasion, the available data do not support the existence of a genuine clonal antigenic variation process involving these proteins" [469]. Within another novel gene family, one member has been found to encode a protein localized to the caveolae–vesicle complex and it elicits an antibody response during natural infections [346].

Even when armed with the genome sequence of *P. vivax*, formidable gaps in our knowledge remain. "The trigger for activation of hypnozoites is not understood, though stress seems to play a part, and distinct patterns of relapse related to local mosquito seasonal abundance suggest a Darwinian genetic process at work to ensure transmission and propagation of the species" [469]. The problem of relapse in vivax and ovale malarias is complex and poorly understood: microsatellite markers identified from the vivax genome sequence have revealed that relapse often occurs as a result of activation of parasites that are different from those causing the primary infection. Will further studies of the vivax genome be able to predict relapse? What is the function and role of Schüffner's dots and which genes are responsible for this remodeling of the red cell? No genes associated with the high flexibility of the *P. vivax*-infected red cell were identified in the genome sequence so what is the basis for this? To date it has not been possible to determine how *P. vivax*-infected reticulocytes are able to move through the spleen without removal nor whether the spleen plays a role in the expression of the variant antigens. Through longitudinal studies in different populations and age groups with variable intensities of transmission of *P. vivax*, it may be possible to uncover the nature of acquired immunity. Such studies are yet to be undertaken. Identification of vaccine candidates for vivax malaria is sorely needed. Will the genome sequence be able to provide more sensitive vivax-specific diagnostics? As first-line therapies for the radical cure of vivax malaria have not changed in half a century and there is evidence of increased failure rates due to parasite resistance to the presently available drugs, will a functional analysis of the vivax genome be able to provide insight into the mechanisms of resistance? Although it is well known that *P. vivax* infections cause fever even with very low numbers of parasites in the blood and with very high levels of inflammatory cytokines, little is known about the parasite molecules that trigger such a response.

For effective intervention and control of vivax malaria, there are technical impediments — ones that lie outside its genome sequence — that still need to be addressed. One critical need is for the development of a continuous *in vitro* culture system for *P. vivax* blood and liver stages. A better understanding of the ecology of the 25 anopheline species that act as vectors for *P. vivax*, many of which tend to have outdoor biting habits and are less attracted to human blood than those anophelines that are the main transmitters of falciparum malaria in sub-Saharan Africa, will be required for effective control where the disease is endemic and transmission high.

Chapter 10

The *Anopheles* Genome and Transmission Control

In an article on 10 June 2010, *Time* magazine described a village in Uganda: "Lake Kwania is more of a giant swamp: shallow, full of crocodiles and choked with lily, papyrus and hyacinth. The malaria parasite loves it here. Kwania's creeks...are perfect for a deadly... mosquito, *Anopheles funestus,* which feeds almost exclusively on humans, with an appetite to shame a vampire. The nearby town of Apac is packed with a living blood bank of people. The average funestus bites humans 190 times a night. And, the average resident is bitten tens of thousands of times a year, including 1,586 bites — four a day — that carry malaria" [515] (http://www.time.com/time/specials/packages/article/0,28804,1995199_1995197,00.html?artId=1995199?contType=article?chn=specials).

Since the 1897 report by Surgeon-Major Ronald Ross, based in Secunderbad, India, it has been known that some kinds of mosquitoes transmit malaria parasites to humans. Indeed, for two years Ross struggled with insusceptible (refractory) mosquitoes, but eventually he was able to rear 20 adult "brown" female mosquitoes (*Anopheles*) from larvae collected in the field and used these to bite Hussein Kahn, a volunteer, whose blood contained crescents of malignant tertian malaria (*P. falciparum*). Over a four-day period, Ross was able to describe the development of the pigmented malaria parasite into pimple-like structures (oocysts) on the surface of the mosquito stomach. The significance of Ross's discovery lies in three areas: malaria transmission by *Anopheles*, epidemiology, and the "identification of what is perhaps the most vulnerable stage in the parasite life cycle for effective intervention" [622]. Ross's work led to the rapid adoption of mosquito control campaigns involving personal protection, house screening to prevent adult mosquitoes from biting, water management to eliminate larvae and, later, the use of insecticides. The earliest of the control campaigns recognized that the most effective means for reducing transmission of the disease was to attack the malaria-carrying mosquito; however, the appearance of insecticide resistance in areas with high transmission (as well as the failure of CQ treatment due to parasite resistance), especially in sub-Saharan Africa, has prompted a search for new control techniques.

In 1991, a meeting on "Prospects for Malaria Control by Genetic Manipulation of its Vectors," sponsored by the MacArthur Foundation and WHO-TDR (the Special Programme for Research and Training in Tropical Diseases) held in Tucson, Arizona, led to a consensus among the participants that the use of molecular approaches to mosquito and disease control should be pursued as a real possibility and not an impossible dream [53]. On the basis of this, the TDR established a 20-year plan for the development of insusceptible (refractory) mosquitoes for malaria. Initial projects were focused on gene mapping, identification of transposable elements, and establishing a database. Discussion by L. Miller (NIAID), F. Collins (Notre Dame University) and F. Kafatos (European Molecular Biology Laboratory) at the 1997 Biology of Disease Vectors course, held in Bamako, Mali, led to a draft plan to fully sequence the genome of *A. gambiae*, the major vector for *P. falciparum* in sub-Saharan Africa. By 2000, modest funding for sequencing was provided by NIAID and WHO-TDR. An *A. gambiae* Genome Summit was convened in 2001 at the Pasteur Institute (Paris) sponsored by TDR and, 18 months later, the full genome sequence was completed by an international consortium of researchers (the Vector Biology Network, Celera Genomics and Genoscope) with funding from NIAID and the French Ministry of Research. It was hoped that the availability of the *A. gambiae* genome sequence, consisting of 278 million base pairs, would "serve as a valuable molecular entomology resource, leading ultimately to effective intervention in transmission of malaria and perhaps other mosquito-borne diseases" [326]. The *Anopheles* genome sequence was envisioned to facilitate studies of: (1) natural populations for their susceptibility and insusceptibility (refractoriness); (2) new targets for control of transmission; (3) ways to introduce suitable traits into natural populations so as to reduce the size of the mosquito population involved in transmission [120]; and (4) the genetic mechanisms responsible for insecticide resistance.

Making mosquitoes immune

Mosquito immunity involves both non-cellular responses that involve the production/secretion of anti-microbial peptides and cellular responses mediated by circulating blood cells, hemocytes, which act to phagocytize and/or encapsulate the invading microbe. In the former category are molecules called defensins that act mainly against Gram-positive bacteria, and cecropins with activity against Gram-negative bacteria. Defensins and cecropins have been identified in *Anopheles*.

One of the most striking of mosquito responses is melanotic encapsulation. The capsule is made of a proteinaceous melanin polymer, and the process begins with the recognition of the microbe's cell wall, followed by activation of serine proteases called prophenoloxidase-activating enzymes, which activate prophenoloxidase and culminates in its cleavage to form active phenoloxidases that oxidize tyrosine to dihydroxyphenylalanine and subsequently into other dopaquinones that serve as precursors of the melanin polymer.

Melanization was first described by Ross in 1898, who described "black spores," corresponding to dead melanotically encapsulated parasites. Ross was, in fact, frustrated by this observation made in several *Anopheles* species, because the mosquitoes concerned did not appear to be efficient malaria vectors. Impressions of the killing mechanism, as studied in refractory mosquitoes, showed that ookinete lysis may or may not be followed by encapsulation with the black melanin pigment [148]. Melanization of *Plasmodium* in the midgut is "one of the best examples of mosquito immune reactions against parasite development;" [448] however, it has been shown that it is actually a post-killing event.

Mosquito species–parasite species combinations do influence the outcome of an infection of mosquitoes, involving the intensity of the immune response as well as the route taken by the ookinete in crossing the midgut [736]. For example, in *A. stephensi*, many more oocysts were found after infection with *P. berghei* than with *A. gambiae*. A knockdown of one serpin gene, the products of which inhibit prophenoloxidase activity, increased melanization and adversely affected the ability of *P. berghei* to invade the midgut and develop into oocysts; however, when this same serpin was silenced in an *A. gambiae* strain from a local population (Cameroon), it did not influence the development of local isolates of *P. falciparum* [453]. Natural populations of *A. gambiae* show significant genetic variation that affects their susceptibility to parasite infection. The genes that control resistance to the rodent malaria *P. berghei* have been localized to a single region on chromosome 3, and this region overlaps with regions previously identified as controling melanization. The region contains 975 genes, among which 35 can be classified as "immune-related," [70] and includes the TEP1 gene encoding a thioester-containing protein 1 [67–69], shown to be involved in phagocytosis of bacteria and that binds to *P. berghei* ookinetes resulting in their lysis. Knockdowns of TEP1 by gene silencing increased the numbers of developing oocysts and silencing of this gene also increased the number of parasites crossing the midgut (or ookinetes) and inhibited their melanization. This suggests that killing precedes melanization. Indeed, in a particular strain of *A. gambiae*, ookinetes were melanized only after crossing the midgut.

In *A. gambiae*, two related genes, leucine-rich repeat immune protein 1 (LRIM1) [493] and *Anopheles–Plasmodium*-responsive leucine-rich repeat (APL1) [553], are important players in parasite killing. These proteins circulate in a complex with TEP1 [228, 526] and silencing of these genes resulted in an increase in oocyst numbers. LRIM1 is upregulated in the mosquito after infection with bacteria and *Plasmodium* and, when silenced in *A. gambiae*, there was a substantial increase in the number of oocysts. These studies show that TEP1 and LRIM1 block *Plasmodium* development in the mosquito; however, they are not normally sufficient to fully block parasite transmission, suggesting that there is an evasion mechanism. Indeed, two sugar-binding proteins, called lectins, protect the parasite from LRIM1/APL1/TEP1 killing. Thus, development of oocysts in the midgut of an adult female mosquito is the result of a dynamic interplay between positive (e.g. the lectins, C-type lectin CTL4, and C-type lectin mannose CTLMA2) and negative (TEP1, APL1, LRIM1) factors and is more complex than previously realized. TEP1 and LRIM1 are regulated by REL1 and REL2 [237].

A recent microarray analysis identified 650 genes that are transcribed upon ingestion of a blood meal by *A. stephensi* infected with *P. berghei* and, in a parallel analysis using RNAi, 11 genes were identified as either agonists or antagonists [747]. A genome-wide characterization of circulating hemocytes from female *A. gambiae*, using microarrays, identified 1,485 transcripts with enriched expression in these cells. Of 119 transcripts with increased hemocyte expression levels, 16 encode known immunity-related proteins and six of these have been shown to have direct or indirect anti-parasite activity. Of 12 genes tested for RNAi effects on parasite survival, seven had a significant effect on either numbers of oocysts or ookinete melanization [519]. However, there is a downside to these studies, in that the convenient *P. berghei* model may not be a reliable predictor of the mosquito immune response against *P. falciparum* in the field. What is clear is that no single gene is sufficient on its own to prevent *Plasmodium* transmission. Further, targeting or overexpressing genes of the immune system would most likely have a detrimental effect on mosquito development as has been demonstrated with the serpins. "Future studies will need to concentrate on natural interactions occurring in the field between *P. falciparum* and its vectors.... to identify genes showing unequivocal phenotypes impacting on the transmission of human malaria and to develop efficient ways to manipulate their function" [120]. Some of these reports provide a measure of encouragement for novel methods to target the immune system of *Anopheles*; however, most of the strategies are still many years from practical application.

Making mosquitoes insecticide resistant

For the foreseeable future, the control of malaria-bearing mosquitoes will rely on the use of insecticides through spraying or insecticide-impregnated bednets. However, the shortage of licensed insecticides coupled with insecticide resistance provokes continuing investigations on the identification of novel insecticidal targets and understanding the mechanism of resistance. Genomics may provide clues for the identification of the genes responsible for insecticide resistance, i.e. those enzymes involved in insecticide detoxification. Through the use of microarrays it has been possible to identify a short list of candidate genes. However, among these, only the glutathione S-transferase on chromosome 3 of *A. gambiae* has been shown to be increased fivefold in DDT-resistant strains over the susceptible strains and the recombinant protein has high DDT dehydrochlorinase activity [539].

The identification of genes implicated in insecticide resistance, it has been argued, will provide an early warning mechanism for incipient resistance and will enable the assessment of the efficacy of insecticide-malaria control strategies and allow for making informed decisions as to the choice of alternative insecticides where resistance has been found. Indeed, "molecular assays to detect mutations in the sodium channel gene, the target site of pyrethroid and DDT insecticides are already being employed in many malaria control programs" [120].

Replacement and/or reduction

A strategy for the reduction of mosquito populations is the sterile insect technique. It is based on the field release of mass numbers of sexually active but genetically sterile insects, always males, over large areas. The sterile males mate with females and they produce inviable progeny. In theory, if release is repeated it will eventually result in the eradication of the local population. The technique has been successful in the eradication of the New World screw-worm from the Southern US, Mexico and all of Central America and with other insect agricultural pests; however, most programs involving mosquitoes have been unsuccessful except for the sterile release of *A. albiminus* in El Salvador. (It may be a suitable technique for islands.) Some of the problems associated with sterile release are the loss of male competitiveness owing to the sterilization procedure (irradiation or chemosterilization), and the unavailability of a genetic sexing mechanism to ensure that only males are released. The latter is essential as female mosquitoes, even when sterile, would still be able to act as transmitters of disease.

A variation of the release of sterile males may be the use of transgenic mosquitoes. The identification of developmentally regulated promoters and genes essential for fertility that could be manipulated to develop efficient sexing mechanisms, or to induce sterility in males, will be essential to the success of the use of the sterile technique or other strategies in which a mosquito carrying a lethal gene, the expression of which can be controled during early development, is used. As envisioned by molecular entomologists, after the transgenic male mosquitoes mate with wildtype females, some or all of the offspring will die as a consequence of inheriting and expressing the lethal genes; this would result in a decline in the population. Another approach would be "death-on-infection" where mosquitoes carry a conditional lethal gene that is activated by the presence of the parasite. This would result in the selective death of the infected mosquito and would diminish the possible ecological effects associated with population reduction while, at the same time, diminishing the rate of transmission. Both of these techniques require mass release, an effective system for driving the gene into wild populations, and identification of genes that confer resistance. A *sine qua non* for transgene expression in mosquitoes is that, to be useful in control strategies, too great a genetic load must not be imposed on the transmission competency of the mosquito population [127].

Population replacement strategies are designed to substitute susceptible mosquitoes with refractory ones and do not require changes in the mosquito population densities. For this to work there must be a gene drive to introduce and establish in a population a replacement gene or genes. Initially it was expected that the means for driving a refractory gene construct quickly and efficiently would involve transposable elements, a technique that was successful in the fruit fly, *Drosophila*; however, this approach did not work with *A. gambiae*. A recent and more promising approach has been to use Medea — a selfish genetic element first discovered in the flour beetle *Tribolium* — that spreads rapidly in the fruit fly populations. Medea encodes a maternally expressed toxin and an antidote expressed in the offspring. The result is that the toxin causes the death of all progeny lacking Medea and the antidote rescues the Medea-bearing offspring. The death of all females that do not inherit the gene is the reason for its acronym — maternal-effect dominant embryonic arrest — with reference to the mythological Greek figure who murdered her own children. It is through this mechanism that the Medea-bearing individuals increase in each generation and it is hoped that, by attaching a refractory gene, resistance to malaria will come along for the ride [429]. An active effort to construct Medea systems for mosquitoes is underway in several laboratories; however, as yet no such system has been made.

The first requirement of any transgenic mosquito control project is the discovery of genes that confer resistance to infection. In a recent study, transmission of malaria was blocked by a genetically modified *Aedes aegypti* (not *Anopheles*) with coexpression of cecropin and defensin. In these transgenic mosquitoes infected with *P. gallinaceum* (not *P. falciparum*), the number of oocysts was dramatically reduced on the mosquito stomach and no sporozoites were found in the salivary glands upon blood feeding; feeding on naïve chickens produced no infection [377]. Although it has been shown that cecropins and defensins are not a natural mosquito defense against *Plasmodium,* infection of *A. gambiae* did not produce activation of defensins and cecropins, and RNAi of cecropins and defensins did not affect the *in vivo* development of *Plasmodium* in the described system, the effect is different in that activation of these anti-microbial peptides is linked to blood feeding. It remains to be seen whether the system will work with *Anopheles* and *P. falciparum.*

Smell plays a crucial role in host-seeking and host-feeding behavior by mosquitoes, and this in turn determines the effectiveness in transmission as well as in mating behavior. The *A. gambiae* genome sequence [326] identified 79 candidate odor receptors and 76 taste receptors. Although regarded as promising novel targets for mosquito control, none has made it to field trials. Other suggested genetic approaches for mosquito control now on the drawing board are: ablation of genes essential for parasite survival, production of knockouts that would target mosquito receptors necessary for parasite invasion into host cells, and expression of antibodies and/or small peptides that target salivary gland sporozoites [552].

Malaria control with transgenic mosquitoes remains a challenge but it may be that in the foreseeable future — 10–15 years from now — there may be available refractory genes and gene drive systems for *P. falciparum* in *Anopheles*. It is unlikely, however, that transgenic mosquitoes will provide an all-in-one solution. Transgenic mosquitoes should be considered in the context of an integrated vector management strategy that should also include insecticide-treated bednets, indoor residual spraying with insecticides, and treatment of infected individuals with antimalarial drugs. There may be other problems; genetic driving mechanisms may not be reliable where there are multiple and unique populations of *Anopheles*. For example, *A. gambiae* typically breeds in small temporary rain-dependent pools and puddles, whereas *A. funestus* exploits large permanent or semi-permanent bodies of water containing emergent vegetation. It attains maximal abundance in the dry season after populations of *A. gambiae* decline. To be successful, mosquito control must take into account the unique biology of the ~50 *Anopheles* species [589] that are transmitters of malaria. Finally, ethical concerns with the release of genetically modified mosquitoes will have to be resolved. These include questions of informed consent in communities that are largely illiterate, unfamiliar with genetic modification and sometimes uneducated on the role of mosquitoes in disease transmission. Further, there is also the possibility of untoward consequences of release, namely an increase in the transmission of non-target diseases [429].

Chapter 11

The Monkey's Tale

For the detection of genetic changes that allow malaria parasites to become drug resistant and to understand the mechanisms of immune evasion, it is essential to know their evolutionary history. Because malaria parasites lack bones or other semi-permanent structures they have left no traces in the fossil record. Without evidence of fossilized malaria parasites is it possible to create a "time machine" that would be able to reveal past encounters of the various kinds of *Plasmodium* with our ancestors? The short answer is: Yes. It is possible to trace the evolutionary history of human malarias using a "time machine" known as the molecular clock. It has no hands for hours, minutes or seconds, and it does not tell the time of day; nevertheless it is a timepiece able to put a series of evolutionary changes by *Plasmodium* into a chronological order. The remarkable feature of the molecular clock is its ability to run backwards and, in so doing, it is able to trace the history *of Plasmodium* and allow a family tree to be drawn.

The construction of the molecular clock begins in 1962 with an idea proposed by Linus Pauling (1901–1994) and Emile Zuckerandl (1922–) when they studied the favorite food of *Plasmodium* — hemoglobin. Zuckerandl was born in Vienna, Austria, to an eminent Jewish family [465]. As World War II loomed, the Zuckerandl family fled first to Paris and then to Algiers in order to escape the Nazi death camps. After the war, Zuckerandl studied biology at the Sorbonne in Paris and then received a Master's degree from the University of Illinois (carrying out research at the Marine Biological Laboratory in Woods Hole, Massachusetts, under the tutelage of the physiologist E. Ladd Prosser). Returning to France, he worked at the marine laboratory at Roscoff where he studied the relationship of the blood pigment hemocyanin on the molting cycle of crabs. In the 1950s, Zuckerandl, as a biologist with an inclination toward molecular problems, arranged to come to the US to work with Linus Pauling at CalTech. Pauling, a physical chemist, had studied proteins since the 1930s and, in 1949, he used the new technique of electrophoresis to show that sickle hemoglobin moved differently in an electric field than did normal hemoglobin. He concluded that sickle hemoglobin must have a different electric charge and, on this basis, he coined the term "molecular disease" to describe sickle cell anemia. (The basis for the charge difference is that a mutation in the globin gene replaces the charged amino acid glutamic acid in normal hemoglobin by the neutral charged valine in sickle hemoglobin.) In 1954, Pauling received the Nobel Prize in Chemistry "for his research into the nature of the chemical bond and its application to the elucidation of the structure of complex substances," i.e. proteins. In 1959, when Zuckerandl arrived at CalTech, he expected to continue his work on hemocyanin, but Pauling said, "You know this subject of yours on hemocyanin...I think the results are going to be difficult to interpret and I think you would do better to work on a protein about which much more is known...why don't you work on hemoglobin?" (This is reminiscent of Hoppe-Seyler's advice to Miescher to study pus cells.)

Using the newly developed biochemical techniques of digesting proteins with enzymes and then analyzing the amino acids in the resultant peptides, Zuckerandl began to compare the hemoglobins from different animal species. Looking at the differences, the idea came to Pauling and Zuckerandl that using the number of amino acid substitutions in a hemoglobin molecule would allow for an estimation of the time of divergences between them. They wrote [465]: "It is possible to evaluate very roughly and tentatively the time that has elapsed since any of the hemoglobin (chains) present in a given species...diverged from a common...ancestor." For example, if there were 18 amino acid differences between horse and human hemoglobin and from fossil evidence the common ancestor of man and horse was 100–160 million years ago, then there would be nine mutations in that time period, or 11–18 million years per amino acid substitution among the 150 amino acids in hemoglobin, with a median time of 14.5 million years. Using this value they calculated, using this molecular clock, the time of divergence between man and gorilla to be 11 million years, a time consistent with that in the fossil record. Early reaction to Pauling and Zuckerandl's evolutionary timepiece by anthropologists, paleontologists, and geneticists was uniformly hostile, and only later — especially when it was possible to sequence DNA — would it become accepted.

In 1967, Allan Wilson (1934–1991) and his doctoral student Vincent Sarich studied the changes in serum albumins using antibody, and claimed that, by immunological evidence, a time scale for human evolution could be established. Wilson, a native New Zealander, had initially trained and received a Ph.D. (1961) with the molecular biologist Arthur Pardee at UC Berkeley, and followed this with a post-doctoral fellowship at Brandeis University (Waltham, Massachusetts) with the biochemist Nathan O. Kaplan (later Craig Venter's doctoral thesis advisor at UC San Diego). At the time of Wilson's fellowship, Kaplan was using lactic dehydrogenase as a tool for determining the evolutionary relationships between various kinds of animals. (Indeed, he was shocked when his analysis found that the horseshoe crab [an arachnid] was not a close relative of crabs and lobsters [crustaceans].) I first met Wilson in 1961 when he was visiting the New York Zoo to collect the carcasses of dead birds in order to compare the lactic dehydrogenase activity in the muscles of flightless and flying birds. He knew of my interest in lactic dehydrogenase, having had access to the sample of *P. lophurae* I had provided to Kaplan's Brandeis colleague Julius Marmur for studies of nucleic acid base composition (see p. xiv). Later that year, I was invited to give a seminar at Brandeis, and Wilson said, "be certain to bring a fresh sample of *P. lophurae* since the other frozen one has lost all enzymatic activity." During my brief stay at Brandeis it was discovered that the lactic dehydrogenase activity of *P. lophuare* was distinctly different in having a preference for a particular and unusual cofactor (acetylpyridine nucleotide) for its activity [593, 595]. Later, this would be found to be the case with all other malarias; measurement of lactic dehydrogenase activity with the acetylpyridine analog has become the basis for the commercially available diagnostic test MalStat for vivax and falciparum infections (see p. 179).

Wilson joined the faculty at UC Berkeley in 1965 and, as noted earlier, by 1967 he had shaken the human family tree when he challenged the accepted wisdom of human evolution using the Zuckerandl–Pauling protein-based molecular clock. Later, in the 1980s, Wilson turned his attention from using proteins to estimating the evolutionary time by a comparison of DNA sequences. This genetic molecular clock was based on the fact that the process of DNA replication is not error-free and nucleotide substitutions (point mutations or small nucleotide polymorphisms, called "snips," and written as SNPs) as well as replication slippages (producing microsatellites) occur continuously and are fed into the gene pool. Those genetic changes that are harmful are eliminated by natural selection; however, others — neutral mutations — have little or no effect on survival and these accumulate at a steady rate of two to five per gene per million years. The DNA sources for identifying these mutational changes have been varied: nuclear genes, the genes for ribosomal RNA and the genes for mitochondrial products. The mitochondrial genes are of particular interest because the sperm mitochondrion remains outside the egg when fertilization occurs and hence its genetic material is lost. Therefore, mitochondrial DNA, Wilson reasoned, could be used to trace maternal inheritance. Using mitochondrial DNA, Wilson and his students were able to show universal human ancestry through maternal lines, from a woman who they claimed lived ~200,000 years ago. They hypothesized that all modern humans evolved from one "lucky mother," dubbed Mitochondrial Eve, and this was hyped in the popular press as the "Garden of Eden Hypothesis." This claim was even more controversial than his earlier one, and again many paleontologists relying principally on evidence from fossils rejected Wilson's molecular clock-based conclusions.

In 1991, when Wilson, aged 56, died while being treated for leukemia, further evidence to support the Mitochondrial Eve hypothesis was at hand; however, there were others who contended that Wilson and co-workers had overinterpreted their data. A later analysis of the original work by Wilson and colleagues (in a paper published posthumously) challenged both the methods and conclusions, and weakened the earlier conclusions somewhat [5, 295]. But, to complicate matters further, a more recent analysis calls into question the use of mitochondrial DNA for studies of human evolution, stating that it provides "no evidence for a bottleneck during recent human evolution" [161].

A molecular clock can provide insights into all living creatures, including the non-living viruses for which there are gene sequences. As noted, this is particularly valuable when there is no fossil record or when that record is patchy in time and space [89]. In some cases, the molecular clock has been able to resolve family tree conflicts. For example, the molecular clock has been used to date the last common ancestor of HIV in the 1930s and to counter the claims that the virus was spread through contaminated polio vaccine in the 1950s (allegedly manufactured using chimpanzee tissue infected with the simian immunodeficiency virus [89]). Regrettably, the same resolution of a time scale has not been established with certainty with the human malarias.

In 1998, based on the sequences of 10 falciparum genes, it was hypothesized that about 3,000 to 5,000 years ago there was a severe population restriction of *P. falciparum* — a bottleneck — and that all extant forms were evolved from this stock. Following Wilson's work it was named the Malaria's Eve hypothesis [548]. In a subsequent study after sequencing several non-coding gene sequences in eight *P. falciparum* isolates, this hypothesis was supported; it gave an estimate of common ancestry from chimpanzees of 3,200–7,000 years ago and went on to claim that falciparum malaria became widespread as a human pathogen only after the spread of agriculture and concomitant with an extension of the range of *Anopheles* mosquitoes [716].

Shortly after the genome of *P. falciparum* had been sequenced, a study using SNPs from 204 genes on chromosome 3 estimated the time to the most recent common ancestor to be 100,000–180,000 years, a time believed to be coincident with the expansion of the human population and the movement of our ancestors out of Africa [467]. Then, using the variation in 100 worldwide mitochondrial DNA sequences, the age of *P. falciparum* was estimated to be even older than that of the Malaria's Eve estimate — ~50,000–100,000 years and "perhaps during the Pleistocene expansion in humans was then followed by migration out of Africa 40,000 to 130,000 years ago" [353]. Other studies, using substitutions in the cytochrome b gene (a mitochondrial gene), estimated that the divergence of *P. falciparum* from *P. reichenowi*, described by Eduard Reichenow between 1917 and 1920, in blood obtained from wild chimpanzees and gorillas in the Cameroon was between 1.2 and 2.5 million years ago. If this estimate were accurate then falciparum malaria infected our human ancestors for several million years and "likely was relatively benign through much of that period" [550]. Then, it is hypothesized, a more recent expansion of *P. falciparum* coincident with our "settling down" to become farmers took place [597], thereby increasing the density of the human populations and facilitating large-scale migrations.

For many years, *P. reichenowi* was considered to be the malaria of the common chimpanzee (*Pan troglodytes*) and was represented by a single isolate collected by Reichenow. Using the ribosomal RNA from this single isolate, Escalante and Ayala [211] established a malaria parasite family tree that estimated the divergence for *P. reichenowi* from *P. falciparum* to a time when chimpanzees diverged from humans ~5–7 million years ago. A recent analysis of the gene sequences from cytochrome b, an apicoplast protease and ribosomal RNA has confirmed these dates [338].

A search for additional isolates of *P. reichenowi* in 18 blood samples from 91 African apes has led to the claim that *P. falciparum* did not originate with the chimpanzee (as previously believed) but rather evolved in the closely related bonobos (*Pan paniscus*) and that the extant populations of *P. falciparum* originated by a single host transfer of *P. reichenowi* when our human ancestors encroached on the dwindling forest habitats where the bonobos lived [380]. Thus, it has been speculated that two populations of *P. falciparum*, one in bonobos and the other in humans, diverged from one another ~1.0–3.0 million years ago, a time when bonobos diverged from the common chimpanzee ancestor. It is estimated that falciparum evolved as a species in bonobos 0.4–1.6 million years ago, before it switched to humans [380]. Based on SNPs in two housekeeping genes (an ATPase and an enzyme in the pyrimidine biosynthetic pathway, adenylosuccinate lyase), it was estimated that modern humans were infected with *P. falciparum* prior to their expansion and migration out of Africa some 50,000–60,000 years ago, and this malaria followed our ancestors into other tropical areas with the exception of South America where it arrived more recently *via* the slave trade [662].

Support for the millions of years old origin of *P. falciparum* comes from another quarter: red blood cell susceptibility and the molecules involved in merozoite invasion (see Chapter 7). *Plasmodium reichenowi* infects chimpanzees and bonobos but not humans, whereas when *P. falciparum* has been used experimentally to infect chimpanzees it only causes a brief and mild infection. It has been postulated that the reason for this differential susceptibility has to do with the sialic acid molecules in the red blood cell surface that serve for merozoite docking via EBA-175 (see p. 115). The principal sialic acid on the human red cell is Neu5Ac, a precursor of N-glycolylneuraminic acid (Neu5Gc), which they lack. By contrast, red cells of chimpanzees have just the opposite condition: an abundance of Neu5Gc and a paucity of Neu5Ac. Obviously, a major change must have occurred in the human lineage after divergence from our common ancestor with chimpanzees and bonobos. It is hypothesized that *P. reichenowi* preferred Neu5Gc over Neu5Ac and the loss of binding for Neu5Gc in the human lineage occurred after the one was converted to the other ~3 million years ago; one consequence of this may have been to provide our emerging *Homo* ancestors with temporary relief from this form of malaria [431]. However, with the loss of Neu5Gc there was an increased expression of its precursor Neu5Ac and, through selective evolution of EBA-175 and its preferential binding to Neu5Ac, *P. falciparum* became the human parasite we know today [195, 547].

The biological and ecological differences between *P. falciparum* and *P. vivax* have been recognized for more than half a century. Unlike *P. falciparum*, *P. vivax* infects reticulocytes, is distinctive in having Schüffner's dots, forms caveolae–vesicle complexes, with few exceptions does not invade Duffy-negative red blood cells, develops chronic infections when the "sleeping" hypnozoites are aroused, and is very rare in Africa but quite prevalent in Southeast Asia where more than 80% of the 80–250 million cases of vivax occur. Today, about 40% of the world's population is at risk from vivax malaria. Until recently, it was endemic throughout North America and Europe and presently is responsible for 70% of infections in South America. In 2002, to jump start gene discovery in *P. vivax*, the remaining funds from the *P. falciparum* genome project were used to begin to sequence the Salvador I strain of *P. vivax*. After a halt of more than a year due to insufficient funds, the genome sequence was completed in 2005 [108].

The Salvador I strain of *P. vivax*, isolated from a patient with a naturally acquired infection in the La Paz region of El Salvador, was chosen for sequencing. Since its isolation this strain has been passed through human volunteers and monkeys by mosquito and blood transfer. Similar to the status of the 3D7 clone as the reference strain of *P. falciparum*, Salvador I is often regarded as the standard reference strain of *P. vivax*. Genomic DNA was obtained from splenectomized squirrel monkeys (*Saimiri boliviensis*) infected with blood stages of *P. vivax* [107]. The chromosomes were separated by pulse-field gel electrophoresis and then sequenced using the shotgun method (see p. 49). The sequencing of the *P. vivax* genome has provided much information about the genetic differences between it and falciparum: the vivax genome with ~27 million bases and an average G + C content of 42% is the richest in G + C of any *Plasmodium* genome sequenced to date [110].

Studies of the mitochondrial genes of *P. vivax* and its closest relative *P. cynomolgi* suggested that, unlike *P. falciparum*, these malarias belong within a group parasitizing Old World monkeys including various kinds of macaques [402]. Following a comprehensive analysis of mitochondrial DNA [468], it was postulated [153] that malarial parasites radiated about 3.5–4.7 million years ago during the Pliocene when there was an increasing number of geographically overlapping primate populations. (Other researchers suggest that this might have occurred more recently in the late Pliocene to Middle Pleistocene ~0.7–2.5 million years ago [212].) Using the older dates and an estimate of between 3.2 and 4.9×10^{-9} mutations per nucleotide per year, the divergence of *P. vivax* from *P. cynomolgi* was estimated to be between 1.2 and 1.6 million years ago. Precisely when and where the host switch from *Macaca sp.* to humans took place is not known as the specific lineages of the malaria parasites of non-human primates that gave rise to *P. vivax* are not known. As, during this period, it is likely that primates were spread out across most of Eurasia and Africa, the switch could have occurred at any place on the land masses where humans and the monkey hosts overlapped and when there were suitable conditions for malaria transmission. Using mitochondrial genomes from the older and more diverse populations in Asia, it suggests that the world-wide populations of *P. vivax* came into existence as a species distinct from *P. cynomolgi* between half a million and 1–2 million years ago somewhere upon the land masses of Eurasia and Africa [153]. Examination of the sequences of two nuclear genes (β-tubulin and cell division cycle 2) and an apicoplast gene (elongation factor Tu) was used to support an "out of Asia" origin for *P. vivax* as a human parasite. In this study, the estimate for the time of origin was relatively recent (~45,680–81,607 years) and again it claimed that there was a switch from *Macaca* to humans [212]. This time frame includes the accepted estimates for the migration of *Homo sapiens* into Southeast Asia.

What, then, may we conclude, however tentatively, concerning the time and location of the origin of *P. vivax*? There has been no scientific dissent concerning the close genetic relationship between human *P. vivax* and the malaria parasites of Asian monkeys. Moreover, that there was a host switch between the ancestors of these monkeys and humans of the line leading to modern *P. vivax* has not been contested. There is, however, some confusion in the literature concerning the significance of a particular undoubted fact about *P. vivax* and modern human populations in Central and Western Africa.

A gene, the Duffy antigen gene, which allows *P. vivax* to infect humans, is absent in virtually all people of pure Western and Central African descent. Such individuals cannot become infected with *P. vivax* malaria. Nowhere else in the world do people lack the gene for the Duffy antigen, except with extreme rarity. Nowhere else in the world, as far as is known, are people completely refractory to infection with *P. vivax* malaria. It has been postulated that these facts imply that *P. vivax* malaria could have been the selective force for the presence of Duffy negativity in Africa, driving the Duffy negativity gene, which is apparently without harmful effects, to be present almost throughout these populations, and thus causing the parasite, *P. vivax*, almost to disappear from them. Unsurprisingly, *P. vivax* is, indeed, virtually unknown — some say it is completely absent — from Central and West Africa.

Because it has been postulated that *P. vivax* selected for Duffy negativity in West and Central Africa, this has sometimes been understood to be a statement that *P. vivax* "originated" in Africa. The high frequency in Africa of the Duffy-negative blood group that protects against *P. vivax* infections is NOT, however, evidence that *P. vivax* "originated" in any phylogenetic sense in Africa. It IS, however, a case for postulating that a, or some, population(s) of *P. vivax* spent a lot of time in Africa probably within the last 100,000 years. That case — that *P. vivax* widely infected humans in Africa over much of the last 100,000 years — is further supported as follows.

Some 15 million years ago, Africa, Europe and Asia were a single land mass allowing for an exchange of primate malarias. Approximately 2–3 million years ago, by most molecular clock evidence, *P. vivax* appears to have diverged from other monkey malaria parasites presently found in monkeys of Southern Asia and the Western Pacific. Based on the rather incomplete fossil record, populations of early humans (long before the appearance of *H. sapiens* in Africa around 200,000 years ago) were also dispersed across Eurasia and Africa. By around one million years ago, marked changes in climate, i.e. glacial cycles, had begun to take place. These glacial cycles, commonly called Ice Ages, usually last ~100,000 years, and are interrupted by warmer periods lasting 5,000–20,000 years; currently we live in one of these warmer interglacial periods that began ~10,000 years ago. When the most recent Ice Age began ~120,000 years ago, ice sheets covered most of the Northern hemisphere with ice covering Southern Europe, Central Asia, Northern China, and Far-Eastern Siberia. Permafrost would have extended deep into North America to the Great Lakes region, as well as encompassing all of Iceland, Britain and Scandinavia. During this time, only sub-Saharan Africa and Southern and Southeastern Asia remained warm and humid, and in Southern Europe and on the shores of the Mediterranean the climate was cooler, humid and more temperate. It has been speculated [113] that, under these climatic conditions, many primate species retreated into two regions: sub-Saharan Africa or Southern and Eastern Asia and the Western Pacific. It was in these warm humid regions (called refugia) that the primates, with their malarias, would have remained, whereas in the much colder — ice-bound — places both malaria parasites and humans would have perished.

Thus, by around 60,000 years ago, when all of our modern human ancestors, i.e. *H. sapiens*, still lived in Africa, the ancestral stock of modern *P. vivax* may have survived only among these human populations in Africa. This would be consistent with the proposition of the "*P. vivax*/Duffy negativity" hypothesis outlined above. Moreover, two human population genetic analyses suggest that selection for negativity in the Duffy antigen began in African human populations at least 5,000–10,000 years ago and certainly within the past 100,000 years. Hamblin and Di Rienzo [289] estimate between about 6,500 and 97,000 years ago with an average estimate of about 33,000 years for time of fixation or start of selection; and the estimates of Seixas *et al.* [588] range from 4,350 to 31,800 years for the date of "fixation" of Duffy negativity, implying that the selection force was in play from before these times. If Duffy negativity was selected for in Africa by the presence of *P. vivax* during this period, then, because Duffy negativity is so rare in human populations outside Africa, it also implies that the ancestors of these populations must have left Africa before the gene for Duffy negativity had been selected to anything above negligible frequencies [117].

All of this is consistent, therefore, with *P. vivax* having been present as a significant selective force amongst the ancestors of West and Central Africans for a significant period within the past several tens of thousands of years. It does not, however, by any means exclude a long and ancient association of *P. vivax*-type parasites with pre-modern humans, now extinct, who inhabited the Asian and African land masses for well over a million years before and around 100,000 years ago. And it says nothing at all about where, upon the whole African and Eurasian landmass, *P. vivax* began to diverge from the stocks of malaria parasites that infected the monkeys of these continents.

At some point within the past few tens of thousands of years, but most probably during the interglacial period of the past 10,000 years, *P. vivax* would have spread back out again from its African refuge, across the Arabian Peninsula into India and Central Asia and then into the Far East. Dispersion of malaria into Asia is supported by written accounts of the disease in China 4,700 years ago and in India 3,500 years ago.

Plasmodium vivax in humans in the Americas is genetically distinct from that in Asia and Africa [406]. If, as is almost certain, the source for the introduction of *P. vivax* [113] from the Old World into the New World involved colonists and explorers, then the European vivax must have differed from that of Asia and Africa. This might have occurred, it is suggested, by *P. vivax* persisting as a parasite of those Neanderthals who had been able to escape the lethal effects of the Ice Age by occupying the warmer regions of the Mediterranean, where it came to be genetically isolated from the vivax stocks in sub-Saharan Africa and Southeast Asia. Some time after ~40,000 years ago, the Neanderthal-type vivax would have infected modern humans as they migrated into this region. Following the extinction of the Neanderthals, ~30,000 years ago, vivax malaria would continue to persist in modern human European populations. If this scenario is true then why is there no Duffy negativity in Europe? The answer may lie in the size of human populations and the manner of transmission. In the Mediterranean of the Ice Age, transmission would have been patchy among the small mobile groups of hunter-gatherers and thus there would have been transitory contacts with vivax, a situation unlike that in tropical Africa where transmission would have been widespread and affecting significantly larger human populations. As a consequence, in Southern Europe, selection for Duffy negativity "might have been a non-starter in southern European populations in the Ice Age" [113].

There is yet another conundrum concerning the evolutionary relationship of *P. vivax* to the genetically similar *P. simium* present in New World monkeys. The question raised is simple: how did "Old World" vivax malaria manage to reach the "New World" continent of South America? It has been speculated that it may have occurred by pre-Columbian migrations of humans from Asia along the cold northern route across the Bering Straits. However, if it did indeed arrive from Asia, it is much more likely to have been achieved *via* a sea passage from the Asian mainland or the Western Pacific. There is, however, no good evidence of vivax as a human infection in South America until recent, post-Columbian, times. Perhaps, as has been hypothesized, the humans in South America were unable to sustain malaria transmission because of lifestyle. However, it is as likely, if not more so, that an Asian source, *P. vivax*, was, as already suggested above, introduced by Europeans from European vivax stocks following the arrival of Columbus in the Americas in 1492. As the susceptible South American monkeys appear capable of sustaining their own transmission of *P. vivax*, a straightforward host switch, this time from humans to monkeys, would be a sufficient explanation for the presence of *P. vivax*, also known as *P. simum*, in these monkeys [113]. In this way, *P. vivax* became *P. simium* [402].

Much less is known, or at least speculated, about the origins of the other two *bona fide* human malarias, *P. malariae* and *P. ovale*, and because they pose less of a major public health problem that will probably remain so. It has been hypothesized that these malarias, as well as *P. falciparum*, may have diverged from one another 100 million years ago, a time long before "the lines leading to distinct mammalian, let alone primate, orders of today" [117]. The ancestral *P. malariae* was, and is today, a natural parasite of African great apes, and it probably cross-infected with humans as the two diverged ~5 million years ago. *Plasmodium malariae*, unlike other human malarias, can remain for decades within the human body and is therefore potentially infectious for mosquitoes throughout that time. It may have been maintained in human populations at very low infection rates among the sparse and mobile hunter-gatherer populations and was best adapted to the conditions that preceded the Agricultural Revolution. In addition, there is an indistinguishable form as a natural parasite of chimpanzees in West Africa and *P. brazilianum*, found in New World monkeys in Central and South America, is morphologically indistinguishable from *P. malariae*. Molecular genetics has failed to distinguish differences between the two and it has been contended that *P. brazilanum* is a form of *P. malariae* that was introduced only recently into New World monkeys by humans [117]. It is probable that the evolutionary roots of *P. ovale* lie in Africa, originally in apes and later in humans where its prevalence increased in the pre-agricultural populations living in Africa. Unlike its tertian malaria cousin *P. vivax*, *P. ovale* has been unable to spread much beyond its place of origin (save for isolated populations in New Guinea and the Philippines) and, to this day, it remains a warm-climate African parasite [117].

The genetic examination of malaria parasite populations in the field using SNPs and microsatellites is not simply an academic exercise practiced by "gene hunters" interested in using the latest tools of molecular biology. Genome sequencing provides more than a window on the ancient origins of malaria parasites, it can be a key to a present-day understanding of the dynamics of disease transmission and the spread of antimalarial drug resistance, and it will certainly be a critical and invaluable adjunct in the design and testing of control measures. Genome sequencing may give insight into the effects of immunity on the rate of parasite evolution and the mechanisms of immune evasion, as well as providing a basis for a more complete understanding of what triggers relapse to establish long-standing and incapacitating infections as well as death.

Chapter 12

A Not So Sweet Solution

During World War II as American and Allied forces engaged in battles in North Africa, Asia and the Pacific, the troop losses due to malaria were sometimes as great as those due to bullets and bombs. Further, with the fall of Java (today Indonesia) to the Japanese, the sources of quinine, the only effective antimalarial, became unavailable and so the Allies began to use Atabrine (also named quinacrine or mepacrine). Atabrine was a marginally effective antimalarial that turned the skin a bright yellow, caused gastroenteritis and, most disturbingly, occasionally caused temporary insanity. Atabrine was hardly the ideal antimalarial but there were no other drugs available and it did prevent death among the troops.

It was the search for the mode of action of Atabrine (and, to a lesser degree, quinine) that stimulated the earliest biochemical studies on malaria parasites. The first such study was conducted in the UK at the London School of Hygiene & Tropical Medicine by Sir Rickard Christophers and James D. Fulton using *P. knowlesi*, a malaria discovered in 1932 in a Malayan kra monkey that had been sent to the Calcutta School of Tropical Medicine *via* Singapore. Benign in the kra monkey, the malaria infection was virulent in the rhesus monkeys then available in the UK.

Fulton (1899–1974), trained in chemistry and medicine (M.B.1934), began his work on anti-protozoan chemotherapy with Warrington Yorke at the Liverpool School of Tropical Medicine and, under the aegis of the MRC, was appointed to the staff of the London School of Hygiene & Tropical Medicine where Sir Rickard introduced him to the malaria parasite. From 1939 until retirement in 1963, Fulton worked at the NIMR on chemotherapy, physiology and the immunology of all of the major protozoan diseases of humans. From 1938 to 1945, Fulton and his associates, using the available analytical tools of Barcroft–Haldane manometry (to measure gas utilization) and colorimetry found that: malaria parasites did not store glycogen, malaria-infected red cells rapidly depleted the medium of glucose, and respiration stimulated by the presence of glucose was inhibited by 0.001 M cyanide. Parasites removed from the red cell behaved similarly [239].

In 1944, the U.S. Office of Scientific Research and Development was created with a Board for the Coordination of Malaria Studies "to better understand the mode of action of antimalarials in order to protect U.S. troops in Southeast Asia, North America and the Pacific." A member of that Board was W. Mansfield Clark (1884–1964), Professor of Physiological Chemistry (1927–1952) at the Johns Hopkins School of Medicine. His colleague in the department was Leslie Hellerman (1896–1981), whose work focused on metalloenzyme functions. Hellerman and his student Marianna Bovarnick (1911–1995) involved themselves in the testing of the biochemical effects of Atabrine. Another malaria project was sponsored at Harvard under A. Baird Hastings, Chairman of the Department of Biological Chemistry and also a member of the Board. Eric Ball (1904–1979) was a graduate student and then a faculty member in Mansfield Clark's department at Johns Hopkins (1930–1939), and spent 1937 to 1938 with Otto Warburg in Berlin (1883–1970) where microdetermination of oxygen had been developed using the gas uptake (manometric) technique of Barcroft and Haldane (1902). In Berlin, as a colleague of Otto Meyerhof (1884–1951), Ball demonstrated that xanthine oxidase contained flavin as a cofactor, and measured the oxidation–reduction potentials of the cytochromes. Ball was recruited by Hastings to join the Harvard faculty and, after arrival in the fall of 1940, became acting head of the department (1943) when Hastings was heavily involved in the Committee on Medical Research, as secretary to the Panel on the Biochemistry of Antimalarials. The team at Harvard embarked on a project to cultivate the malaria parasite *P. knowlesi in vitro* and to study its metabolism. Ball, in turn, recruited to the malaria project another Harvard faculty member, Christian Anfinsen (1916–1995), who had received his Ph.D. under Hastings in 1943. The group was joined by Ralph W. McKee (1912–1992), who had joined the Harvard faculty in 1940 after receiving a Ph.D. in Biochemistry from St. Louis University where he had isolated vitamin K from fish meal, and Quentin M. Geiman (1904–1986), an Assistant Professor in the Department of Comparative Pathology. Geiman, trained as a parasitologist, received his Ph.D. from the University of Pennsylvania in 1934 working on amebas.

Manometry of intact malaria-infected red cells, malaria parasites removed from infected cells or cell-free extracts showed that malaria parasites contained cytochromes and increased flavin adenine dinucleotide levels [35]. Ball *et al.* [34] found that several 2-hydroxy-3-alkylnaphthoquinones, which were active against both *P. lophurae* and *P. knowlesi*, strongly inhibited the respiration of these parasites and suggested that they acted by inhibiting the cytochromes in its mitochondrion. However, because the effect of nine other drugs (several being sulfas, as well as quinine and Atabrine) on the respiration of malaria parasites *in vitro* did not correlate with the *in vivo* responses, Coggeshall *et al.* [140] wisely cautioned that "inhibition of respiration...alone should not be depended upon to furnish an index of chemotherapeutic efficiency but should be used as an adjunct to *in vivo* experiments."

During this same period, work on the biochemistry of bird malarias (*P. gallinaceum, P. lophurae*) was initiated at the Johns Hopkins University and the University of Chicago under contracts with the Board for the Coordination of Malaria Studies. *Plasmodium gallinaceum*, described by Emile Brumpt in 1935, naturally infects jungle fowl, but it can be maintained in chickens, and in this host it was established in laboratories across the globe. *Plasmodium lophurae*, isolated in 1938 by L. T. Coggeshall from a fireback pheasant, *Lophura igniti*, living in the New York Zoo, was maintained by blood passage in chicks and ducklings.

The malaria research group at the University of Chicago School of Medicine was formed under the leadership of Earl A. Evans Jr.

(1910–1999), who had worked as an undergraduate in Mansfield Clark's department at Johns Hopkins, received his Ph.D. from Columbia University in Hans Clarke's laboratory (one of the earliest centers for radioisotope techniques), and joined the Chicago faculty in 1937. In 1939, he worked with Hans Krebs (1900–1981) on CO_2 fixation and the newly formulated (citric) tricarboxylic acid cycle. James W. Moulder, who received his Ph.D. in 1946, was Evans' graduate student, as was Joseph Ceithaml, who received his Ph.D. in 1941, and John F. Speck. Moulder joined the faculty in the Department of Microbiology at the University of Chicago and remained there until his retirement in 1986. Thus, it was the "connected" background and training of these biochemists that inspired and influenced their work on the metabolism of malaria parasites (and, as might be expected, it also affected their interpretations).

In the 1920s and 1930s, the enzyme reactions by which glucose, in the absence of oxygen, was fermented to lactic acid or to ethyl alcohol and carbon dioxide were described in extracts made from liver, muscle, and yeast. Much of this work was due to the German biochemists Otto Embden, Gustav Meyerhoff, Jakob Parnas, and Otto Warburg. What they discovered was that, when sugar was dismembered into smaller molecules, the energy stored in chemical bonds of the sugar was turned into the cell's energy currency, the molecule ATP. After the cell took up glucose, the soluble enzymes in the cytoplasm, in the process of glycolysis (literally breaking down of glucose), split it into two molecules of pyruvic acid (or lactic acid). These transformations could occur in the absence of oxygen. With malaria parasites, glucose, found to be indispensable for *in vitro* growth, was rapidly consumed by the infected red cells. Its breakdown was by glycolysis and, in both the monkey (*P. knowlesi*) and bird (*P. gallinaceum*) malarias, the end product was lactic acid (reviewed in [240, 445, 466]). In short, the malaria biochemists had followed a well-trodden path and their findings did not deviate from the then current views on energy metabolism.

By contrast, in the early 1930s, very little was known about the stages through which sugar (and pyruvic acid) was completely oxidized to carbon dioxide and water. This became the focus of research of Hans Krebs, who would receive the Nobel Prize in Physiology or Medicine in 1953 for his discoveries of how "fragments of our foodstuffs become incorporated in the so-called Krebs cycle where they will be able to act as the fuel of life." Krebs was born in Germany, the son of a physician, and received his M.D. at the University of Hamburg in 1925. In 1926, he was appointed assistant to Professor Otto Warburg at the Kaiser-Wilhelm Institute at Berlin where he remained until 1930. For the next three years he worked in German hospitals but in 1933, with the rise of the National Socialist (Nazi) movement and virulent anti-Semitism, Krebs, nominally Jewish, had his appointment terminated. He was invited by Sir Frederick Gowland Hopkins to join the School of Biochemistry at Cambridge and was appointed Demonstrator in 1934. In 1935, he was appointed Lecturer in Pharmacology at the University of Sheffield and later Lecturer-in-Charge. In 1945, he became Professor and in 1954 was appointed Professor of Biochemistry at the University of Oxford, where he remained until his death in 1981.

Krebs noted in his Nobel banquet speech that the "research I had been doing — studying how foodstuffs yield energy in living cells — does not lead to the kind of knowledge that can be expected to give immediate practical benefits to mankind; however I am convinced that an understanding of the process of energy production will eventually help us in solving some of the practical problems of medicine."

Krebs found that, unlike glycolysis, the enzyme-catalyzed oxidative reactions could not be obtained in cell extracts made from pigeon muscle and liver by mincing and grinding the tissues as the oxidative reactions in the extracts rapidly deteriorated. It appeared to him that the enzymes involved in oxidative reactions must be particulate rather than soluble as was the case with the glycolytic enzymes. He persisted in extracting the particles and eventually found that there was a periodic formation of a number of dicarboxylic and tricarboxylic acids. He noted that it was possible to formulate a complete scheme of glucose oxidation that he named the citric acid cycle. In this cycle, pyruvate or a derivative of pyruvate (later described by Fritz Lippmann, with whom he shared the Nobel Prize, as acetyl-CoA) condensed with oxaloacetate to form citrate and, by a sequence of reactions in which cis-aconitate, isocitrate, α-ketoglutaric, succinic, fumaric, malic, and oxaloacetic acid were intermediates, one acetic acid equivalent was oxidized and the oxaloacetic acid required for condensing was regenerated. He also noted that not only was there production of energy, but the reactions of the cycle could also supply intermediates necessary for the synthesis of cytochromes and hemoglobin. In keeping with the biochemical conservatism of the time, Krebs' announcement of a citric acid cycle in 1937 was met with skepticism. Later, however, when similar findings were made in almost all living things from bacteria to yeasts to plants and animals, his former critics became supporters of the proposed cycle. Indeed, research by Speck with *P. gallinaceum* showed that pyruvate was oxidized to CO_2 and H_2O by infected red cells and free parasites "by a cycle like that proposed by Krebs" [637]. Also, parasites removed from the confines of the red blood cell oxidized the Krebs cycle intermediates succinate, fumarate, oxaloacetate, and α-ketoglutarate at rates equal to that of pyruvate. When Moulder [466] observed that quinine increased the rate of glucose consumption and the formation of lactate, he concluded that the drug acted by inhibiting the oxidative utilization of pyruvate, a view supported by the earlier work of Bovarnick *et al.* [80]. However, it should be emphasized that, in these studies, evidence for a Krebs cycle was drawn from oxygen utilization not substrate utilization and, as such, the correlations may have been unwarranted.

All of Krebs' work on the citric acid cycle was done using conventional and the then current biochemical techniques of manometry and spectrophotometry. However, developments in nuclear energy post-World War II added two other tools: the production of radioactive compounds and methods for their detection. The first radioactive isotopes for biochemical use were spin-offs following construction of cyclotrons in the US and the UK. By the 1940s, studies using radioactive phosphorous (^{32}P) were possible and later radioactively labeled carbon (^{14}C) compounds were introduced.

The use of paper and ion-exchange chromatography (much of it developed at the NIMR) provided a simple and effective means for the separation of products from substrates. The first study on malaria parasite respiration using radioactive glucose was that of Fulton and Spooner [241] with *P. berghei*. When P. T. Grant, who had completed a Ph.D. at NIMR and had collaborated with Fulton, moved to the Department of Biological Chemistry at Aberdeen University, it was logical that one of his students (I. B. R. Bowman) would study the effects of Atabrine on the metabolism of *P. berghei* growing in rat reticulocytes, using uniformly labeled ^{14}C glucose as well as glucose labeled in the C-1 and C-6 positions [81]. The reason for the use of C-1- and C-6-labeled glucose is that, by measuring $^{14}CO_2$ from glucose labeled in the C-1 position, it is possible to determine the activity of the pentose pathway as the CO_2 released is derived from the decarboxylation of the C-1 carbon of glucose-6-phosphate, whereas CO_2 released from glucose labeled in the C-6 position represents that from glycolysis. Bowman et al. [81] found that the major end product from glucose was lactate with small amounts of citrate, fumarate, malate, oxaloacetate, and α-ketoglutarate. Atabrine depressed glucose catabolism but did not affect the ratio of CO_2 from $(1-^{14}C)$ or $(6-^{14}C)$, suggesting that the drug did not affect the pentose pathway. Further, as the percent $(U-^{14}C)$ glucose converted to $^{14}CO_2$ was unaffected, it suggested that the drug acted on glycolysis; based on the redistribution of products, they suggested the most sensitive enzyme to be 6-phosphofructokinase and, to a lesser extent, hexokinase. (Earlier studies by Bovarnick et al. [80] with *P. lophurae*-infected red cells claimed that Atabrine interfered with some phosphorylation reaction and this was consistent with the findings of Speck and Evans [636] who found hexokinase to be most strongly inhibited.) The selective inhibition of 6-phosphofructokinase by Atabrine suggested to Bowman et al. that "the 6-phosphofructokinase of this parasite differed from that of the host cell" and this possibility was supported by Bueding's finding [95, 97] that the 6-phosphofructokinase of the blood fluke *Schistosoma mansoni* was 80 times more sensitive to tetravalent antimonials than the enzyme in mammalian host tissues. Indeed, the prescience of Bowman *et al.* [81] would be borne out when multiple molecular forms of enzymes (isoenzymes) with similar but nonidentical catalytic functions were discovered in my laboratory using *P. lophurae*.

Isoenzymes, the Achilles' heels of *Plasmodium*

In 1960, when I started my post-doctoral tenure at the Rockefeller University with William Trager, he had just returned from a conference in Florida where one of the other speakers was Ernest Bueding (Johns Hopkins). As early as 1955, Bueding (1910–1986) and his co-workers had shown, using immunologic methods, that the "Einheit" in biochemistry did not hold: there was enzyme heterogeneity in blood flukes (schistosomes) as well as in the tissues of the host [96, 305]. Of course, in this he was at odds with some of the most influential biochemists of the time [94]. During the Florida meeting, Bueding asked Trager if he thought that malaria parasites had enzyme heterogeneity. Returning to Rockefeller, Trager suggested that I concern myself with the possible heterogeneity of lactate dehydrogenase (LDH) in *P. lophurae* and its duckling host.

At the time, Markert and Møller [426] had already demonstrated the presence of isoenzymes in various vertebrate tissues using an overlay method after electrophoresis in starch gels and, in 1960, in the Rockefeller laboratory of Alexander Bearn, Elliot Vesell was separating LDH isoenzymes by starch gel electrophoresis and then directly visualizing the enzyme in the gel. With the help of Vesell and another Trager post-doctoral investigator, Philip D'Alesandro, I was soon able to separate lophurae LDH from that of the red blood cell [593].

In the 1960s, Nathan Kaplan and his colleagues at Brandeis University were concentrating on isoenzymes that required nicotinamide adenine dinucleotide (NAD) as a cofactor. The Kaplan laboratory was actively synthesizing analogs of NAD and had found that the ratio of enzyme activity with NAD and its various analogs (especially acetylpyridine NAD) was a very sensitive measure of the differences of various dehydrogenases (especially

LDH) in different species and in different organs. When Kaplan heard of my work with lophurae LDH, he invited me to Brandeis to give a seminar. During that visit it was discovered that *P. lophurae* LDH had an exceptionally high affinity for the acetylpyridine analog of NAD [595]. Subsequently, this proved to be the case for other species of *Plasmodium* and, today, LDH activity with acetylpyridine-NAD serves as the basis for the diagnostic tests OptiMal and Malstat [129, 422] for human malaria infections and also as an immunodiagnostic [672]. It has also been used to quantitatively measure the adherence of *P. falciparum*-infected red cells *in vitro* [531] and sequestration in comatose children [590].

The LDH gene from *P. vivax* has been cloned and sequenced [691] and shown to have >90% homology to that from *P. falciparum*. The crystal structure of LDH from *P. vivax* with nicotinamine adenine dinucleotide (NADH) and reduced acetyl pyridine adenine dinucleotide (APADH) has shown the active sites and cofactor-binding pockets to be similar to those of *P. falciparum* and distinct from the human enzyme [122]. In addition, the LDH genes from all four human species of *Plasmodium* have been cloned, sequenced, the kinetic properties of the proteins compared [93], and a series of azole-based compounds that inhibit the LDH of *P. falciparum* have been described [102].

In addition to LDH, we studied the heterogeneity of several other dehydrogenases including malic dehydrogenase (MDH) and glutamic dehydrogenase (GDH). We found the MDH isoenzymes from *P. lophurae* and *P. berghei* to differ from those of the host [594]. We speculated that MDH might play a role in reoxidizing NADH formed at the triose phosphate dehydrogenase step of glycolysis. Later, MDH from *P. falciparum* was purified [394] by conventional biochemical methods and, through the use of an antiserum, was shown to be cytoplasmic and not mitochondrial [395]. The falciparum MDH gene, located on chromosome 6, has been cloned and expressed [690] and has high homology with bacterial LDH and MDH. In support of our earlier work with *P. lophurae*, the *P. falciparum* MDH did not use NADP, but was NAD/NADH-specific and used the acetylpyridine analog of

NAD; it was speculated that MDH could complement the function of LDH in *P. falciparum*.

In other studies, we found that glutamate was one of the products of carbon dioxide fixation[2] by *P. lophurae*-infected red cells and it could be metabolized [605]. This prompted us to investigate the enzyme GDH [601]. The lophurae GDH was in the cytoplasm and was NADP-specific. We suggested that, in the conversion of glutamate to α-ketoglutarate, "a product of the reaction that may be of considerable significance is the formation of reduced NADP...which can be used in reductive synthesis." (Our other hypothesis that α-ketoglutarate might be a supplemental source of energy by feeding into the Krebs cycle would subsequently turn out to be false.) Later, Vander Jagt *et al.* [704] showed the falciparum enzyme to be parasite-specific and thus a useful tool for monitoring the contamination of sub-cellular fractions (intended for use as antigens). Vander Jagt *et al.* [703], using conventional biochemical methods, found GDH purified from *P. falciparum* to have a Mr of 230 kDa and to be capable of producing reduced nicotinamide adeninine dinucleotide phosphate (NADPH) at about 10% of the capacity of the red cell with activity increasing throughout the growth of the parasite. The falciparum GDH was not inhibited by CQ.

They went on to suggest that, as the isocitric dehydrogenase had higher activity, it could offer the parasite an additional pathway for NADPH synthesis,

Further characterization of the falciparum GDH has come from cloning and expression in *E. coli*. It is a homohexamer with a subunit Mr of 49.5 kDa; its 470 amino acids show 23% sequence homology with the human enzyme and 50% homology with that from lower eukaryotes and bacteria. The enzyme has been crystallized and the most obvious differences between the human and malaria enzyme are the sub-unit interfaces of the hexameric protein and the unique N-terminal extension of the falciparum GDH [718, 729].

---

[2]In CO_2 fixation, CO_2 is initially added to pyruvic acid; the endproducts are alanine, malate, citrate, aspartate, and glutamate with oxaloacetate and α-ketoglutarate as intermediates.

Although these investigators did not directly demonstrate that the source of glutamate was glutamine and not carbon dioxide fixation, their proposal receives support from the following observations: there is a 100-fold increase in glutamine accumulation by *P. falciparum*-infected red cells [389] and the addition of glutamine to RPMI 1640 medium enhances the *in vitro* growth of *P. falciparum*. According to their proposal [718], the glutamine would be converted to glutamate by the parasite synthase, glutamine amide α-ketoglutarate amidotransferase: glutamine + 2 α-ketoglutarate + NADPH- → 2 glutamate + NADP. Although the glutamine amide α-ketoglutarate amidotransferase reaction consumes one NADPH, it produces two molecules of glutamate that can yield via GDH two NADPH that can, in turn, serve as an electron source for the antioxidative enzymes glutathione reductase and thioredoxin reductase, and glutamine could also play a key role in the Krebs "cycle."

Most of the enzymes of glycolysis found in the bird and monkey malarias using conventional biochemical methods have been found in *in vitro*-grown *P. falciparum*-infected red cells [561]. With the completion of the *P. falciparum* genome there was further evidence for the presence of glycolytic enzymes and, by 2007, almost all of the genes for the isoenzymes of glycolysis had been cloned and sequenced [745].

Tackling the Krebs "cycle"

The DNA of the *Plasmodium* mitochondrion is much smaller than that of most other eukaryotes — 6 kb in length — and is linear rather than circular. It was first isolated in 1987 [696] and its sequence was characterized as mitochondrial by the Vaidya and Wirth laboratories two years later [10, 695].

Akhil B. Vaidya (1947–) received his Ph.D. in 1972 from the University of Bombay, India. He was trained as a retrovirologist at a time when molecular biology was moving into its recombinant DNA phase. His laboratory at Drexel University was involved in constructing genomic libraries and fishing out endogenous retroviruses. William Weidanz, a colleague in the department who worked on rodent malaria parasite immunology, suggested that he construct genomic libraries of malaria parasites to identify antigen genes. A new graduate student, Prema Arasu, did just that and, at Vaidya's suggestion, checked the quality of the library by screening it with a probe made from labeled total parasite DNA; the idea was that repeated sequences in a eukaryotic genome would be represented frequently in a good-quality library and be detectable by the labeled total DNA. Arasu's library was quite good and many clones lit up, apparently containing the repeated sequences. These clones were put aside for a future leisurely look. In 1988, the sequence of one of these clones was obtained. Surprisingly, the 6 kb tandemly arrayed "repeated" sequence encoded mitochondrial electron transport genes. Vaidya had literally stumbled onto the mitochondrial genome of malaria parasites.

The discovery of the smallest mitochondrial genome, its unusual gene content and organization altered Vaidya's scientific career. He gave up his funded retrovirus program, canceled the subscription to the *Journal of Virology*, and focused his laboratory's interests on these unusual mitochondria. Vaidya found that the sequence of the electron transport chain proteins was significantly divergent from that of their mammalian counterparts, explaining the selective toxicity of antimalarials targeting the parasite mitochondrion.

The *P. falciparum* mitochondrion contains 30–100 copies of the genome yet it encodes only three proteins (cytochrome b and subunits I and III of cytochrome oxidase) and ribosomal RNA. The ribosomal RNA genes are fragmented and scrambled; it does not encode any tRNAs, ribosomal proteins or ATPase sub-units. As such, the *Plasmodium* mitochondrion "denotes an endpoint in the slimming down process whereby the ancient progenitor of all mitochondrial genomes began the slow march to near total dependence on host nuclei" [738].

In eukaryotes, the electron transport chain consists of four integral membrane complexes localized to the inner mitochondrial membrane: complex I (NADH–ubiquinone oxidoreductase), complex II (succinate–ubiquinone oxidoreductase), complex III (ubiquinol–cytochrome c oxidoreductase), and complex IV (cytochrome c oxidase), plus coenzyme Q (ubiquinone) and cytochrome c. As first shown by Fry and Beesley [238], the malaria parasite electron transport chain is different in lacking complex I; however, a single sub-unit NADH dehydrogenase is present and is homologous to that found in plants, bacteria, and yeast but not animals [384, 693, 694, 700].

Electron flow in the mitochondrion of *Plasmodium* involves the substrates NADH, succinate, malate, dihydro-orotate (DHO), and glycerol-3-phosphate, located within the intermembrane space or the matrix, being acted upon by the NADH dehydrogenase (rotenone-insensitive [alternative] complex I), succinate dehydrogenase (complex II), malate–quinone oxidoreductase, dihydro-orotate dehydrogenase (DHODH), and glycerol-3-phosphate dehydrogenase present within the inner membrane. The electron acceptor of all of these dehydrogenases is ubiquinone, oxidized by complex III (cytochrome bc_1 complex), the complex that transfers electrons to cytochrome c. The reduced cytochrome c is oxidized by complex IV (cytochrome c oxidase) that transfers electrons to oxygen. It is the oxidation of cytochrome c (first described by Ball in the 1940s) by molecular oxygen that explains why malaria parasites require small amounts of oxygen for their reproduction.

However, is it possible that malaria parasites also have a functional and conventional Krebs cycle? Working with *P. lophurae* and finding significant increases in the liberation of $^{14}CO_2$- and ^{14}C-labeled intermediates from labeled glucose (and still operating under the influence of earlier findings/suggestions that the bird malaria *P. gallinaceum* has a Krebs cycle), we claimed that this was evidence for a canonical Krebs cycle in *P. lophurae* [606]. Similarly, Shakespeare *et al.* [591] at the NIMR suggested that, in *P. knowlesi*, the production of CO_2 from $6-^{14}C$ glucose "was the result of breakdown of pyruvate catalyzed by reactions of the Krebs tricarboxylic acid cycle." Indeed, gene expression profiles of blood-stage parasites have indicated an apparent coordinated expression of genes encoding Krebs cycle enzymes; however, as the expression is usually so low, it is understandable why earlier biochemical assays failed to detect such activities. As plasmodial mitochondria lack pyruvate kinase, the conventional generation of acetyl-CoA within the mitochondrion cannot occur and, as a consequence, a block exists in the step necessary for initiation of the Krebs cycle. In retrospect, it now appears that Shakespeare *et al.* and Sherman *et al.* were mistaken in attributing CO_2 production from glucose to be indicative of a conventional Krebs cycle.

The movement of electrons across the inner mitochondrial membrane drives the synthesis of ATP by a multiprotein enzyme complex, F_0F_1 ATP synthase (Complex V). The genome of *P. falciparum* lacks the genes for encoding sub-units a and b in the F_1 ATP synthase and hence the enzyme cannot function to generate ATP [697, 700]. (This confirms earlier work that suggested that the mitochondrion contributes little to the ATP pool of the parasite [238].) Because of this it is unlikely that, even in the insect stages of malaria, there is a conventional Krebs cycle. Perhaps a key point against the presence of a canonical Krebs cycle in all asexual stages of malaria is that found more than 50 years ago: essentially all of the glucose consumed by *Plasmodium* ends up as lactate.

As the Krebs cycle enzymes do not appear to function for ATP production, what role might they play? As had been suggested by Krebs, they function biosynthetically, providing succinyl-CoA for heme biosynthesis. In this pathway, succinyl-CoA is conjugated to glycine by δ-aminolevulinate synthetase to form δ-aminolevulinate (δ-ALA), the first substrate in the heme synthetic pathway. δ-ALA synthetase has been localized to the mitochondrion; however, as all of the other enzymes in the heme biosynthetic pathway are found in the apicoplast, it poses a problem as to the mechanism for shuttling substrates between these two organelles. A way around this may be the findings of Varadharajan *et al.* [709], in which ferrochelatase was localized to the apicoplast. Surprisingly, as the cytoplasm of the parasite contains all of the heme biosynthetic enzymes (which are host-derived), it is possible that, when provided with the mitochondrial-formed δ-ALA, heme assembly takes place there. Although such a scheme has attractive aspects, it is not free from problems: How are the reactions regulated? How are the apocytochromes chaperoned? How is free heme maintained prior to import into the mitochondrion? What mechanisms serve for heme transport? Finally, yet to be discovered is the source of iron used by ferrochelatase.

Although the *Plasmodium* mitochondrion, with its limited number of cristae, is not a significant source of ATP, its electron transport system is still critical for parasite survival. Its primary metabolic function, as shown in a series of elegant experiments from Vaidya's laboratory, is to regenerate (through complex III) the oxidized ubiquinone required as the electron acceptor for DHODH, an enzyme essential for the *de novo* synthesis of pyrimidines [500]. Although mitochondrial electron transport can be dispensed with by providing a cytosolic bypass for orotate synthesis, mitochondrial membrane potential is still required for other metabolic steps relegated to the mitochondrion.

For more than half a century, the malaria parasite Krebs cycle has been a black box. Blood plasma glucose provides the rapidly dividing asexual stages of *Plasmodium* with a readily available source of energy by glycolysis. As malaria parasites do have mitochondria, albeit with fewer cristae, why does *Plasmodium* rely exclusively on anaerobic glycolysis for generating ATP rather than the more efficient mitochondrial pathway of oxidative phosphorylation? Further, the *P. falciparum* genome sequence has revealed that all the genes encoding enzymes for the complete oxidation are present and these are actually transcribed during blood-stage growth of the parasite.

Using mass spectrometry of carbon-labeled glucose and glutamine coupled with the information provided by the sequencing of the *P. falciparum* genome, "the fundamentally different from any described" pathway has been elucidated [489]. Indeed, although *Plasmodium* synthesizes all of the Krebs cycle metabolites, the pathway that it employs is unique: instead of a cycle, the pathway is two-branched. Both pathways begin not with glucose (as would be expected from the work of Krebs, Shakespeare *et al.* [591] and Sherman *et al.* [606]) but with the amino acid glutamine, which is rapidly taken up by the malaria-infected red cell from the blood plasma; once inside the parasite, this is enzymatically deamidiated to glutamic acid and then to α-ketoglutarate. Enzymes in one branch convert the latter into isocitrate and then to citrate; the citrate, in turn, is cleaved into a two-carbon fragment that is converted into acetyl-CoA and oxaloacetate, and the latter is then reduced to malate by MDH. The malate is excreted from the parasite. The glutamine-derived acetyl-CoA is used for histone acetylation. In the other branch, the α-ketoglutarate is oxidized to malate to generate both reducing power and succinyl-CoA, an essential precursor of the biosynthesis of heme. Again, malate is excreted. Typically, the isocitric dehydrogenase of the Krebs cycle uses the cofactor NAD; however, in the *P. falciparum* mitochondrion, the required cofactor is NADP and this is generated by GDH (an enzyme that we [601] described in *P. lophurae* four decades ago!).

It has been hypothesized that this novel branched pathway is "an evolutionary trade-off in which metabolic flexibility is lost to optimize growth" [489] of the parasite within the red blood cell. The abundant supply of glucose from the blood plasma ensures that the parasite is provided with a constant supply of energy (*via* fermentative enzymes in its cytoplasm), whereas the higher levels of glutamine are a source of molecules five carbons in length that drive the production of ubiquinones, succinyl-CoA and acetyl units in the mitochondrion. The significance of this genome-related work is that it provides a rational basis for the addition of glutamine for good *in vitro* growth of *P. falciparum* and it allows for a better understanding of mitochondrial electron flow, heme biosynthesis and histone acetylation, all of which are current or suggested targets for yet to be discovered antimalarials.

Chapter 13

To Search and Find

To understand a disease, the illness must be described objectively in a reproducible way. Since the time of Laveran it has been recognized that an objective description of the disease malaria means consistently finding the causative agent, *Plasmodium*. It is estimated that presently worldwide there are millions of cases of malaria. With massive increases in international travel, malaria imported from these regions can pose a significant and ever-increasing problem for those living in developed countries. Indeed, it is reported that ~30,000 travelers from industrialized countries contract malaria annually and, despite treatment, between 1% and 4% of those who acquire *P. falciparum* will die. Accurate and practical diagnostic tests are critical for control and treatment, and it is estimated that a sensitive and specific tool for diagnosis (requiring minimal infrastructure) could potentially avert over 100,000 malaria-related deaths and ~400 million unnecessary treatments [297]. There is a pressing need to circumvent the limitations of light microscopy — still the gold standard for diagnosis.

One of the earliest of the molecular techniques for diagnosis (often referred to today as rapid malaria tests or rapid diagnostic tests [RDTs]) was the finding of a parasite-specific enzyme lactic dehydrogenase (LDH) in infected blood samples. However, when I (in collaboration with the Brandeis University laboratory of Nathan Kaplan) discovered that the bird malaria *P. lophurae* had a distinctly different LDH, little did I realize that this would be useful for the diagnosis of human malarias.

Kaplan received his Ph.D. in 1943 with David M. Greenberg at UC Berkeley working on the metabolism of phosphate in rat liver using ^{32}P, and then did post-doctoral work with Fritz Lippmann (who had apprenticed with Otto Meyerhoff in Germany where he was able to determine the structure of coenzyme A, thereby establishing its role in the Krebs cycle). Kaplan's laboratory at Brandeis University (1957) found that the ratio of enzymatic activity with NAD and its various analogs (especially the acetylpyridine NAD) was a very sensitive measure of the differences of various dehydrogenases (especially LDH). When a fresh preparation of packed frozen *P. lophurae* was taken by me to Brandeis and analyzed in Kaplan's laboratory, it was discovered that lophurae LDH had an exceptionally high affinity for the acetylpyridine analog of NAD [595]. This was later found to be the case for other species of *Plasmodium* and, today, LDH activity with acetylpyridine–NAD serves as the basis for the diagnostic tests OptiMal and Malstat [129, 422] for human malaria infections and also as an immunodiagnostic [672].

Rapid diagnostic tests

As plasmodial LDH (pLDH) utilizes acetylpyridine adenine dinucleotide (APAD) ~200 times more rapidly than human LDH or, in our studies, duck LDH, this activity can be used to measure the presence of parasites in fresh blood samples. Usually the detectable range is between 0.2% and 10% of infected red cells; however, the degree of infection cannot be determined. Another limitation, in addition to sensitivity — especially in the field — is the necessity to have live parasites for enzyme activity. To counter this limitation, the test has been modified from one that is strictly enzymatic to one that is immunologic. For a targeted parasite antigen such as LDH, two sets of antibodies — one a capture antibody and the other a detection antibody — are used. The capture antibody is sprayed as a strip onto a nitrocellulose membrane to which it binds. These fixed antibodies serve to extract and bind parasite antigen from a sample of lysed blood when it is drawn up the strip (as in paper chromatography). The second antibody, the detection antibody, is conjugated to a dye and this will bind to the parasite antigen that has been captured on the nitrocellulose strip, producing a stained line. Another antibody specific for the detection antibody is bound to the nitrocellulose and acts as a control line. One of the first commercially available immunochromatographic tests using LDH was OptiMal. Since that time, other manufacturers have developed variations on the theme of pLDH using other parasite antigens such as aldolase or *P. falciparum*-specific HRP-2, a water-soluble protein found in the cytoplasm and on the surface of the infected red cell, which diffuses into the plasma. The amount of HRP-2 increases during asexual development. However, due to sequestration in *P. falciparum* infections, the detection of antigen may not correlate with the actual parasite biomass.

In 2006, the TDR and the Foundation for Innovative New Diagnostics launched an evaluation program to assess the performance of the commercially available malaria RDTs. Forty-one products — all based on pLDH, HRP-2 and aldolase — from 21 manufacturers were evaluated for sensitivity of performance, stability, and specificity. Each product varied in quality, stability, and utility, and the WHO is currently developing measures to ensure lot-to-lot quality controls before these can be used in lieu of the "gold standard" — light microscopy.

The advantage of the RDTs is that they require little technical experience and minimal equipment; however, the disadvantages are cost (~50 cents/test), low sensitivity (~100 parasites/μl) and, in the case of HRP-2, the persistence of antigen in the blood plasma when parasites are no longer present in the red cells.

Nucleic acid-based tests

The RDTs described above were not significantly improved upon by the Malaria Genome Projects; nor were most of the nucleic acid-based tests that were developed as diagnostics in the early 1990s using the PCR. In the conventional PCR assay — often developed by targeting the multi-copy small sub-unit 18S ribosomal gene found in all species of malaria — the 18S RNA is amplified using primers that anneal to nucleic acid sequences that are highly conserved among the species. Next, the PCR products are subjected to a second round of amplification using four different primers, one for the gene for each species. Products are then analyzed by gel electrophoresis or species-specific probe hybridization or dot blots. The advantages of using PCR are that the assay can be automated using a plate reader and is highly sensitive (detecting 1 parasite/μl); however, it is not quantitative, is labor intensive, time consuming and requires expensive equipment.

To obviate some of these limitations, real-time PCR assays have been developed. This method is used to amplify and simultaneously quantify a targeted DNA molecule. It enables both the detection and quantification (as an absolute number of copies or relative amount when normalized to DNA input or additional normalizing genes) of one or more specific sequences in a DNA sample. The reaction is prepared as usual, with the addition of fluorescent double-stranded DNA dye, and the reaction is run in a real-time PCR instrument. After each cycle, the levels of fluorescence are measured with a detector; the dye only fluoresces when bound to the double-stranded DNA (i.e. the PCR product). With reference to a standard dilution, the double-stranded DNA concentration in the PCR can be determined. The latter is necessary because, over the 20–40 cycles of a typical PCR, the amount of product reaches a plateau determined more by the amount of primer in the reaction mix than by the input of the template/sample. Relative concentrations of DNA present during the exponential phase of the reaction are then determined by plotting fluorescence against cycle number on a logarithmic scale (so an exponentially increasing quantity will give a straight line). A threshold for detection of fluorescence above background is then determined. The cycle at which the fluorescence from a sample crosses the threshold is called the cycle threshold. The quantity of DNA theoretically doubles every cycle during the exponential phase and relative amounts of DNA can be calculated. As all sets of primers don't work equally well, one has to first calculate the reaction efficiency. Thus, by using this as the base and the cycle difference cycle threshold as the exponent, the precise difference in starting template can be calculated. (Amounts of RNA or DNA are then determined by comparing the results with a standard curve produced by real-time PCR of serial dilutions [e.g. undiluted, 1 : 4, 1 : 16, 1 : 64] of a known amount of RNA or DNA.) As mentioned above, to accurately quantify gene expression, the measured amount of RNA from the gene of interest is divided by the amount of RNA from a housekeeping gene measured in the same sample to normalize for possible variation in the amount and quality of RNA between different samples. This normalization permits accurate comparison of the expression of the gene of interest between different samples, provided that the expression of the reference (housekeeping) gene used in the normalization is very similar across all of the samples. Choosing a reference gene fulfilling this criterion is therefore of high importance, and often challenging, because only very few genes show equal levels of expression across a range of different conditions or tissues.

In real-time PCR, the sensitivity and specificity are high and, with a closed vessel system, cross-contamination is low and the turnaround time can be high. An advantage of real-time PCR is that it can also be used to detect drug resistance genes so that emerging drug resistance can be monitored. However, the high cost of this method precludes its use in developing countries and the need for highly trained technicians as well as quality control and equipment maintenance limit its deployment in resource-poor settings. To circumvent the limits of large-scale application of conventional and real-time PCR, i.e. cost and human resources, pooling of samples prior to diagnostic testing (i.e. if a pool test is positive then the individual samples in the pool are evaluated) has been employed [663]. This method can obviate >90% of individual tests.

As noted earlier, because RDTs are less sensitive with parasite numbers below $100/\mu l$, individuals who live in endemic areas and are asymptomatic (but who may serve as a reservoir as their blood is infectious for mosquitoes) may be missed. As a consequence of this lack of identification they may introduce malaria into regions that have been freed of disease. Because with conventional and real-time PCR the sensitivity is 0.5 parasites/μl and the threshold of sensitivity is 20 times more sensitive than light microscopy and 100 times more sensitive than OptiMal, these methods are suitable for endemic areas. In a study conducted with specimens from Brazil, it was estimated that, were RDTs to be used exclusively for diagnosis, 30% of the false negatives would have been missed [245].

As the PCR-based methods require expensive equipment, are time consuming in sample preparation and run, and the reagents used are costly, at present it has not been developed for high throughput for thousands of samples. A recent study [650] used a *Plasmodium*-specific conventional PCR based on the 18S rRNA gene followed by dot-blot detection using species-specific probes and CYTB based on the cytochrome b gene, followed by detection using SNP analysis of 11 previously published sequence alignments. In both methods, the blood was spotted onto filter paper, air dried, stored at 2°C and then processed for PCR. The PCR products were transferred to a Bio-Dot apparatus for detection using digoxigenine-labeled sequences. Because the copy number of the rRNA gene varies from four to eight and the copy number for the cytochrome b gene is estimated to be 30–100 per parasite, the latter has greater sensitivity. The dot 18S and CYTB methods detected 0.75 and 0.075 parasites/μl, respectively. The authors state: "Many of the phases of the current processes are amenable to adaptation to pipetting robots including DNA extraction and detection by real-time PCR" and "could be transferred to a microarray format." This "should allow for detection and species identification of thousands of samples in a mass screening program in less than three days of laboratory work." Perhaps, by the selection of highly specific and sensitive probes using sequences from hypothetical genes and the development of novel and faster extraction techniques including more rapid PCR methodologies, the time for diagnosis as well as cost could be reduced so that it could be applied in endemic areas [245].

Loop-mediated isothermal amplification

Some investigators suggest loop-mediated isothermal amplification (LAMP) to be a more appropriate nucleic acid-based technique for diagnosis in endemic areas [210]. A specialized polymerase amplifies sequences from double-stranded DNA such that the products fold into looped structures, causing the reaction mixture to become turbid. Turbidity can be detected by eye or in a turbidimeter. This assay bypasses DNA isolation by using heat-treated blood supernatants. It does not require DNA denaturation and thus eliminates the need for expensive PCR thermal cyclers. Sensitivity is high (10–100 gene copies can be detected), the method is rapid, relatively inexpensive (35 cents/reaction), semi-quantitative, and the equipment used (a $5,000 turbidimeter) is much less expensive than the equipment required for PCR. However, the reagents must be kept cold and clinical trials need to be undertaken to validate the feasibility and clinical utility of LAMP [210].

Erdman and Kain [210] note: "While investments must be made to harness the power of high-tech molecular tools for application to this threatening disease, it is also critical to invest in the reality of malaria — that is to improve the quality, availability and affordability of existing malaria diagnostics and surveillance techniques currently used in the developing world, including microscopy." Indeed, the advantages of the use of the information provided by the Malaria Genome Project for diagnosis are yet to be fully realized.

Chapter 14

The Elusive Vaccines

In the late 1800s, the great microbe hunter Robert Koch was dispatched by the German Foreign Office to German New Guinea (now Papua New Guinea) to study the pattern of disease of those living in the regions where malaria was endemic. Koch observed that malaria was more apparent in the blood of young children and almost completely absent in adults, that individuals constantly exposed to falciparum malaria develop resistance to the disease, and there may be an outbreak of disease when an area receives an influx of malaria-naïve individuals. Koch concluded that immunity to malaria requires constant exposure over a number of years and sterile protection never (or almost never) occurs. Clinical malaria, i.e. the absence of disease in the presence of blood parasites, is characterized by a reduction in parasite densities with a concomitant reduced risk of illness. "The slow acquisition of anti-parasite immunity is often thought to reflect the need for long-term exposure to the parasite" [645] as well as the age of the individual. Koch suggested that immunity against one *Plasmodium* species offers little protection against other species. Despite these century-old descriptions of acquired immunity, not only are the mechanisms of anti-parasite immunity unclear but so too are the targets (antigens). Indeed, despite intense research over many years, there is still no licensed malaria vaccine. Without identification of the protective antigens, the malaria vaccine is likely to remain elusive. Simply put, it is taking so long to develop a vaccine against malaria because the relevant antigens from the many life cycle stages (see p. 22) are yet to be identified, isolated, purified, and tested for efficacy in humans.

Antigens before the genome

In the early1940s, Wendell Gingrich showed that, for canaries, immunization with a dozen intravenous injections of large numbers of formalin- or heat-killed *P. cathemerium*-infected red blood cells was protective. If the dose of infected red cells was reduced, however, or if the number of immunizing doses was fewer, the vaccine failed to influence the severity of the infection; despite this, most birds did survive (eight out of 11). Jacobs [345] showed that ducklings vaccinated with insoluble extracts of *P. lophurae*, when mixed with a staphylococcus toxoid (that had been used to make non-antigenic materials more antigenic), showed increased protection in four out of six ducklings against challenge. He concluded that the antigens were either too insoluble or partially antigenic, or both, and that the parasite material did not efficiently provoke antibody production. This suggested to Jules Freund, whose name today is remembered most for the practical advantage of his adjuvant (from the Latin word *adjuvare* meaning "to aid") techniques for immunization, that a vaccine might be possible if the potency of the malarial antigens could be enhanced. Freund and co-workers [235] incorporated formaldehyde-killed *P. lophurae* combined with a lanolin-like substance, paraffin oil, and killed tubercle bacteria to vaccinate two-month-old white Pekin ducklings. (Older ducks weighing 2–3 kg were used because the experiment required a period of at least two months from the time of first vaccination to challenge and, as these were probably age-immune, the challenge dose had to be high, i.e. 1–5 billion parasitized red blood cells to produce a lethal infection in control birds.) After three injections given four weeks apart and challenged with blood-stage parasites by intravenous inoculation one month after the last immunizing injection, seven out of eight vaccinated ducks survived in contrast to 50% mortality in the control birds. The survivors, however, had a few parasites in the blood.

In a more extensive study, Freund *et al.* [669] found that when vaccinated ducks received two injections of vaccine none died of malaria; they also found, however, that four out of five developed a low-grade blood infection and that one had no detectable parasites in a blood smear. With a single vaccination there was also evidence of protection, and this seemed to persist for several months. Although protection could be achieved without adjuvant, it was of relatively short duration. The work of Freund and co-workers had shown that it was possible to produce immunity to malaria without a prior infection. However, immunization did not prevent the animal from becoming infected altogether. In addition, inoculations had to be done intramuscularly, and protection appeared to be related to adjuvant-induced local tissue reactions, which at times were severe. The intramuscular lesions produced by Freund's adjuvant preclude its use in humans; it is clear from Freund's experiments, however, that vaccination against malaria could be achieved (but only when a suitable and potent adjuvant was used).

In 1932, Sinton and Mulligan isolated a malaria parasite from a long-tailed Malayan kra monkey (*Macaca irus* = *M. fascicularis*) imported from Singapore to Calcutta. It was maintained at the Malaria Institute of India (in Delhi) in rhesus monkeys and named *P. knowlesi*, after the Director of the Institute Robert Knowles [623, 734]. In 1932, Knowles and Das Gupta succeeded in experimentally transmitting *P. knowlesi* from monkey to human and then from human to human by inoculation of infected blood. In the early 1930s (before the advent of penicillin), this monkey malaria was widely used for the treatment of general paralysis of the insane, tertiary syphilis being one of the main reasons for admission to neuropsychiatric institutions. However, it soon became apparent that this infection could become uncontrollable and, after several fatalities, its use was discontinued in favor of the less virulent *P. vivax* (see p. 198).

Eaton and Coggeshall [199], working in the Rockefeller Foundation Laboratories, found that it took between one and five *P. knowlesi*-infected red blood cells to produce an infection and, once the infection was established, it was almost invariably fatal unless treated. Further, immunization with parasites killed by heat, freezing and thawing, formalin or drying produced no resistance to challenge. By contrast, repeated reinfection of chronically infected monkeys enhanced the potency of immune serum [143] and, when injected into animals with an acute infection, it had a variable but generally depressing effect on the course of infection if administered in daily doses; relatively large amounts of immune serum given shortly before or at the time of injection of parasites, however, had only a minor effect [142]. The protective effect of immune serum was more marked when the challenge was with 10 *P. knowlesi* parasites than with 1,000 parasites, and protection was more effective when immune serum was incubated with the parasites before injection and given daily during the incubation period and the first stages of infection; if administered after parasites appeared in the blood, however, it was more difficult to protect the monkey in this way [141]. Thus, in protection experiments, survival or death of the rhesus monkey was dependent on both the amount of immune serum and the number of *P. knowlesi* parasites used for challenge.

Armed with Emil von Behring's demonstration that immunity stemmed from antibodies to toxins in the serum, and with the passive transfer of immune serum by Coggeshall [143], Michael Heidelberger was optimistic that a malaria vaccine could be developed for humans. In 1942–1945, as a Professor of Immunochemistry at Columbia University and under a contract recommended by the Committee on Medical Research funded by the Office of Scientific Research, he and a graduate student, Manfred Mayer, prepared a vaccine from *P. vivax*-infected red cells obtained from malaria-infected troops who had returned to the US from the South Pacific and volunteered to donate blood [303]. They used *P. vivax* because it was "safe" — it usually did not cause serious or fatal infections and it had been used for decades in the treatment of those with late-stage syphilis (see p. 197). Blood containing large, hemozoin-laden parasites were used as the starting material; this was fixed with dilute formaldehyde, lysed by freezing and thawing and "purified" by centrifugation. It took 15–20 hours to process 500 ml of blood! Using this material, Heidelberger conducted the first controled active immunization studies for vivax malaria. Some 200 volunteer patients suffering from chronic relapsing vivax malaria were divided into three groups. One group received routine therapy, a second group received normal red cell membranes, and the third group received the vaccine, consisting of 2–4 billion formaldehyde-killed parasites administered by intracutaneous, sub-cutaneous and intravenous routes in divided doses over a period of four to five consecutive days. The relapse rate in the three groups was the same [302]. A subsequent experiment with late-stage syphilitics found no protection from challenge [304]. They also administered formaldehyde-killed sporozoites to healthy volunteer inmates, and again this vaccine had no effect on susceptibility to infection by the bite of *P. vivax*-infected mosquitoes. Heidelberger concluded the experiments to be "a complete failure" [304] and he never again worked on a vaccine for malaria.

Heidelberger's disappointing findings for a malaria vaccine were a direct contrast to those of Freund and co-workers, who had reported [235] that monkeys could be immunized against malaria with formaldehyde-killed parasites combined with paraffin oil and killed tubercle bacilli (Freund's complete adjuvant [FCA]). In a preliminary monkey trial [233], each dose was divided into three or five equal portions, injected into the sub-cutaneous tissues of the neck, armpits, and groin, and then the monkeys were challenged with infected blood. In one of seven vaccinated monkeys, no parasites were seen in blood smears and in the others there were fewer than 10 parasites per 100 red blood cells; in the latter case, the numbers declined until none could be seen. In a more extensive study [234], the vaccinated animals developed low-grade infections, and none died from malaria when challenged; the low-grade blood infection was of short duration and there were no relapses for as long as six months. It was not possible to substitute for the killed tubercle bacteria, however, and no protection occurred when peanut oil was used instead. When killed parasites in saline were injected or when tubercle bacteria were replaced by alcohol–ether extracts of the bacilli, cholesterol or lecithin, there was no protection. The authors stated [234] that "Because of the possible application of vaccination to man it seemed desirable to prepare antigen by a method which Heidelberger and associates found successful in obtaining *P. vivax* parasites from human blood." Of the three monkeys vaccinated, two died of malaria and the course of the disease was similar to controls; the third monkey developed a low-grade infection lasting 28 days, and death occurred on day 34.

The conclusions to be drawn from using bird and monkey malarias are inescapable: protection could be achieved after vaccination with the various stages of the parasite although, in most instances, this required the use of an adjuvant. Protection against reinfection was stage-specific and more often than not was non-sterilizing. And, in spite of the promise of malaria vaccination, the protective antigens had still not been identified.

Up until 1960, serious reservations about a malaria vaccine for humans were raised because of clinical, logistical, and economic considerations. Because (as had been first noted by Koch) immunity to malaria in humans develops slowly and incompletely, the assumption was that vaccination would not improve on the immunity developed by repeated severe infections. There was also the belief that problems would arise with the acceptability of a vaccine for children within a target population. And, finally, some contended that a malaria vaccine would not only be costly to develop, it could serve only as an adjunct to the inexpensive and effective insecticides and antimalarial drugs that were the basis for the boastful proclamation of "man's mastery of malaria." Full of hubris in 1955, the Eighth WHO Assembly meeting in Geneva, Switzerland, endorsed a policy of global eradication of malaria with reliance on CQ treatment and DDT spraying. No mention was made about conducting research that might lead to a protective vaccine. However, by the 1960s, many parts of the world where eradication had once seemed possible were experiencing a resurgence of malaria; in other places, economic constraints forced premature relaxation of surveillance and, in some countries, the control programs were curtailed because of political turmoil. There was also widespread mosquito resistance to insecticides and parasite resistance to the cheap and once effective antimalarial drugs. It became apparent, even to the WHO, that the Global Eradication of Malaria Program was a failure and that eradication could not be achieved, and so there was renewed interest in a vaccine.

In 1956, Ian McGregor and co-workers [442] reported that newborn infants resident in the Gambia who had received weekly CQ possessed significantly lower concentrations of γ-globulin than did a group of unprotected children. (The protected group never showed parasites in the blood, whereas the unprotected group was infected by the second year of life.) A follow-up investigation showed that, although the γ-globulin concentration fell for the first three to six months after birth, subsequently there was a progressive rise with age similar to that of Europeans, but the level in the African children was always higher. These findings convinced McGregor that there was an association of malaria with enhanced levels of serum γ-globulin; at the time, however, there was no proof that the raised levels reflected a specific antibody response or that the γ-globulin response was protective and responsible for effective immunity. McGregor recognized that to conduct definitive experiments would require collaboration with an immunologist. He recruited Sydney Cohen at Guy's Hospital (an expert on γ-globulin) to the project. When Cohen studied the synthesis of γ-globulin in Europeans and adult Gambians exposed to malaria, the former group was found to synthesize up to 80% less than the latter, suggesting that the Gambians were making more γ-globulin as a protective response to malaria [144]. McGregor arranged to collect a pool of serum from healthy Gambian volunteers and sent this to London, where Cohen carried out the fractionation. The 7S γ-globulin from adult Gambian serum, as well as the purified 7S γ-globulin fraction and serum minus the 7S component from adult Gambians, was provided to McGregor. In addition, the 7S fraction from serum of UK blood donors was prepared as a control. The therapeutic effect of these fractions was assessed in young Gambian children suffering from acute clinical *P. falciparum* and *P. malariae* as well as in untreated children. The fractions were administered intramuscularly at 8–24-hour intervals for three days with a total dose equivalent to 10–20% of a child's own γ-globulin. By the fourth day, the parasite density had not increased and, by the ninth day, parasites were not seen in the blood in eight out of 12 cases. Protection by passive transfer was limited, however, lasting only three months. Unlike the other two 7S fractions, only the 7S fraction from immune adults reduced both the levels of parasites in the blood (asexual but not sexual stages) and clinical illness. Here was the first reliable experimental data to support the view that humans repeatedly exposed to malaria-infected mosquitoes could develop an immunity capable of restricting clinical illness and parasite blood density, and that this immunity could be passively transferred to non-immunes (children) via γ-globulin [146]. These observations led to the proposal that, at least in theory, vaccination against malaria could be feasible. And, when McGregor and Cohen showed that the 7S fraction from adult Gambian serum had the same therapeutic effect in Tanzanian children with *P. falciparum*, it suggested that West and East African strains of malaria had antigenic similarities and a vaccine prepared against parasites from one region of Africa might be effective against parasites from other regions. This emboldened them to begin a hunt to identify the blood-stage antigen responsible for protection against falciparum malaria.

In 1965, the WHO invited a group of scientists and health educators to its headquarters in Geneva to suggest innovative ways to rescue the failing Global Malaria Eradication Program. Paul Silverman (of the University of Illinois) was an attendee, and although at the time he had no "hands on" experience with malaria he pressed his case for the feasibility of a malaria vaccine. One of the other attendees, Lee Howard, the Head of the Health Division of the U.S. Agency for International Development (USAID), listened to Silverman's arguments and was intrigued. At the meeting's end Silverman rearranged his flight home so that he could sit next to Howard, continue the discussion, and persuade him of the benefits of a malaria vaccine. Indeed, Silverman went so far as to say that, by lifting the burden of disease through a protective vaccine, USAID would be able to reduce and perhaps even eliminate its continuing economic support programs in areas where malaria was endemic [176, 611]. Silverman's proposal was for a combination vaccine that would target malaria transmission and pathology to two stages: the sporozoite and the asexual blood forms. At this time, there were formidable obstacles against achieving these goals. There was no method for continuously growing the blood stages of *P. falciparum* (or, for that matter, any other species infecting humans) in the laboratory for use in tests for protection, none of the human malarias had been adapted to infect primates, and there was no practical way to mass produce sporozoites of sufficient purity to be used as a vaccine. Silverman dismissed these hurdles as being "simply technical problems" that could be overcome by a major, well-funded research program. Several weeks later, Silverman was invited to Washington, DC, and presented his plans to the Health Division of the USAID. The organisation, having no in-house scientific expertise, sent Silverman's proposal to an *ad hoc* panel of malaria researchers. Their judgment was that it was not feasible to develop a practical malaria vaccine for humans and consequently USAID should continue its business of distributing food and insecticides, providing advice, and assisting with economic programs rather than investing in vaccine research. The division ignored the expert advice and, in 1966, elected to support a $1 million contract with the University of Illinois under the direction of Silverman. The project was designed to determine the feasibility of developing a vaccine against human malaria and involved testing of sporozoite and erythrocyte antigens and the *in vitro* cultivation of these stages.

In 1972, when Silverman became the Vice President for Research at the University of New Mexico, the USAID contract moved with him and a sub-contract was established with Rush Memorial Institute to develop a monkey (*P. knowlesi*)–human malaria model system. Following a 1974 Malaria Vaccine Workshop where the current status of research on vaccines was reviewed, a "road map" for more specific research approaches and priorities was established as well as the means for increasing interest and cooperation among scientists in the area. The priorities were: a continued emphasis on erythrocyte antigens including testing in monkeys, use of adjuvants, the continuation of basic studies of sporozoite immunization but with a de-emphasis on sporozoite antigens *vis-à-vis* red blood cell antigens, and consideration of attenuated strains. In an attempt to simulate the immune status of children, the Silverman group vaccinated juvenile rhesus monkeys with purified protein factor (PPF) and Freund's adjuvant (first dose with complete FCA and second dose with Freund's incomplete adjuvant [FIA] given at four-week intervals and followed by challenge four weeks later), and six out of nine monkeys survived [621]. Even when the antigen was freeze-dried, it gave 75% protection when used with an adjuvant such as Bacillus Calmette–Guerin (BCG) or FCA [582].

In 1975, Cohen's group [461] reported a spectacular finding: of six monkeys vaccinated twice intramuscularly with freshly isolated *P. knowlesi* merozoites emulsified in either FCA or FIA and challenged with the same variant, no parasites were found in the blood in three and the remainder had a low-grade infection (maximum 1.5%) that persisted for 6–11 days. Six other monkeys challenged with a different variant from that used in vaccination with FCA showed a low-grade infection that terminated in less than two weeks, and the remaining two animals died. After initial challenge, all surviving immunized animals were resistant to challenge up to 16 weeks. Merozoite vaccination using FCA was required for resistance to challenge with a different variant but not for the same variant. Inoculation of blood from vaccinated animals after clearance of blood parasites into naïve monkeys did not result in an infection, indicating that the vaccination had induced sterilizing immunity. Merozoites frozen in liquid nitrogen provided comparable protection, as did freeze-dried merozoites stored for up to 20 weeks at 4°C; the latter, however, gave somewhat less consistent protection. The work was hailed in a 1975 *British Medical Journal* [23] article entitled "Malaria Vaccine on the Horizon."

By the late 1970s, the claims and counterclaims of successful vaccination on both sides of the Atlantic, i.e. principally between the Cohen and Silverman groups, led to questions as to whether the schizont or merozoite was better as an antigenic source. A competition was contemplated despite the fact that, in all of the previous vaccine trials with blood-stage vaccines in which some degree of protection was shown, the toxic FCA was required, the "pure" preparations were still not free of host cell contamination, the degree of protection was less than desirable for use in human trials, and in most cases the vaccinated animals became infected. At the time, Cohen, the Chairman of the WHO Scientific Group on the Immunology of Malaria, was asked by USAID to visit New Mexico to assess progress. It was suggested by a group of consultants that, to resolve the differences between the American and British vaccines, there should be a direct comparison of the antigens in a single laboratory. By the time of this trial (1977), the USAID–University of New Mexico malaria vaccine program was under the direction of Karl Rieckmann, a physician who had been taken on earlier by Silverman in the hope that he would conduct vaccine trials with human subjects. Without a suitable vaccine for humans, however, Rieckmann was to supervise a trial using rhesus monkeys, not human subjects. A total of 32 monkeys were used in the study. Eight controls and 24 rhesus monkeys were divided into groups of eight to test three antigenic materials: a frozen and thawed *P. knowlesi* schizont preparation, a freeze-dried preparation, and fresh *P. knowlesi* merozoites. Each preparation was emulsified in FCA, monkeys were inoculated intramuscularly twice at six-week intervals, and then they were challenged with a variant different from that used for vaccination. All controls developed severe blood infections and died within 12 days. Of those that received the frozen schizonts, six died, whereas the surviving monkeys had a blood infection that lasted 10–14 days with 0.01% and 2.7% parasites in the blood. Four of the eight monkeys that received the freeze-dried preparation died and the remaining animals had low-grade blood infections (0.07%, 0.4%, 0.9%, 1.2%) that persisted over a 9–15 day period; only two monkeys out of eight that had received the merozoite preparation survived.

Clearly, none of the vaccine preparations was as successful as reported previously. Silverman blamed the poor showing of his vaccine on storage problems and countered that a fresh preparation might have produced better results. The British claimed that, like good wine, their vaccine did not "travel well" and the prolonged period over which the merozoites were collected to obtain sufficient antigen for the second immunization may have contributed to the lower degree of protection. In a subsequent study, a fresh merozoite preparation was prepared at Guy's, sent to New Mexico in a frozen state, and then stored in liquid nitrogen until use; of eight monkeys vaccinated, four survived a challenge with blood parasites and the remainder died. The results of this head-to-head competition were presented at the NMRI/USAID/WHO Workshop on the Immunology of Malaria held in Bethesda on 2–5 October 1979. Upon peer review of these experiments, the USAID-supported University of New Mexico group lost support from the scientific community and its USAID contract was terminated.

In 1964, Quentin M. Geiman (1904–1986) at Stanford University (Palo Alto, California), encouraged by the U.S. Army Medical and Research Command, attempted to cultivate *P. falciparum*. Despite his lack of recent experience with malaria, Geiman was the only one from the World War II Harvard project who was available and interested in the cultivation of malaria parasites. However, unlike *P. knowlesi* (which Geiman had used at Harvard) he found it difficult to grow *P. falciparum in vitro*. Further, apart from nine cases of human malaria and two samples of infected blood, there was no steady supply of falciparum parasites available in the San Francisco Bay Area and storage and freezing of blood caused changes in the parasite before *in vitro* culture began. It was the 1966 report of Young, Porter and Johnson [752] from the Gorgas Memorial Laboratory that prompted Geiman to try to circumvent the shortage of human cases and to investigate the possibility of adapting *P. falciparum* to night owl monkeys. Night owl (*Aotus*) monkeys obtained from a local dealer were infected with blood from a soldier who had returned from Vietnam and, later, blood from a female patient who had contracted an infection in Uganda was used to infect a monkey. Infections were then transferred from these owl monkeys to other owl monkeys by inoculation of infected blood [259]. At Stanford, the former line was named FVO, for Falciparum Vietnam Oak-Knoll, and the latter FUP, for Falciparum Uganda Palo Alto.

To assist him in the cultivation of *P. falciparum*, Geiman recruited Wasim A. Siddiqui. Siddiqui had received a Ph.D. in 1961 from the University of California at Berkeley working on amebas, returned to India briefly, and then spent 1963–1966 working at the Rockefeller University with William Trager studying the nutritional requirements of *P. lophurae*. At Stanford, and with support from a U.S. Army Department of Defense contract, Siddiqui used commercially available media for the short-term cultivation of *P. falciparum* in *Aotus* red blood cells ostensibly to produce high yields of malaria-infected red cells for isolation of antigens to be used in vaccination studies [502]. In 1970, Geiman retired from Stanford and shortly thereafter Siddiqui moved to the University of Hawaii. Beginning in 1975, a USAID contract with the University of Hawaii (Principal Investigator Wasim Siddiqui) was dedicated to developing a malaria vaccine and continuing to improve on methods for *in vitro* cultivation of *P. falciparum*. Taking clues from the fberghei studies, Siddiqui [612] used blood-stage antigens obtained from *Aotus* infected with *P. falciparum* (FUP) and later used a merozoite-enriched preparation from short-term *in vitro* cultures [614]. In the latter experiment, two doses of vaccine (2.73 mg) emulsified with FCA were administered intramuscularly at three-week intervals and, three weeks following the second vaccination, the animals were challenged by intravenous injection of 6.2×10^5 parasites. All of the control animals died within two weeks with fulminating infections, whereas three vaccinated monkeys survived; however, all had low-grade blood infections. Subsequently, Siddiqui *et al.* [616] were able to show cross-resistance in monkeys vaccinated (using FCA) with FUP and challenged with FVO.

Siddiqui recognized that a barrier to the development of a protective malaria vaccine was the need for a suitable replacement for

FCA. He wrote [613]: "The ultimate objective of all malaria vaccine studies is to develop a vaccine that can be used to immunize and protect man, not monkeys or rodents. Therefore, the development of an immunologically satisfactory and pharmacologically acceptable adjuvant is imperative in the development of a malaria vaccine acceptable for use in man." Siddiqui tried muramyl dipeptide (MDP), a substance reportedly able to replace whole tubercle bacteria in FCA, and which had already been shown to enhance the immunological response of animals against an antigen when injected with FIA (mineral oil). However, vaccination trials of *Aotus* with merozoites in MDP in peanut oil or mineral oil were unsuccessful.

Siddiqui continued to pursue alternatives to FCA. He tried stearoyl-MDP adjuvant with carrier liposomes (cholesterol plus lecithin). The vaccine contained a crude antigen (50–60% schizonts with merozoites and the reminder other developmental stages) with 2.86 mg protein emulsified in the adjuvant. Animals were vaccinated twice at four-week intervals and, 17 days after the last dose, the monkeys were challenged with 750,000 parasites (FUP) obtained from an ongoing blood infection in an *Aotus* monkey. Two of the controls died within two weeks of challenge and, remarkably, the third control, despite having 25% of its red cells infected, survived. All four of the immunized monkeys survived challenge, two developing low-grade infections that lasted one week, two developing infections ranging from 5 to 15%. A limited number of owl monkeys became negative for blood parasites after a month; however, as blood from these animals was not injected into naïve monkeys it was not known whether the animals still harbored small numbers of parasites undetectable by microscopy. The results were reported in a *Science* article [617], "Vaccination of experimental monkeys against *Plasmodium falciparum*." A possible safe adjuvant was considered to be important and significant (by Siddiqui) as the stearoyl-MDP adjuvant and liposomes did not produce an inflammatory reaction at the injection site and "an effective, safe malaria vaccine may be possible."

Siddiqui immunized *Aotus* monkeys with a crude vaccine made from cultured *P. falciparum* mixed with FCA and, although a few of the vaccinated monkeys were protected from challenge, all had low numbers of parasites in the blood. The vaccine could not be used with humans; however, USAID publicized the findings as if a human vaccination trial was surely to come in the near future. Press conferences were called, the State of Hawaii legislature expressed gratitude to Siddiqui, and the University of Hawaii honored him with its most prestigious award for excellence in research, stating that this was for his "finding the first promising candidate for a malaria vaccine." USAID was also wildly enthusiastic: "Technically speaking a vaccine may be available for human testing as early as 1985" [176]. In August 1984, USAID announced a major breakthrough in the "development of a vaccine... and the vaccine should be ready for use around the world, especially in developing countries, within five years."

To its credit, USAID was one of the first agencies to recognize the need for a malaria vaccine and it funded such research. However, those who were responsible for the USAID malaria vaccine program were neither patient nor realistic, and they did not want to countenance the fact that vaccine development can be a slow and deliberate process. Seeking to ensure that federal funds would continue to be provided, USAID did not produce a realistic timetable of 15 or 20 years or perhaps longer — something they feared the public and government would not like to hear — and so they promised a "magic bullet" in just a few years. The significance of the vaccination results was frequently exaggerated and given undue prominence. Within USAID there was a lack of critical self-assessment and accountability, a disregard of the advice of professionals, and an unbridled desire to be the first developer of a protective vaccine. The researchers in the USAID network had become an exclusive club whose members were its cheerleaders, particularly the project director.

In the 1970s, several discoveries were made to indicate that the landmark investment by USAID in a malaria vaccine might pay off. By 1974, flush with self-delusion, USAID sponsored a malaria workshop though the National Academy of Sciences to review the current status of research, to provide the agency with a more specific focus of approaches and priorities for development of a vaccine, and to suggest ways of stimulating increased interest and cooperation among scientists working in the area. In 1975, a panel of experts proposed that USAID expand its support to more than one laboratory. In 1975, USAID advertised for contractors to develop the means for the *in vitro* cultivation of *P. falciparum*. This resulted in funding projects at Rockefeller University, the University of Hawaii, Parke-Davis Co. and New York University. Within a year the Rockefeller University project made a significant breakthrough in culture. The scientific consultants who had advised USAID that a malaria vaccine would require the cultivation of *P. falciparum* had been correct, but they could not have foreseen how serendipity would play a role. William Trager, who had spent a lifetime studying the nutrition of malaria parasites, was specifically invited by USAID to apply for funding.

In his proposal to USAID, Trager specifically asked for support of a post-doctoral fellow experienced in the cultivation of intracellular parasites. James B. Jensen (1943–), who received his Ph.D. in 1975 from Auburn University (Alabama), was the person selected. During February 1976 they tested the suitability of commercial media using *P. falciparum*-infected red cells removed from an *Aotus* monkey infected with the FVO strain of *P. falciparum* (obtained from Trager's former post-doc Wasim Siddiqui who, at the time, was working at Stanford with Geiman); the cells were washed, diluted with human AB red cells, suspended in 15% human serum, placed in "flow vials" and a variety of tissue culture media were screened. The newly developed RPMI 1640 medium was found to be superior to all others tested. They also changed the gas mixture from 5% CO_2 + 95% air to one high in carbon dioxide and low in oxygen: 7% CO_2 + 5% O_2 + 88% N_2. Under these conditions, and with a settled layer of red blood cells in the flow vials, it was possible to maintain the parasites for 24 days by adding fresh uninfected red cells every three or four days [348, 681–683]. One day Jensen decided to take some of the infected red cells and place them into small flat glass dishes invented by Koch's young co-worker Julius Petri. When he placed the Petri dishes containing the infected red cells into a 5% CO_2/air incubator the parasites died out after two or three days. It was then that Jensen decided to employ a candle jar instead of the CO_2 incubator — a method he had used to grow the cells for the cultivation of various coccidian parasites (and when the CO_2 cell culture incubators in the virology laboratory at Utah State were unavailable to him). Jensen located a large glass desiccator, placed his Petri dish cultures inside and, after lighting a candle, closed the stopcock when the flame went out; this was then incubated at human body temperature for several days. At first Trager was dismayed to observe Jensen's use of a 19th century technology, not far removed from the days of Robert Koch; however, when he was shown the Giemsa-stained slides, Trager was convinced Jensen was on to something. In the summer of 1976, Milton Friedman, a graduate student in the Trager laboratory who was working in Ian McGregor's MRC laboratories in the Gambia, arranged for a sample of human blood infected with *P. falciparum* to be sent to New York. This was diluted with RPMI 1640 (the best of the commercial media), placed in Petri dishes, put into a candle jar and incubated. The line grew very well (much better than FVO) and became the FCR-3 strain. Later, other lines would be established using similar methods and the impact of continuous cultivation of *P. falciparum* would be phenomenal, allowing for the sequencing of the *P. falciparum* genome, identification and isolation of blood- and insect-stage antigens, and providing a means for identifying vaccine candidates.

The successful culture of *P. falciparum* by Trager and Jensen encouraged USAID [683] to advertise for contractors to improve on the cultivation of the red blood cell stages of malaria with the goal of purifying the antigens produced (with special attention to removal of red cell membranes), culturing different strains, developing methods to produce gametocytes, and providing methods for harvesting parasite antigens and testing their immune potential. There were 25 responses and "after intensive review" two additional projects were selected, bringing the "collaborating network on malaria immunity and vaccination" to 10 projects, including Rockefeller University, University of Hawaii, WRAIR, GML,

Parke-Davis and Company, and the University of New Mexico to work on the antigenic nature of *in vitro*-grown parasites and New York University to conduct sporozoite immunity research.

Trager recruited a team to carry out vaccine research. Robert Reese, an immunologist, would study the isolated antigens. Susan Langreth, an electron microscopist, would monitor the purity of the isolated antigenic fractions. Harold Stanley's expertise in tissue culture would improve the *in vitro* growth conditions and synchrony, and Araxie Kilejian, a biochemist, was to isolate and purify antigens. To support the USAID program, Trager was given a large budget, and a number of laboratory rooms to accommodate the increased number of investigators. However, before vaccinations of *Aotus* were begun, it was learned that there were nine different kinds of chromosomal patterns (karyotypes) and each had the potential to respond in an independent fashion to challenge with *P. falciparum* [543]. Indeed, preliminary studies indicated that monkeys of the type VI karyotype (*A. boliviensis*) were more resistant than karyotypes II or III (*A. griseimembra*). To minimize the variability in immune responsiveness, *Aotus* type II was used. Schizonts and mature trophozoites, harvested from Trager–Jensen cultures of *P. falciparum* (FCR-3 strain), were used to prepare merozoites by saponin lysis and this served as the vaccine. The first experiment employed six monkeys; three received merozoites emulsified in FCA administered intramuscularly and a booster vaccination was given three weeks later consisting of the same number of merozoites, but this time emulsified in FIA. The controls received only an injection of the adjuvant. (A more suitable control would have been uninfected red cells; however, that was not done.) At three weeks after the second vaccination, the animals were challenged by intravenous inoculation of one million parasitized red cells of the same strain derived from an *Aotus* monkey. Within two to three weeks after being challenged, all of the monkeys were dead with blood infections ranging from 10% to 55%.

In the belief that the amount of antigen was more important than the manner of administration, the protocol was altered so that three intramuscular injections were given at three-week intervals, with the first two containing merozoite antigen emulsified with MDP (a synthetic derivative of N-acetylmuramyl-L-alanyl-D-isoglutamine) and mineral oil (FIA), and the third containing solely the antigen in mineral oil [543]. (The basis for using MDP was that it had been found [461, 463] to have some protective effect, i.e. two out of six monkeys survived challenge when vaccinated with *P. knowlesi* merozoite antigen emulsified in MDP and FIA. This result was in contrast to FCA, where five out of six vaccinated animals survived challenge.) Although the number of *P. falciparum* merozoites in the vaccine given to each monkey had been increased and the challenge injection reduced to 500,000 infected red blood cells, two out of three vaccinated monkeys died. One died one week later than controls, a second died two weeks after the controls, and the third monkey had less than 0.5% of its red blood cells infected. All of the "protected" *Aotus* monkeys showed anemia and, in those that succumbed, this was assumed to be the cause.

Reese *et al*. [543] claimed the work to be "significant for two major reasons:" merozoites from *P. falciparum* cultured *in vitro* for over a year were still able to induce immunity, that is they had not lost their antigenic "punch," and synthetic MDP could substitute for FCA when a sufficient amount of antigen was used. In 1980, Reese moved his laboratory across the country to the Scripps Research Institute (La Jolla, California) where immunology was in the forefront. At Scripps, Reese and co-workers studied the synthesis of merozoite proteins [330], complement activation [647], developed an assay for measuring the effects of immune sera on *in vitro* parasite maturation [648], prepared monoclonal antibodies from *Aotus* monkeys and used these to identify a merozoite surface antigen [649], constructed a cDNA library and screened this with serum from immune *Aotus* to identify three merozoite surface antigens [26, 27]. When Reese did not gain tenure at Scripps, the USAID project was moved to the Agouron Institute (also in La Jolla). There the Reese laboratory carried out immunologic modeling [549] and peptide mapping. They identified a high-molecular-weight phosphoprotein [329], and suggested it to be involved in parasite transport [646], as well as rhoptry antigens [328]. However, for the most part, these "significant finds" produced nothing of any importance toward the development of a protective malaria vaccine. Indeed, in 1987, after a negative review of the project by a group of scientific consultants, the USAID contract to the Agouron Institute was terminated. Shortly thereafter, Reese left the field of malaria vaccine research.

Beginning in 1972, Siddiqui had set his sights on conducting vaccine trials with a human parasite using *Aotus* monkeys as surrogates for a related primate, i.e. us. In 1988, acting on "information received," the Office of Inspector General of USAID began investigating Siddiqui's and the University of Hawaii's handling of research funds [21]. There was evidence to support the allegations that the Principal Investigator (Siddiqui) apparently diverted funds to his and his secretary's personal use, and monies were used to refurbish his offices at the University with these construction costs charged to the grant as consultant payments. On 14 September 1989, the Grand Jury in Hawaii indicted Siddiqui and his administrative assistant Susan Lofton with theft in the third degree and criminal conspiracy. The Deputy Attorney General of Hawaii charged that some of the USAID money was siphoned off between 1984 and 1987 through illicit accounting tricks and a kickback arrangement with a Honolulu travel agency that Siddiqui had engaged to run a 1985 Asia Pacific Conference on Malaria. The USAID conference check was deposited with the Research Corporation of the University of Hawaii and, at Siddiqui's direction, was to pay the bills for the conference. There was $100,000 as an advance payment for services to be used given as a deposit to the Pacific Hotel where the conference was to be held. In March 1985, a month before the conference, Siddiqui instructed the travel agency to begin paying him directly $1,260 per month and to pay his secretary $1,000 per month. These salary supplements were to continue for the next two years. Between December 1984 and April 1986, the travel agency paid Siddiqui $17,400 and Lofton $16,000. USAID was sent a bill for $35,425 for services and departmental rental related to the conference; they paid the bill, but the indictment disputes that the money was used for that purpose [427]. This was only the beginning of

USAID's troubles. The Senate Appropriations Committee on Foreign Operations, which has jurisdiction over the USAID budget, launched a General Accounting Office (GAO) investigation of the entire $8.5 million malaria vaccine program budget. On 18 October 1989, the *Washington Post* reported [76, 342, 439, 487] that the GAO had found irregularities at the University of Hawaii.

In spite of the troubles at the university, Siddiqui and his associates continued their vaccine-related studies. They used serum from vaccine-protected *Aotus* monkeys to identify 13 protein antigens [356] and then, in a final vaccine trial using *Aotus*, were able to protect against challenge using a single merozoite surface coat protein (Pf195 = MSP-1); protection against challenge was achieved only when the antigen was emulsified in FCA [615]. During a decade of research (1987–1996), with generous support from a USAID contract, Siddiqui and his collaborators were able to show that serum from animals vaccinated with Pf195 inhibited the *in vitro* growth of *P. falciparum* [339] and that B30-MDP (a lipophilic muramyl dipeptide derivative) and LA-15-PH (a synthetic equivalent of monophosphoryl lipid A) could replace FCA in inducing growth inhibitory antibodies in rabbits to Pf195 [340]. A protective malaria vaccine, however, was far from at hand.

In addition to the problems in Hawaii, there were others. In 1983 or 1984, USAID decided that a large number of candidate vaccines suitable for testing in humans would be available and it was believed that before this could occur they would have to be tested in South American owl (*Aotus*) and squirrel (*Saimiri*) monkeys from Peru, Colombia, and Bolivia. Between 1984 and 1988, approximately $6.7 million of USAID monies would be obligated to acquire, house, and maintain monkeys. A GAO investigation [21] found indications of fraud in a USAID-funded project at the National Institutes of Health in Bogota, Colombia, which was to supply and house the monkeys. James Erickson, USAID's chief technical officer for the malaria program, was also involved in monkey business. In April 1985, USAID decided to purchase up to 600 Bolivian monkeys, using unexpended funds from the suspended American Institute for Biological Sciences (AIBS) sub-contract with the

Colombian National Institutes of Health. AIBS, the management arm of the vaccine program, was to arrange for the purchase of the monkeys and a person named George Diaz would be given a contract to find and acquire the monkeys. Diaz and Erickson then contacted Worldwide Primates based in Miami to obtain 200 owl monkeys at $475 each and 400 squirrel monkeys at $375 each for a total of $245,000. Diaz then told AIBS that the monkeys would be obtained from Gerrick International for $336,000 — $630 for each *Aotus* and $520 for each S*aimiri*. It turned out that Gerrick International was a shell corporation for Diaz and Erickson. On 12 September 1985, Erickson had ordered stationery with the letterhead Gerrick International and Diaz opened a bank account in the name of Gerrick International. AIBS issued a check for $168,000 to Gerrick International in partial payment for monkeys and Gerrick, in turn, paid Worldwide Primates $122,500. In January 1986, Diaz wrote a check for $8,500 to his brother-in-law and this was then signed over to Erickson. In March, $11,886 was withdrawn from the Gerrick account and the money was used to purchase two cashier's checks made payable to J. Erickson. In October 1985, Gerrick was paid the remainder of the $168,000 and presumably Erickson and Diaz made a profit on this also. The grand jury used this to level charges against Erickson of conspiracy, conflict of interest, and accepting a gratuity. As he paid no tax on the profits, Erickson was also indicted on three counts of submitting false tax returns. By April 1987, Erickson's superiors had reassigned him pending allegations of mismanagement. To this were added allegations of sexual harassment and anonymously supplied information of illegal activities. In October 1987, Erickson was put on administrative leave and he languished for a full year before launching an offensive, claiming that he was the victim of a witch-hunt. He sued AIBS through a whistle-blower statute and, in July 1988, he legally enjoined USAID to demand that it come to a decision to reinstate him or fire him. The court agreed and, as nothing had been proven, USAID docked him a week's pay for bad judgment and personal misconduct. Erickson might have escaped without punishment had it not been for the continuing probe by the Inspector General and the GAO, whose investigations brought down the house of cards Erickson had built. In 1990, after pleading guilty, Erickson was fined and received a short sentence to be served in a half-way house. He served his time and was not heard from again.

The conclusion of the 1989 GAO report [21] was that in the vaccine program there was evidence of fraud and waste of government funds, that sub-projects were selected under questionable circumstances, performance was not subject to adequate evaluation, and poor quality research may have been tolerated. The USAID malaria vaccine program illustrates the problem with so-called push programs [379], i.e. those that pay for research inputs rather than results and where funds are committed before a product is developed. In push programs, applicants tend to exaggerate the prospects that their approach will succeed and, once funded, researchers may divert the resources to other products and to pursue avenues that will more rapidly advance their careers. Indeed, even when it was clear that there were problems in finding a protective malaria vaccine using monkey models, researchers kept requesting more funding and AID administrators kept approving it. In the development of a malaria vaccine it is imperative that the perils inherent in push programs be recognized and avoided.

In 1992, USAID elected to shift emphasis from support of research to the development of promising vaccine candidates and to a testing of investigational vaccines. They formed the Malaria Vaccine Development Program (MVDP), effectively replacing Malaria Immunology and Vaccine Research (MIVR), to build a pipeline from early pre-clinical vaccine development, through the regulatory process and field-testing of vaccine candidates. The major target products of the MVDP are blood-stage vaccines, particularly MSP-1 and AMA-1, for children living in areas where malaria is endemic.

Two blood-stage vaccine candidates

The concept of a blood-stage vaccine is based on a fundamental principle: mimic the immunity that occurs in nature.

In malaria-endemic areas, natural acquired immunity is induced by continued exposure to the bite of infected anophelines and, although blood parasites persist, successive bouts of clinical disease are diminished due to reduced parasite multiplication and growth. The assumption of proponents of a blood-stage vaccine is that it will accelerate the acquisition of acquired immunity, inhibit parasite invasion into red blood cells, and affect parasite multiplication to reduce the parasite burden, thereby limiting disease severity and death. This would "offer enormous benefit to the public health, particularly infants and children living in endemic areas who suffer the most...due to malaria" [179].

The majority of blood-stage vaccines are clustered around two hypothesized mechanisms of protection: block red blood cell entry by the merozoite and/or attack the development of the malaria parasite within the red blood cell, directly or indirectly. Leading blood-stage vaccines include MSP-1 and AMA-1.

The road to merozoite surface protein

In 1977, the recently appointed head of the Division of Immunobiology at the Wellcome Foundation, the immunologist James Howard, was intent on rebuilding and reorientating the Department of Immunochemistry to better reflect emerging molecular approaches to microbiology and parasitology. George Cross first met Howard at a meeting of the Royal Society of Tropical Hygiene and Medicine early in 1977 and, after successive meetings, was offered a job at Wellcome to build a group that would include a prominent commitment to malaria immunology/vaccinology. Although he had no formal training with malaria, he had a free hand to learn "on the job." Cross began to hire smart, experienced and keen young scientists to spearhead malaria research. The first was a talented young immunologist, Robert (Robbie) Freeman, who had recently finished his Ph.D. on the immunology of the rodent malaria *P. yoelii*. Freeman arrived in 1978 and persuaded Cross that *P. yoelii* was a better model (for human) malaria, existing as it did in virulent and less virulent clones — and was capable of being scaled up for biochemical studies. With an excellent supporting staff, especially Avril Trejdosiewicz, who had been hired to run the monoclonal facility, the first monoclonal antibodies to the asexual blood stages of rodent and human malaria parasites were made. One of Freeman's first contributions was to show that the best way to identify useful relevant monoclonal antibodies was to screen, from the beginning, by the time-consuming but more informative approach of immunofluorescence microscopy. The next step was to see if any of the monoclonals would suppress a mouse malaria infection. The first experiment involved two monoclonals: one that clearly reacted with the merozoite surface, which could be recognizing a protective antigen, and the other, which recognized a merozoite internal structure, and was unlikely to be protective. In fact, both monoclonals were strongly protective in "passive immunization" experiments [232].

From this point, the Immunochemistry group moved rapidly to purify the target antigen by affinity purification, for which the monoclonal worked well. It was at this point that Cross persuaded Anthony Holder to get involved with the malaria project and contribute his skills in protein purification and analysis. In one experiment, using parasites harvested from 300 mice, they were able to purify at least 1 mg of protein. This enabled true vaccination experiments to be done [319].

Cross left Wellcome for his present position as a Professor at the Rockefeller University in 1982, but was retained as a consultant from 1982 to 1987. Robbie Freeman left Wellcome in 1985, to pursue a different career and lifestyle, and died tragically by his own hand eight years later. In 1988, Holder moved to head the Division of Parasitology at the NIMR, but the Wellcome malaria program continued until 1990, when it was terminated because Wellcome had moved from being owned entirely by a charitable foundation to being partly a public company. This triggered a review of the company's position and future direction, and resulted in the sale of its vaccine business to Medeva, and a subsequent reorganization of its molecular biology resource that terminated all work on vaccines and other biological therapies at the UK research site.

A year after Holder joined the Wellcome group, the monoclonal antibodies were used to identify a *P. yoelii* merozoite-specific protein with an Mr of 235,000 [319]. The protein (named Py235) could be purified and prepared in large amounts. After immunization with this protein in FCA, the mice survived challenge, whereas controls died. But, Holder and Freeman wondered, was this of any relevance to *P. falciparum*? They answered their own question upon finding a cross-reaction between the *P. yoelii* antigen and a merozoite antigen from *P. falciparum*, and suggested [320] that "this 195,000 mol wt *P. falciparum* protein, and its form on the merozoite surface, are...of considerable interest for malaria vaccine development." They had discovered the first and most abundant MSP, named MSP-1. (MSP-1 has also been called, at various times, precursor to major merozoite surface antigens, a *P. falciparum* 195,000 molecular weight antigen [Pf195] and falciparum merozoite protein 1 [FMP1].)

The initial studies on MSP-1 were rapidly expanded to other *Plasmodium spp.* including sequencing of its gene in *P. falciparum*. How could they be sure it was the right gene? Interestingly, it was protein sequence from a fragment of MSP-1 released into culture supernatants that provided the definitive proof. Studies on the processing of MSP-1 followed, particularly by M. J. (Mike) Blackman, and the discovery that a monoclonal antibody isolated by Jana McBride in Edinburgh directed against one end of the molecule inhibited merozoite invasion of red blood cells focused attention on this part of the molecule. Human studies, with Eleanor Riley in the Gambia and Roseangela Nwuba in Nigeria, supported the view that this part of the molecule was important as a target of protective immunity and a vaccine candidate. With Irene Ling and Sola Ogun, in parallel to work being carried out by Carole Long, a recombinant ~100 amino acid fragment (MSP-1_{19}) of *P. yoelii* MSP-1 was shown to be effective in protecting mice against a parasite challenge infection.

MSP-1 is encoded by a single gene on chromosome 9 in *P. falciparum* and is synthesized late in the blood-stage developmental cycle. It has not been possible to "knock out" the gene for MSP-1, suggesting that it is essential for invasion and parasite survival. Sequence analysis shows that MSP-1 contains conserved regions interspersed with polymorphic regions. MSP-1 undergoes extensive processing during merozoite formation and, soon after merozoite release, the resulting four fragments remain as a complex on the merozoite surface with the 42,000 Mr protein (MSP-1$_{42}$) being attached *via* the GPI anchor. A final processing step, mediated during invasion, cleaves this fragment into 33,000 and 19,000 fragments (= MSP-1$_{19}$); the larger fragment is shed from the merozoite surface along with the remainder of the MSP-1 complex, whereas the smaller MSP$_{19}$ remains on the merozoite surface during invasion.

Full-length MSP-1 has been suggested to bind to red cells in a sialic acid-dependent manner, or it may form a complex that binds to band 3 protein (see p. 119). Antibodies directed against MSP-1$_{19}$ block invasion and prevent processing of MSP-1 [66]. As immunization of *Aotus* with recombinant MSP-1$_{42}$ in the presence of FCA partially protected against challenge, and appeared to be as effective as whole merozoites, it was suggested to be a blood-stage vaccine candidate. It was clear, however, that the immunogenicity of the recombinant protein needed to be improved [321]. Antibodies directed against MSP-1$_{19}$ in mice and monkeys elicited protection against challenge and similar results were reported with antibodies raised against MSP-1$_{42}$ [627]. Immunization of *Aotus* monkeys with the recombinant protein MSP-1$_{42}$ produced in *E. coli* resulted in high-titer antibodies and led to protection against an ordinarily lethal challenge with *P. falciparum* [627]. However, with recombinant MSP-1$_{42}$ and MSP-1$_{19}$, protection of *Aotus* was seen only when high antibody titers were obtained using formulations with FCA [654].

So what were the results of a Phase 1 (safety) vaccination trial in adult human volunteers in Kenya using MSP-1$_{42}$ (also called FMP1) in combination with AS02A (an oil-in-water emulsion containing 3-deacylated monophosphoryl lipid A and QS21, a saponin-like adjuvant) or Imovax rabies vaccine? Slightly more recipients of the three intramuscular vaccinations at zero, one and two months of FMP1/AS02A experienced more pain than with the

Imovax group but otherwise the two groups were equal. The Kenyan volunteers already had high pre-existing antibody titers to FMP-1 and the titers increased after each immunizing dose (as was the case with malaria-naïve subjects in the US [652]). The vaccine appeared to be well tolerated and immunogenic, but its efficacy in adults and in children was disappointing and further work on it as a vaccine has been abandoned.

The road to apical membrane antigen 1

The demonstration that antibody from rhesus monkeys immune to *P. knowlesi* inhibited the multiplication of the parasites *in vitro* [145], and that passive transfer of immune monkey serum also confers protection, provided a rationale for using sera from immune hosts to characterize malaria parasite antigens and identify those with putative protective activity. As it had already been shown that vaccination with merozoites with FCA induced some degree of protection (see p. 212), a search for the protective antigens of the merozoite began. Two of the 11 antigens recognized by immune serum were found to be from merozoites [169]. But were these protective and how could sufficient amounts be obtained for vaccination purposes? Sydney Cohen's group at Guy's Hospital, now including Judith Deans and Alan Thomas, elected to apply the recently developed monoclonal antibody techniques to an antigenic analysis of *P. knowlesi* merozoites to identify putative protective antigens and to produce these for vaccination.

In 1982, supported by funds from the MRC and the malaria component of the United Nations Development Programme/ World Bank/WHO Special Program for Research and Training in Tropical Diseases, Deans and Thomas prepared two rat monoclonal antibodies that inhibited the *in vitro* multiplication of *P. knowlesi* and both reacted specifically with a single protein of Mr 66,000 [170]. Later, it was found, using an antiserum prepared against the 66,000 Mr antigen, that it did not prevent merozoites from attaching to erythrocytes, but did inhibit their orientation [464]. As saponin had been shown to be effective with a vaccine containing whole merozoites (see p. 221), rhesus monkeys were immunized with the purified 66,000 Mr antigen with saponin as adjuvant; monkeys produced antibody that inhibited merozoite invasion of red cells and four out of six monkeys vaccinated with the purified antigen were protected [171].

Using the phage technology of the time, Andy Waters, who joined the group at Guy's to do his first post-doc, had been given the task of obtaining the *P. knowlesi* AMA-1 gene, PkAMA-1. The monoclonal antibodies that Thomas and Deans generated gave him the tool to eventually do so, although it was certainly not straightforward [725]. However, Robin Anders' group at the Walter and Eliza Hall Institute had already published on the molecule after identifying it by a different route.

As it has been impossible to delete ("knock out") the gene for AMA-1, it is presumed to be critical for parasite survival. The complete amino acid sequence of AMA-1 from several strains of *P. falciparum* has been determined. Antibody levels to AMA-1 are higher in human populations exposed to natural malaria infections than most other antigens including MSP-1 and the prevalence of such antibodies increases with age. AMA-1 has been considered to have potential as a vaccine candidate. However, the finding (in 1990) that the AMA-1 gene is polymorphic, i.e. different strains may have up to 100 different amino acid sequences and therefore produce the different shaped or sized forms of AMA-1, "has tempered expectations in terms of vaccine-induced broad protective immunity" [544]. A further complication for use as a vaccine is the finding that AMA-1 has to fold correctly to induce parasite-inhibitory responses [574].

There have been several published reports on Phase 1 (safety, immunogenicity, and efficacy in a non-endemic setting) vaccine trials with AMA-1 with the adjuvant Montanide ISA720 (containing squalene, a metabolizable oil). There were no adverse effects with increasing dosages of 5, 20 and 80 μg given intramuscularly at zero, three and six months; however, it showed poor immunogencity in six out of 29 subjects, suggesting a loss of potency [580]. In another study with "Combination 1" containing the AMA-1 protein produced in the yeast *Pichia pastoris* and adsorbed to Alhydrogel and with a similar dose and schedule, it was also well tolerated but again immunogenicity occurred in only four out of 22 subjects [423] and, when AMA-1 was packaged in a pox virus and delivered as NYVAC Pf7, antibody production was poor and the protective effects could not be ascribed to any one component in the vaccine [485]. In the most recent trial, the malaria vaccine FMP2.1/AS02A, consisting of a recombinant protein based on AMA-1 with the adjuvant AS02 (an oil-in-water emulsion with deacylated monophosphoryl lipid A and QS21, a saponin agent derived from the soap bark tree *Quillaja saponaria* and manufactured by GlaxoSmithKline [GSK]), was found to be safe and immunogenic when tested in a Phase 1 and Phase 2 trial (safety, immunogenicity and efficacy determination of dosage in an endemic setting in children in Mali, West Africa, with doses of 25 and 50 µg at zero-, one- and two-month intervals). The sera inhibited parasite growth *in vitro* [665]. As of this time there is no surrogate measure for protection other than testing in human subjects; this continues to hamper evaluation of vaccine responses. The authors of the Phase 1 and Phase 2 trial sagely concluded: "Although these results (with AMA-1) are promising, until a blood stage vaccine demonstrates clinical efficacy, immune correlates of vaccine induced protection and the choice of immunogencity endpoints for clinical development decisions will remain a matter of reasoned conjecture" [665].

A failed promise, SPf66

In 1987, with his announcement that he had developed the first synthetic vaccine [510], the Colombian physician Manuel Elkin Patarroyo stunned the scientific community. He said he had protected monkeys from malaria and now was beginning to test his vaccine on thousands of his countrymen. Patarroyo could have tried several approaches in the development of a protective vaccine, i.e. identify the proteins recognized by immune sera, measure the T cell responses in infected populations to antigens identified by other means (such as antibody), and identify the proteins on the surface of the merozoite or the infected red blood cell. He used none of these approaches, but directly identified antigens of asexual-stage parasites that would protect monkeys against a fatal falciparum infection and then synthesized a 45-amino-acid peptide.

Working in labs housed in three Victorian-era hospital buildings in Bogota and financed by a series of Colombian presidents, the strong-willed and flamboyant Patarroyo seemed to have done what no one else had been able to: make a polymer consisting of three peptides from the merozoite surface, including MSP-1, and then, with a deft stroke of *legerdemain*, link these using a peptide from the sporozoite surface [90]. Patarroyo, after a short period of training at the Rockefeller University with Bruce Merrifield (who in 1984 received the Nobel Prize for his "development of methodology for chemical synthesis [of peptides] on a solid matrix"), was leading a team of 60 scientists — chemists, molecular biologists, and computer scientists — that had prepared a multi-antigenic, multi-stage synthetic peptide vaccine, named SPf66. Initially, SPf66 was reported to be able to protect four out of eight *Aotus* monkeys. After repeating the monkey work several times with larger groups of animals [509], he was planning to go forward with humans. This leap forward to human trials was undertaken despite the fact that Socrates Herrera, a former student of Patarroyo, was unable to replicate the work, and the Centers for Disease Control (CDC) (Atlanta, Georgia), at the request of the WHO, had tested two preparations in *Aotus* monkeys and found an inconsistent immune response and a lack of protection against challenge. Patarroyo in response claimed the CDC had failed to couple the peptides properly.

In 1988, in a preliminary trial with human volunteers, Patarroyo claimed that two out of the five immunized volunteers were protected. In a field trial involving 1,548 human volunteers in Colombia, the overall protection rate was reported to be 39%, and in adults was 67% with immunity lasting one to three years [437, 699]. Patarroyo was not interested in why the vaccine worked: he simply held news conferences. Like Rumplestiltskin, the little man in the Grimm brothers' fairy tale who could spin straw into gold, Patarroyo, with a twinkle in his eye and a smile of satisfaction on his lips, showed the assembled reporters a jar of the new vaccine and boasted that the "three series dose costs thirty cents and that's less than the price of Coca-Cola. It's a chemical product, it's completely reproducible, it's completely pure. There is no possibility of mutation" [90]. Some were enthusiastic, others were skeptical, some were jealous [92].

Patarroyo was hailed as a modern Louis Pasteur and was received as if he were a rock star. By 1993, SPf66 had been tested on 41,000 people in Brazil, Colombia, Ecuador, and Venezuela, where it was claimed the vaccine gave 40–60% efficacy. Critics charged that these field trials had been faulty in design [428]. The trials in South America had not included a placebo control, the identities of those immunized with SPf66 were not hidden from the vaccinators, and those who had received the vaccine knew they were not getting a placebo. Louis Miller of the NIH said the trials were unethical as volunteers were placed at an excessive risk of a dangerously high parasite load. A further concern was that no one outside Patarroyo's group had examined the data in his laboratory notebooks. Another criticism was that the vaccine had not been produced by good manufacturing practices. Undeterred, Patarroyo donated a license for SPf66 to WHO and WHO responded by backing field trials.

In a 1994 field trial (backed by the Swiss Tropical Institute) in Tanzania with 586 children vaccinated, the incidence of disease was reduced by 31% — falling far short of earlier claims of much higher efficacy [12, 269]. In a 1993–1995 U.S. Army-sponsored trial among 1200 children in Thailand, said to be the most expensive ($1.5 million) and the best designed to date, there was no efficacy whatsoever. In another trial supported by the MRC and held in the Gambia, 630 infants (6–11 months in age) were immunized with three injections but the vaccine had little effect on protection against first or second attacks of malaria and a delay or prevention of malaria was seen in only 8% of those vaccinated. During a second year follow-up [77] there was no significant protection. In 1999,

Acosta *et al.* wrote: "Given the modest protection previously documented in older age groups and the lack of efficacy in younger infants this vaccine in its current alum-based formulation does not appear to have a role in malaria control in sub-Saharan Africa" [4].

Earlier, Patarroyo, in a fit of hubris, said, "Why should we wait so long? The technology is now ready to be used, why not use it right now? And that's what happened — we made it." At the time of Patarroyo's announcement there was optimism, but by 1996 and after exhaustive field trials, W. Ripley Ballou of WRAIR (who was now involved in developing the competing vaccine RTS, S) concluded that the SPf66 vaccine "does not protect against clinical falciparum malaria and the... results are so disappointing that further efficacy trials are not warranted" [428]. Nick White, of the Oxford Tropical Medicine Research Programme of Wellcome Mahidol University in Thailand, put it more gently: "We have not reached the end of the journey towards an ideal vaccine against falciparum malaria...but we are still on track" [733].

Transmission-blocking vaccines

"The use of antimalarial drugs for transmission control is, however, by and large not very effective compared with those that directly affect the human–mosquito interaction. The reason is that it is virtually impossible, using drugs alone, to ensure that a malaria-infected individual has received the necessary treatment at the time when he or she is infectious to mosquitoes. In endemic areas drug cover can only be intermittent...unless permanent drug prophylaxis is employed across an entire endemic human population...and in most situations, would be neither practical or affordable" [112]. An alternative to the use of drugs to suppress human infectivity for mosquitoes, some have suggested, is to develop and deploy vaccines designed to limit the transmission of human malaria infections to mosquitoes — a TBV.

During the 24 hours of development within the mosquito midgut, malaria parasites are outside cells and are bathed with the constituents of the blood meal. Thus, the midgut stages are accessible to the immune factors of the vertebrate host, i.e. primarily antibody with or without the activation of the complement cascade. This situation has been exploited by inducing antibodies in the vertebrate host against the pre-oocyst stages in the midgut stages, i.e. gametes and ookinetes. (Because of the difficulty of intact antibody crossing the mosquito midgut, the oocyst and sporozoite are unlikely to be affected.) These antibodies, when ingested by the female mosquito as a part of the blood meal, may lead to sterilization of the infection in the mosquito. As these antibodies have the potential to suppress the transfer of malaria parasites from human to mosquito it suggested the possibility for the development of a mosquito-stage vaccine that would block transmission by preventing infection in the vector. A malaria TBV would not prevent disease in newly infected individuals, but would contribute to "herd immunity" in that although it would not directly influence the course of infection in the vaccinated individual, it would affect the potential for infection in others [114, 361, 578, 579].

This was an idea first investigated in the 1940s by Don E. Eyles of the U.S. Public Health Service. Eyles had shown, using the chicken malaria *P. gallinaceum*, that serum associated with a peak in blood parasites was detrimental to the development of gametocytes in the mosquito [213, 214]. Eyles suspended gametocyte-carrying chicken red cells in a variety of chicken sera and fed these to mosquitoes through an artificial membrane. After a period of time, Eyles carried out Ross-type mosquito dissections and counted the number of oocysts on the stomach wall. Compared with sera from normal birds, that taken from chickens one day after the peak of a blood-induced infection caused a reduction in oocyst numbers, indicating suppression of infectivity of the gametocytes to mosquitoes. However, normal serum as well as serum taken from peak blood infections that had been sporozoite-induced had no effect.

Little was done on TBVs until the late winter of 1975 or early 1976, when Richard Carter, then a post-doctoral fellow in Miller's NIH laboratory, proposed the injection of extracellular gametes of *P. gallinaceum* into chickens to see whether there was any effect upon the infectivity of subsequent blood infections of *P. gallinaceum* to mosquitoes through the effects of antibodies on the gametes as they emerged into the immune blood meal. Carter and colleagues made semi-purified preparations of male and female gametes of *P. gallinaceum*, X-irradiated them to eliminate their infectivity to the chickens, and inoculated them directly, without adjuvant, into the chickens by the intravenous route. There were six such immunized chickens and one control. After three such weekly injections, all of the chickens were then infected with live blood stages of *P. gallinaceum*. Normal, or almost normal, blood infections ensued in all of the immunized chickens and in the control. *A. aegypti* mosquitoes were fed upon the chickens daily throughout their infections and the numbers of oocysts counted in the infected mosquitoes. The normal chicken produced the expected hundreds of oocysts throughout the infection. The immunized chickens produced either none at all at any time, or a very occasional one. The overall reduction in oocysts in the immunized chickens over the full course of their infections was 99.99–100%. Their sera clumped (agglutinated) and immobilized male gametes very effectively [115]. It was found that, up to at least six months after immunization and at a time when anti-gamete antibodies had ceased to be detectable in the circulation of an immunized chicken, a new blood infection dramatically boosted these antibodies back to effective transmission-blocking levels even before the first blood-stage parasites were detectable in the blood. In short, the chickens "remembered" almost instantly how to do transmission-blocking immunity as soon as they were confronted with a new malarial blood infection.

Robert Gwadz, a colleague of Richard Carter, immunized chickens with formalin-killed gametocytes of *P. gallinaceum*. Using the membrane feeding apparatus, he showed that the gametocytes from the immunized chickens entirely recovered their infectivity when they were washed free of their own plasma and resuspended in normal chicken serum, whereas gametocytes from non-immunized chickens, when resuspended in serum from the immune chickens, were almost completely unable to infect mosquitoes. Then, finally, he took the gametocyte-infected blood from the immune chickens and examined it for exflagellation under the microscope. There, for the first time, he observed the "microgamete immobilization reaction" [279]. Gwadz's work revealed the site and effectively the nature of a form of immunity that suppressed infectivity of gametocytes to mosquitoes. After his work there could be little doubt of the reality of anti-gamete infectivity-suppressing immunity.

The next series of studies that Carter and colleagues undertook were devoted to characterizing the protein antigens on the surface of male and female gametes and zygotes, and their developmental successors in the mosquito midgut, the ookinetes, of *P. gallinaceum*. Labeling the surface of gametes with radioactive iodine and then precipitating the now radioactive proteins using transmission-blocking monoclonal antibodies that reacted uniquely with either of these two specific protein antigens of the gametes identified two antigens. These particular monoclonal antibodies also efficiently blocked gamete fertilization and transmission of the parasites to mosquitoes, and this was effective proof that the antigens that they identified were the actual target antigens of the transmission-blocking immunity.

The equivalent antigens of *P. falciparum* were soon to be identified by the same methods [713]. These proteins of *P. falciparum* were named Pfs230 and Pfs48/45, based on their apparent molecular weight from migration through a gel under an electric field. As was to become evident, these two antigens, Pfs230 and Pfs48/45, were expressed on the surface membrane of gametocytes as they developed inside the human bloodstream. Once the parasites were taken up in the blood meal by a mosquito they both remained associated with the surface of the emerged gamete for several hours. The presence of these antigens in the gametocytes is the probable reason why these stages induce transmission-blocking immunity almost as effectively as the gametes themselves. The anti-gamete activity of certain antibodies against these antigens, and especially against Pfs230, is dramatically effective [739] in blocking transmission.

Meanwhile, still working with the *P. gallinaceum* system in chickens, Carter and his colleagues identified two additional antigens, Pgs25 and Pgs28, on the developing zygotes of the parasites appearing respectively at around five and 12 hours after emergence [527]. Thereafter, the antigens continue to be present on the surface of the zygote throughout its development as an ookinete. Thus, unlike Pfs230 and Pfs48/45, these antigens are expressed only during the mosquito phase of the life cycle, and are not found at all in the gametocyte stages in the blood circulation. These proteins, too, were shown to be targets of very effective transmission-blocking antibodies. The mechanism of the effect of these antibodies remains unclear to this day. One thing is certain, however: they are not involved in preventing fertilization as this event is long past by the time their target antigens appear. As had been the case with the gametocyte and gamete surface antigens, homologous protein antigens were soon identified in *P. falciparum* [714] and these were named Pfs28/Pfs25.

The gene for Pfs230 has been located on chromosome 2. When the gene for Pfs230 is deleted, the number of oocysts is reduced but not abolished [739]. At least one of its functions is probably in the fertilization mechanism itself, performing some essential action on the male gamete, whereas the Pfs48/45 class of protein has a role on the female gamete. It may also be that Pfs230 is involved in protection of the parasites from the digestive enzymes, nitric oxide, or against anti-microbial peptides within the blood meal. Pfs230 was originally cloned from a *P. falciparum* cDNA library. In the efforts to make synthetic forms of the Pfs230 and Pfs48/45 proteins that could be used for a vaccine, there have been significant problems. These mainly have to do with the fact that each of these proteins is folded together by a large number of molecular bonds called "disulphide bonds." Disulphide bonds turn out to be extraordinarily difficult to reproduce accurately by synthetic means. So far, only a small region of Pfs230 has been expressed in *E. coli* and is able to induce antibodies with some transmission-blocking effect [739].

Ps48/Ps45, of course, is the other identified target antigen or transmission-blocking antibodies that have their effect at and around the time of fertilization of the gametes of a malaria parasite in a mosquito midgut. Recently, there has been a significant success in expressing Pfs48/45 in an immunogenic form that appears to have solved the riddle of how to recreate successfully the elusive disulphide bonds [494].

One of the great potential strengths of Ps48/45 and Ps230 as antigens for a malaria TBV is that these antigens are expressed in the human host so that there can be a boosting effect. At this point in time, we do not know for how long such immune memory may last. It could be for very many years. If so, it would be of tremendous significance to the deployment of gametocyte/gamete antigen-based TBVs for the containment or even elimination of malaria transmission in endemic regions. Thus, vaccinated populations could have their transmission-blocking antibodies boosted by stray incoming blood-stage infections even many years after most malaria transmission had ceased, thereby locking out their onward transmission and sustaining the low malaria environment.

The likely reality of human immune memory for the gametocyte/gamete surface antigens has been shown in studies in malaria-endemic populations. Patients who experienced more than one *P. vivax* attack within four months had considerably less infectivity to mosquitoes than those experiencing their first attack, suggesting the induction of transmission-blocking immunity by a prior infection [538]. Moreover, antibodies to both Pfs230 and Pfs48/45 have been demonstrated in the sera of humans living in areas where *P. falciparum* is endemic and the presence of these antibodies correlates with transmission-blocking activity in membrane feeding assays [246, 270].

In spite of the manifest potential of the gamete surface antigens of the Ps230 and Ps48/45 type as TBVs, the most advanced of the TBV candidates are the post-fertilization antigens. Identification of the target antigens of anti-gamete immunity was pursued using monoclonal antibodies developed against purified *P. gallinaceum* ookinetes as immunogens. These monoclonal antibodies acted to block transmission [273] and acted as would be expected, i.e. well after the fertilization event. The effect was mediated by antibodies against a protein with an Mr of 26,000 (named Pg26), one of two *de novo* ookinete surface proteins — the other being a protein with

Mr of 28,000 [116, 387]. These proteins are now known respectively as the Ps25 and Ps28 classes of ookinete antigens, and have been found in all species of malaria parasites investigated. The Pfs25 protein, the *P. falciparum* equivalent of the Pg26 ookinete protein, has become the basis of the only advanced human malaria TBV. In 2005, the WHO portfolio of candidate vaccines listed five TBVs in trial. In the December 2010 WHO portfolio, however, there were no TBVs listed.

The sporozoite vaccine revisited

By the early 1940s, with reports of the existence of protective antibodies in the blood of monkeys and birds (see pp. 196–198) and the observation that the enhanced activity of serum used in treating *P. knowlesi* infections was directly correlated with the degree of "stimulation" of the immune system, there was a renewed interest in vaccination for malaria. Paul F. Russell and Major H.W. Mulligan at the Pasteur Institute of Southern India carried out the renewed attempts to produce a protective vaccine using sporozoites. As a firm believer that bird malaria was a reliable indicator for human malaria, Russell in collaboration with Mulligan developed a simple sporozoite-clumping (agglutination) test using chickens and *P. gallinaceum* [564]. They found the agglutinin titer of serum to be elevated when the chickens were in an acute or chronic state. Then they went a step further: when the salivary glands of mosquitoes heavily infected with sporozoites were placed in a shallow dish and exposed for 30 minutes to the direct rays of a mercury arc sun lamp, the sporozoites were inactivated. Chickens vaccinated either intramuscularly or intravenously with these attenuated sporozoites showed a considerable rise in the titer of sporozoite agglutinins. Further, of 14 chickens that received inactivated sporozoites and were then challenged with viable sporozoites (by mosquito bite), 7% did not become infected, 64% recovered spontaneously and 28% developed severe infections and died. Those suffering from the mild infection had a very high agglutination titer. They wrote "It seems fair to conclude, therefore, that repeated injections into fowls of inactivated sporozoites of *P. gallinaceum*, producing an agglutination titer of at least 1/32,000 render such fowls partially immune to the pathogenic effects of mosquito-borne infection with the homologous *Plasmodium*" [470]. In another study, they found that 92% of chickens with an agglutinin titer of 1/32,768 or higher showed spontaneous recovery and 8% died after challenge by the bite of infected mosquitoes, whereas the mortality was 51% in fowls with normal or lower agglutination titers. Significantly, the immune birds did not resist a challenge with intravenous injections of the same strain of blood-stage parasites. They concluded: "This suggests that trophozoites and sporozoites are not immunologically identical." In a subsequent report, they [567] noted "the combined mortality for 19 fowls was 21.1%, which was less than half that of normal fowls similarly infected. This is a significant measure of immunization, although in no case was infection prevented."

Russell returned to the US in 1942. During World War II, he served as a colonel in the Army Medical Corps assigned to General MacArthur's headquarters in Australia. In his 1955 classic work, *Man's Mastery of Malaria* [566], Russell provided a retrospective on the history of malaria and attempts at eradication; however, not a single sentence in the entire book refers to a malaria vaccine. Russell remained an administrator and technical advisor on malaria with the Rockefeller Foundation [565] until he retired in 1959. After his work in India, he was never involved in vaccines again.

During the 1960s, Harry Most, who had worked on the development of CQ while in the U.S. Army during World War II, was Chairman of the Department of Preventive Medicine at the New York University Medical Center (NYU). Most served as Chairman of the Armed Forces Epidemiological Board and Director of its Commission on Malaria, a civilian advisory panel. With the support of the Commission and funded by the U.S. Army Medical and Research Command, a project on the biology of *P. berghei* was initiated at NYU. Most recruited Jerome Vanderberg to the project and, starting in September 1963, Vanderberg immersed himself in what was known about the biology of *P. berghei* in Africa. During the next couple of years, his research was focused on working out the parameters for sporozoite transmission of *P. berghei* to laboratory rodents and the characterization of these infections in laboratory rodents. In 1965, the immunologist Ruth Nussenzweig joined the NYU malaria research group (consisting of Most, Vanderberg, and Meier Yoeli).

Immunization studies with irradiated rodent malaria sporozoites at NYU found the optimal dose of X-irradiation to inactivate the sporozoites dissected out of mosquito salivary glands to be 8–10,000 rad; of 46 mice, each immunized via intravenous injection with 75,000 irradiated sporozoites, only one mouse developed an infection. The remaining mice were challenged 12–19 days later with 1,000 viable sporozoites [479]. The percentage of animals infected varied from 14 to 73% (average 37%), whereas in the controls the percentage of mice infected averaged ~90% and the protection was estimated to be 27–86%. There was no protection against an inoculum consisting of infected red blood cells, confirming the stage specificity observed earlier by Russell. In a subsequent study of 103 mice injected with 75,000 sporozoites irradiated with 8,000 rad, only three developed blood infections, and with 10,000–15,000 rad none of 29 mice became infected. When the blood of the X-irradiated immunized animals was inoculated into other animals it failed to produce an infection and the immunized animals did not develop infections after removal of the spleen; this was taken to indicate that sterile immunity had been induced [708]. Protection remained strong for up to two months and then declined. By repeated intravenous challenge of immunized mice at monthly intervals, there was a boosting effect and protection could be maintained for up to 12 months [492].

One of Vanderberg's most valuable findings was his observation that, upon *in vitro* exposure to serum from immunized mice, an antibody-mediated precipitation reaction was formed around sporozoites, and projected from one end [705]. This reaction was called the circumsporozoite precipitation (CSP) reaction. The terminology for the CSP reaction was suggested by the similar-appearing circumoval precipitation reaction around schistosome eggs incubated in immune serum. The abbreviation, CSP, originally referred to the circumsporozoite precipitation reaction; however, over time it has come, instead, to refer to the circumsporozoite protein. Because of the striking way in which serum from immune animals deformed sporozoites, this was postulated to be the basis for a humoral component of protective immunity against sporozoites. Indeed, incubation of sporozoites of *P. berghei* for 45 minutes with immune serum neutralized their infectivity and did not require complement [139]. The CSP reaction was found to be produced in mice and rats immunized by intravenous inoculation with irradiated sporozoites of *P. berghei* or by the bite of infected irradiated mosquitoes, as well as in animals injected intravenously with viable *P. berghei* sporozoites [139]. After immunization with irradiated sporozoites, there was a reduced circulation time for intravenously injected infective sporozoites [480]. Antibodies to CSP appeared to be critical to protection; however, because protection was never complete in passively immunized animals, it suggested that additional mechanisms might be necessary to bring about complete protection against sporozoite-induced malaria infections.

The discovery of the CSP reaction could have led to an enduring and productive collaboration between the members of the group at NYU but instead it created a rift, the basis of which had nothing to do with the finding but with personalities and priority of authorship. Earlier, Vanderberg and Ruth Nussenzweig had an agreement that research publications on sporozoite "immunity" would have her as the senior author, whereas those on sporozoite "biology" would have him as the lead author. Vanderberg, upon completion of his studies on the CSP reaction, prepared a draft manuscript. This was shown to Nussenzweig, who, despite the fact that she had contributed nothing to the work directly, insisted on lead authorship, claiming that this was her research area. Vanderberg, however, said that the studies were conducted by him alone and concerned sporozoite gliding — a biological phenomenon. Most, as Department Chair, was asked to settle the conflict, and when he failed to support Nussenzweig's position, a stressful and hostile environment was created. Henceforth, they were unable to work together in harmony.

With the initial success in the immunization of mice using X-irradiated sporozoites, Ruth Nussenzweig recognized that it would be important to define the correlation between the *in vitro* detectable, anti-sporozoite antibodies and protective immunity and to characterize the antigens involved. Emboldened by the results of the mouse vaccinations with irradiated sporozoites, she assembled a team of immunologists to carry out investigations to isolate and characterize the antigens responsible for protection, and to extend the rodent malaria immunization studies to monkeys. Vaccination attempts were initiated using *P. cynomolgi* and *P. knowlesi* as in rhesus monkeys the former produces a mild and benign infection, similar to *P. vivax* in humans, whereas the latter is highly virulent and is more akin to *P. falciparum* infections in children. The results were disappointing: an initial study in two monkeys failed to protect against challenge after inoculation of irradiated *P. cynomolgi* sporozoites divided into five immunizing doses over a period of 146 days [139]. In a follow-up experiment, using a dozen monkeys, protection was obtained only after intravenous inoculation of large doses of irradiated *P. cynomolgi* over a period of nine-and-a-half and 13-and-a-half months. However, only two of the animals were totally protected against challenge with 10,000–20,000 infective sporozoites [139]. With rhesus monkeys immunized multiple times with 300–400 million X-irradiated *P. knowlesi*, two developed sterile immunity but not the third animal. Unlike the situation with merozoite vaccination (see pp. 222–223), there was no amplification of the anti-sporozoite protective response with *P. knowlesi* sporozoites emulsified with FCA administered intramuscularly and nor was there protection against sporozoite challenge. Consequently, Ruth Nussenzweig abandoned, at least for the time being, further vaccination work on primates and directed her attention to characterizing the immune mechanisms seen with irradiated sporozoites in mice.

Vanderberg, in contrast to Ruth Nussenzweig, believed it was timely to carry out studies in humans. A collaborative arrangement was set up with David Clyde [491] at the University of Maryland School of Medicine where he was the director of a program studying antimalarial drugs against sporozoite-induced malaria in human volunteers at the Maryland House of Correction in Jessup. The irradiated sporozoite immunizations of human volunteers and subsequent challenge with unirradiated sporozoites proved to be a long and sometimes frustrating effort because the first series of immunizations used X-ray doses that had previously been used with rodent malaria sporozoites. In those studies, sporozoites had been directly obtained from dissected-out mosquito salivary glands; however, these could not ethically be injected intravenously into humans. An alternative approach was suggested from studies on mosquito-borne viruses. Serum surveys, done after epidemics of infections with these viruses, consistently showed that only a very small percentage of serum-positive individuals had actually experienced signs or symptoms of disease. Thus, from an epidemiological standpoint, mosquitoes might be more important as vehicles of immunization than as vectors of disease. The use of *Plasmodium*-infected mosquitoes to induce infection was not entirely novel, its practice dating back at least to 1926 with the treatment of 2,500 patients with late-stage syphilis in England at the Horton Mental Hospital in Epsom. This treatment, practiced in England, the US and other countries, persisted well into the 1950s, but with the widespread use of penicillin to treat syphilis, its usage declined. The human trials first carried out by Clyde [135] and later by Rieckmann [551] would be a reapplication of this method for the purposes of testing the protective efficacy of malaria vaccines.

Accordingly, an initial trial was conducted by Vanderberg with rodent malaria, which used infected, irradiated mosquitoes as substitute "hypodermic syringes" to deliver sporozoites. The argument [706] was: "The technique that we presently use for immunization involves the intravenous injection of infected mosquito salivary glands which have been dissected out, ground up and irradiated. However, this preparation contains considerably more extraneous mosquito debris than sporozoites, and the injection of such material into humans would possibly pose medical risks of embolisms and sensitization. Until sporozoite preparations can be purified it would seem prudent to avoid this. A more reasonable approach for the present would be to X-irradiate infected mosquitoes and then let them feed on volunteers, thus allowing the mosquitoes to inject the sporozoites in a relatively uncontaminated condition. Such a technique would have limited practicality, but it has the advantage of being performable now. If protective immunity could be demonstrated under such circumstances, it might encourage further work on attempts to establish purification procedures for sporozoite homogenates. The injection of irradiated sporozoites by mosquitoes should thus be viewed as an attempt to test the feasibility of vaccination in humans, which if successful could lead to trials using more practical techniques." The results showed that mice so immunized were completely protected from sporozoite challenges that caused blood infections and death in 100% of non-immunized control mice. With the success of this approach, it seemed appropriate to move from mice to men.

The first series of immunizations of volunteers using irradiated sporozoite-infected mosquitoes resulted in some breakthrough blood infections so several months were lost while retooling. A second series was begun with several new volunteers and with higher doses of radiation against the infected mosquitoes [135]. An average of 222 mosquitoes that had been irradiated (17,500 rad) fed on each man and, after six to seven weeks, each man was again fed on by an average of 157 irradiated mosquitoes. This time there were no breakthrough blood infections during the immunizations. The vaccinated individuals, along with unvaccinated control volunteers, were then challenged by bites of infected mosquitoes sufficient to have induced a blood infection in every single volunteer that had ever taken part in prior trials conducted by David Clyde and his associates. Upon challenge by the bite of mosquitoes infected with normal infectious sporozoites, one of three vaccinated volunteers was fully protected, whereas all of the non-vaccinated volunteers developed a blood infection, as expected. After the challenge, sporozoite-infected mosquitoes and serum from the challenged volunteers were brought to New York. When Vanderberg tested these sera for reactivity with live sporozoites, strong CSP reactions were found in the serum of one of the vaccinated individuals (GZ). Vanderberg immediately telephoned Clyde and predicted that GZ would be protected. All were delighted when this prediction turned out to be true. It is perhaps relevant that GZ had taken part in the first vaccination trial, although he was not one of the vaccinees who had experienced a breakthrough blood infection. Thus, he had received an especially extended schedule of vaccinations. They concluded that a sufficient dose of radiation-attenuated sporozoites was necessary to attain a sterile immunity upon challenge.

In 1974, Clyde began experiments on himself to determine whether it was possible to immunize against *P. vivax* as well as *P. falciparum* and whether there was cross-protection [136]. Clyde allowed non-irradiated mosquitoes that were infected with vivax and falciparum malaria to bite him, and when he developed infections to both he knew he was not immune. He described the first attack [13]: "You shake like anything. You are very cold. You have a high temperature and a splitting headache. Then you start vomiting, and that is the most awful part of it. You have about four hours of absolute misery and then it gradually lets off for about another 12 hours. Then it starts again." Clyde went on to see whether different strains had immunological differences and so he allowed irradiated mosquitoes from different geographical regions to bite him. Each time, he received scores of bites. "It was a damn nuisance and very unpleasant to have six cages of 350 mosquitoes hanging on to you but that's part of it." The welts from the bites itched and he applied cortisone cream to relieve the irritation and to prevent himself from scratching. By the end of the experiment he had received over 2,700 bites. To test the efficacy of the "vaccine" he accepted a challenge of being bitten by unirradiated mosquitoes. He was protected. Unfortunately the protection was not long lived — in the case of falciparum three months and for vivax from three to six months [13].

Vaccination studies similar to those reported by Clyde, Vanderberg, and Most were conducted between 1971 and 1975 in a collaboration between the NMRI and Rush–Presbyterian–St. Luke's Medical Center at the Stateville Correctional Center (Joliet, Illinois) under the direction of Karl Rieckmann [551]. Five volunteers were bitten by fewer than 200 *P. falciparum*-infected mosquitoes irradiated with 12,000 rad over a period of four to 17 weeks and, during the immunization, two volunteers developed blood infections (which were quickly cured by CQ). This indicated that the radiation dose was too low to inactivate all of the sporozoites. Of the four men challenged, none was protected. Another three volunteers were selected. One was exposed six times within a two-week interval; mosquito dissection showed that he had been bitten by 440 mosquitoes. At two weeks after the last immunization he was bitten by 13 infected non-irradiated mosquitoes and he did not become infected. A second volunteer was exposed eight times to a total of 984 irradiated mosquitoes and, although the intervals were not exactly two weeks apart, he too was protected against challenge at two to eight weeks after the last immunization; however, he showed no protection when challenged at 17 and 25 weeks after the last immunization. A third volunteer was exposed to 987 irradiated mosquitoes with immunization at irregular intervals during a 38-week period and, when challenged eight weeks afterward with a strain different from that used in immunization, was protected; however, when challenged with the same strain at 18 weeks after the last immunization there was no protection.

Thus, vaccination with irradiated sporozoites was "an encouraging step towards the goal of immunizing man against malaria" (Vanderberg, personal communication). The limited success of these vaccination studies served to establish what had been hoped for, namely, a clear "proof of concept," demonstrating that production of sterile immunity to malaria in humans might be biologically feasible and was deserving of further efforts.

Some of the problems encountered in these early human studies with irradiated infected mosquitoes were that it was difficult to carry them out within a defined interval or with a constant number of mosquitoes as immunization depended on the availability of gametocyte donors, and in addition there was the problem of estimating the number of sporozoites released during feeding. *In vitro* culture and production of falciparum gametocytes on demand, so to speak, solved one of these problems. To better delineate the requirements for irradiation dose and the number of sporozoites introduced, studies were carried out during 1989–1999 [316]. At one hour before immunization, the female mosquitoes were exposed to 15,000 rad of gamma radiation from a ^{60}Co or ^{137}Cs source. (Irradiation that was lower than this dose did not attenuate sporozoites.) The sporozoite dosage was calculated in the following way: 50 mosquitoes were dissected to estimate the percentage of mosquitoes with sporozoites and those with a score of >2 (usually 50–75%) were multiplied by the total number of mosquitoes taking a blood meal to calculate the number of immunizing bites. Using this method of calculation, four volunteers who received >1,000 immunizing bites were protected against seven challenges or rechallenges from five to 10 mosquitoes infected with a different strain of *P. falciparum* than that used in immunization; one individual was not protected despite having the same immunizing dose when challenged by 90 mosquitoes with a different strain. Protection was evident at least nine weeks after primary challenge and for at least 23–42 weeks against a rechallenge. Overall, 33 of 35 challenges within 42 weeks after >1,000 immunizing bites led to protection, whereas only five of 15 challenges with >378 and <1,000 immunizing bites led to protection.

Development of the pre-blood-stage vaccine, RTS, S

Over 90% of adults living in the Gambia, an area of high malaria endemicity, were found to have detectable levels of anti-sporozoite antibodies [139, 473], suggesting that these antibodies may be related to the acquisition of natural resistance by continued exposure to the bites of malaria-infected *Anopheles* mosquitoes. Further, observations with rodent, human, and primate malarias showed that protective immunity mediated by immunization with irradiated sporozoites was associated with antibody induction. A monoclonal antibody raised against the surface of *P. berghei* sporozoites neutralized their infectivity [139]. Passive transfer with the monoclonal antibody protected mice against sporozoite challenge [751]. Sporozoites were neutralized by Fab fragments, suggesting that agglutination was not required to inhibit parasite infectivity, and a surface antigen recognized by this monoclonal was involved in sporozoite penetration of liver cells [525].

The sporozoite antigen of *P. berghei* recognized by the monoclonal antibody was stage- and species-specific, and was named the circumsporozoite protein, CSP [139]. The involvement of CSP in inducing immunity was also suggested by the observation that irradiated oocyst sporozoites, having less CSP, afforded minimal protection, whereas those that had matured in the salivary gland contained larger (10–20%) amounts. Monoclonal antibodies were also prepared against sporozoites of *P. knowlesi* and *P. cynomolgi*; these were used to precipitate similar circumsprozoite surface proteins (CSPs). As with *P. berghei*, sporozoite neutralization was associated with these monoclonal antibodies.

The CSPs from a variety of *Plasmodium spp.* are encoded by one gene (of the 5,300 malaria parasite genes). Ruth Nussenzweig, now joined by her husband Victor, was convinced that the CSP would be an attractive target for the development of a protective vaccine.

In 1981, a practical impediment to the use of CSP as a vaccine was that its only source was the mature sporozoite. This difficulty was overcome by the cloning of the CS gene and the ability to deduce the amino acid sequence of the protein from the DNA base sequence. The first CS gene cloned was from *P. knowlesi* [206]. To clone the gene, several thousand *Anopheles* mosquitoes were raised and fed on a monkey infected with *P. knowlesi* (done in collaboration with Robert Gwadz at NIH). At NYU, the mosquitoes were hand-dissected, the salivary glands separated, and mRNA was extracted from sporozoites by a graduate student (Joan Ellis) in the laboratory of the molecular biologist Nigel Godson. Three cDNA clones were obtained and the region that coded for the immunoreactive region was identified and sequenced. The cloning of the CS gene from *P. knowlesi* was quickly followed by the cloning of the CS genes from several human malarias, i.e. *P. falciparum*, *P. vivax*, and *P. malariae* [163, 207, 393, 440] by a number of laboratories including those at the NIH and WRAIR.

On 12 February 1981, NYU filed a patent application on behalf of the Nussenzweigs and Godson for their cloning of the CS gene. After filing the patent, NYU notified the funding sources, including USAID, NIH, and WHO, and indicated that they were entering into negotiations with a genetic engineering company, Genentech, to produce CSP. When Genentech asked for exclusive licensing to market the vaccine, objections were raised by WHO, which indicated that their support required "public access" and under US patent law USAID held the patent for the work that it supported at NYU. The conflict dissipated in 1983 when the NYU-based research had moved ahead, accomplishing the work that was to be done by Genentech. The bargaining over the market rights was discouraging to the Nussenzweigs who were falsely accused, among other things, of having a financial stake in the patent. The legal wrangling, however, continued for years. In the end, it was resolved and the achievement by the NYU group was heralded in a *New York Times* [75] headline from 3 August 1984: " Malaria vaccine is near." One scientist quoted in the article boldly predicted: "This is the last major hurdle. There is no question now that we will have a vaccine. The rest is fine tuning." In 1989, NYU licensed the CSP patent non-exclusively to GlaxoSmithKline, royalty free; this ultimately would lead to RTS, S.

In early 1984, after the scientists at WRAIR had cloned and sequenced the *P. falciparum* CS gene, it became possible to develop a sub-unit vaccine. WRAIR entered into collaboration with GSK to produce CSP using GSK's recombinant *E. coli* expression system. Although efforts to produce a full-length CSP were unsuccessful, four constructs were expressed, purified, and tested for immunogenicity in animals, and one (R32Ltet32) was selected for clinical development. Combined with alum (as an adjuvant) the vaccine, FSV-1, was tested on volunteers in 1987 [38]. W. Ripley Ballou, then a young U.S. Army physician, and five colleagues (including Stephen Hoffman from NMRI, see p. 251) taped a mesh-covered cup containing five infected *Anopheles* mosquitoes to their arms.

The *Anopheles* were allowed to bite them; afterward, to make certain the mosquitoes in the cup were infective, the heads of the mosquitoes were lopped off in Ross-like fashion using a pair of tweezers and the salivary glands examined with a microscope. Ballou and the other volunteers had been injected a year earlier with FSV-1 and now it was time to be challenged with infectious sporozoites to assess protection. At nine days after the infected mosquitoes had fed, the first unvaccinated control had parasites in his blood and was given CQ to clear the infection. The second control and three vaccinated volunteers also came down with blood infections and on the eleventh day Hoffman, who had traveled to San Diego to give a presentation on the vaccine and had been confident it would work, fell ill with malaria. On the twelfth day, Ballou also succumbed. Only the sixth volunteer, Daniel Gordon, was still healthy and remained so. The efficacy of the vaccine was disappointing; however, for the first time, an individual had been completely protected by a sub-unit vaccine. When the vaccine was tested by WRAIR for safety and immunogenicity in Western Kenya, the majority of vaccinated subjects had antibodies to CSP.

Over the next few years, a series of recombination constructs of CSP ($R32NS1_{81}V20$, $R32NS1_{81}$) were produced by GSK that incorporated the $NS1_{81}$ protein from the influenza virus to stimulate a cell-mediated immune response; when these were tested with volunteers at WRAIR, the immunogenicity of $R32NS1_{81}V20$ was low and further clinical development was not pursued. In parallel, WRAIR tested $R32NS1_{81}$ in combination with the only acceptable adjuvant for use in humans, alum (called FSV-2) and, although it was more immunogenic, this too failed to protect any of the volunteers. In the years that followed, FSV-2, in combination with other adjuvants including monophosphoryl lipid (MPL), or an emulsion of MPL, mycobacterial cell wall skeleton and squalene (Detox, Ribi Immunochem), and cholera toxin were also tested at WRAIR but again the results were disappointing [37].

In 1990, Gray Heppner volunteered for a vaccine trial at WRAIR. Heppner did his undergraduate work at the University of Virginia and, after completing his M.D. at the University of Virginia

Medical School, did an internship and residency at the University of Minnesota where, in the late 1980s, he had assisted a malaria researcher growing *P. falciparum* and, in the process, became intrigued with malaria. Although Heppner joined the Army Reserves while in Minnesota, he didn't sign on for active duty until 1990 and that brought him to WRAIR as an infectious disease officer. At WRAIR, the CSP lacking a central repeat region, called Recombinant *Plasmodium falciparum* (RLF), had been expressed in *E. coli* and when encapsulated in liposomes was found to be immunogenic in mice; further, the anti-RLF serum reacted with the surface of intact sporozoites and was able to inhibit their invasion into liver cells *in vitro* [732]. The safety and immunogenicity of the RLF vaccine were tested in 17 malaria-naïve volunteers, Heppner being one of them. Although RLF formulated with alum or MPL was well tolerated and immunogenic upon sporozoite challenge, all immunized volunteers developed malaria [306].

In 1987, the GSK malaria vaccine program transferred its labs from Philadelphia to its vaccine division in Rixensart, Belgium. Joe Cohen, who had taken over the project at the same time that Ballou and colleagues at WRAIR had been using themselves as human guinea pigs, had another plan for using CSP as an antigen. Cohen, using experience gained from GSK's successful development of a recombinant hepatitis B vaccine, Energix-B, decided to couple CSP with the surface antigen protein from hepatitis grown in yeast (*Saccharomyces cerevesiae*) where, at high concentrations, the protein spontaneously formed virus-like particles, and when used as an immunogen antibody formation is enhanced. Cohen hoped that by fusing the repeat peptide NANP (arginine alanine arginine proline) from CSP to the hepatitis surface antigen (consisting of 226 amino acids) similar particles now festooned with the CSP would be made and be able to provoke antibodies targeted to the sporozoite surface. To overcome the possibility that antibodies alone would not suffice, Cohen added a fragment from the tail end of CSP to stimulate cell-mediated T cell responses. This construct would provide, as Cohen said, a "double whammy" with 19 NANP CSP repeats (R), T cell epitopes (T) fused to the hepatitis B surface antigens (S) coexpressed and self-assembled with unfused S antigen. It was named RTS, S [37]. In 1992, the first clinical trial of RTS, S for safety and efficacy was carried out at WRAIR using volunteers. Malaria-naïve volunteers received RTS, S either with alum or alum plus MPL. Both formulations were well tolerated and immunogenic; however, after challenge with sporozoites, zero out of six in the alum group and two out of eight in the alum–MPL group were protected. These results were considered to be encouraging enough to warrant further improvement in the vaccine to enhance both humoral and cell-mediated immunity [307]. Taking clues from the study, Heppner cooked up a formulation with adjuvants that would produce the right sorts of cell-mediated immune responses. Heppner suggested formulating RTS, S with an oil-in-water emulsion plus the immunostimulants MPL and QS21 (a proprietary GSK saponin derivative from the Chilean soap bark tree, *Quillaja saponaria* [AS0]), and by 1996 — 12 years after the first trials — RTS, S/AS02 was tested in human volunteers and protected six out of seven vaccinees. When the volunteers were challenged six months later, one of five was still protected. Two or three vaccinations were necessary to produce sterile or partial immunity (i.e. delayed appearance of parasites in the blood) in most vaccinees. In liquid form, the RTS, S/AS02 had a limited shelf life and so the RTS, S was freeze-dried and then reconstituted with AS02. When 40 volunteers received the vaccine on a zero-, one- and three-month schedule or on a zero-, seven- and 28-day schedule of vaccinations, protection against sporozoite challenge was seen in 45% and 38% of the vaccines, respectively [643].

In the summer of 1998, trials were held in the Gambia with 250 men receiving RTS, S/AS02 or a rabies vaccine on a zero-, one- and three-month vaccination schedule. The RTS, S showed a 34% reduction in the first appearance of parasites in the blood over a 16-week period. During the surveillance period, 81 of 131 men immunized with RTS, S had parasites in the blood, whereas in the control group 80 out of 119 tested positive. The following summer the men were given a booster and this showed that the vaccine was acting in two ways: protecting against infection and weakening the symptoms in those who became infected [37].

Although GSK was encouraged, the company felt that it needed funding assistance to move RTS, S into trials with infants [37]. Ballou wrote a proposal to the Gates Foundation and it provided $50 million through the Malaria Vaccine Initiative (MVI) via the Program for Appropriate Technology in Health (PATH). Ballou was asked to lead it, but instead took a position with a Washington, DC, biotechnology firm, MedImmune, to work on other vaccines; in 1999 Heppner succeeded Ballou as Chief of the WRAIR Malaria Program.

With the MVI on board, GSK and WRAIR collaborated with Pedro Alonso of the Barcelona Center for International Health in Spain who had developed a research site in Mozambique. Alonso's site would be the biggest RTS, S trial, enroling 2,022 children between the ages of one and four years. By 2003, Ballou had rejoined the effort, having left MedImmune to join GSK at Rixensart. The trials in Mozambique with children showed that the vaccine conferred a 35% efficacy against the appearance of parasites in the blood and a 49% efficacy against severe malaria that was maintained for 18 months after the last vaccination. Another formulation specific for children (RTS, S/AS02D) has undergone tests in 214 infants in Mozambique in preparation for licensing by 2011. Infants were given three doses of RTS, S/AS02D or the hepatitis B vaccine Energix-B at the ages of 10, 14 and 18 weeks [24]. Early on, 17 children in each group had adverse reactions and later 31 had serious adverse reactions, none of which seemed to be related to vaccination. Vaccine efficacy for new infections was 65% over a three-month follow-up after the completion of immunizations. The prevalence of infection in the RTS, S group was lower than in controls (5% vs. 8%). It was not clear why 35% of the children did not respond or why the vaccine protected only 34% of adults and the protection was of shorter duration. Perhaps it stems from problems associated with induction of immunological memory.

Persistance of protection for at least 15 months after vaccination has been shown in Kenyan and Tanzanian infants with RTS, S formulated with AS01 adjuvant. This formulation is also being used in a large phase 3 trial taking place in 11 centers in Africa, with the first results expected in late 2011 [271].

It has been suggested that both antibody and cell-mediated immunity contribute to protection with the RTS, S/AS02 vaccine with the relative importance depending on the pre-existing immune status of the individual and other, as yet undetermined, host factors. There is, however, no immunologic correlate of protection in those vaccinated. RTS, S does induce a strong cell-mediated CD-4 T cell response with the production of interferon-γ; however, there is overlap between protected and unprotected groups. Also, despite RTS, S inducing very high titers of anti-CSP antibodies in vaccines, there is usually some overlap between protected and unprotected individuals and no clear threshold has been identified that defines protection [271]. It has been suggested that perhaps those with the most rapid drop in titer are most at risk of developing an infection. The reason why antibody alone does not eliminate incoming sporozoites is not understood; however, it may be that continuous shedding of CSP-antibody complexes allows the highly motile sporozoites to escape. RTS, S vaccination may reduce but not completely prevent emergence of the merozoites from the liver, so that vaccinated children receive an attenuated low-dose blood-stage infection that allows a more effective immune response to blood stages [742].

Protection using the sub-unit RTS, S vaccine is far from perfect and perhaps with better formulations of adjuvants and with viral vectors [610, 651] there will be enhanced efficacy [365]; however, there remains a nagging question: is immunity actually dependent solely on a CSP construct? Indeed, recently the critical role of CSP in induction of sterilizing immunity has been called into question using rodent malarias. In one instance, using *P. berghei*, sterile immunity was obtained despite there being no immune response specific to CSP [275]; in another study, a mouse strain that was completely immune-tolerant to CSP was found to be fully protected after three immunizations of irradiated sporozoites and challenge with infectious sporozoites [386].

Back to the future

Since the 1980s, sub-unit malaria vaccines have received the most attention and it is sobering to note that there is only a single recombinant protein vaccine on the market for any disease, and there are no vaccines based on synthetic peptides, recombinant viruses, recombinant bacteria or DNA plasmids. In stark contrast to sub-unit vaccine formulations for malaria, protective studies with irradiated sporozoites of rodent and human malarias seem to have fared better. In 1989, faced with the disappointing trials with CSP-based vaccines, gamma-radiation-attenuated sporozoite immunization studies were begun at NMRI and WRAIR using volunteers [209, 316]. After 10 years of clinical experience with these immunizations, Luke and Hoffman [416] concluded: "Immunization with radiation-attenuated *P. falciparum* sporozoites provides sterile protective immunity in >94% of immunized individuals for at least 10-and-a-half months against multiple isolates of *P. falciparum* from throughout the world." They went on to write: "Given the...need for an effective...malaria vaccine...we believe that an attenuated sporozoite vaccine should be produced and tested for safety and protective efficacy as soon as possible." And, echoing the words that Paul Silverman (see p. 202) used decades earlier to convince USAID that a vaccine against blood stages was an achievable goal, i.e. the hurdles associated with a whole parasite vaccine were "simply technical problems" that could be overcome by a major, well-funded research program, Hoffman wrote that "although technically and scientifically challenging such an approach has an enormous advantage over other approaches." Shortly thereafter (2002), Hoffman founded and became the CEO of the only company in the world dedicated solely to developing a radiation-attenuated sporozoite vaccine for malaria. The name of the company Sanaria — meaning "healthy air" — is a clever counterpoint to the Italian word mal'aria, namely "bad air." In 2005, Hoffman filed a patent (200050220822) for "aseptic, live, attenuated sporozoites as an immunologic inoculum" and began the difficult task of putting theory into practice.

Stephen L. Hoffman (1948–), currently the Chief Executive and Scientific Officer of Sanaria, is a lean, green-eyed, exuberant optimist, heralded in the 11 December 2007 *New York Times* as "the soul of a new vaccine." Impatient and intolerant of negativity, he is following the century-old precepts of Louis Pasteur to turn a crippled malaria parasite into a wonderful protective weapon against itself. By 2001, Hoffman had to come to the conclusion that it would take many, many more years to develop a highly effective malaria vaccine based on recombinant DNA technology. He retired from the Navy and in 2001 joined Celera Genomics as Senior Vice President of Biologics with the goal of utilizing genomics and proteomics to produce new biopharmaceuticals, especially immunotherapeutics against cancer. While at Celera he organized a Keystone Symposium on malaria vaccine development, and left the meeting frustrated by the realization that biotechnology was unlikely to produce a highly effective malaria vaccine for at least another 25 years. However, in putting together data from 10 years of work immunizing people with radiation-attenuated sporozoites, he came to the conclusion that there already was a way to make a malaria vaccine. The approach had been pioneered decades earlier by the NYU group (Nussenzweig and Vanderberg) using rodents and later tested with human volunteers in Maryland and Illinois (Clyde and Rieckmann). Believing that this approach would provide a highly protective malaria vaccine, he left Celera in 2002 and founded Sanaria with a single goal in mind: to develop a malaria vaccine for infants, young children, and women using a disarmed version of the whole parasite. Malariologists had always considered radiation-attenuated sporozoites as the "proof of principle" that a malaria vaccine could be developed, but none thought that a vaccine composed of radiation-attenuated sporozoites that met regulatory and cost-of-goods requirements could be manufactured. The accomplishments of Sanaria during the succeeding years have shown that an irradiated sporozoite vaccine can be manufactured. Clinical lots have been produced, and the PfSPZ vaccine has now been shown to be safe in pre-clinical animal studies.

Hoffman is no newcomer to malaria vaccines. In the spring of 1987, as a Commander in the U.S. Navy, he was part of a team of physicians involved in the test of the CSP-based sub-unit vaccine, FSV-1. After vaccination, he allowed himself to be bitten by 3,000 infected mosquitoes. Ten days later he went to a medical conference in San Diego. The morning after he landed he was already shaking and feverish, and shortly thereafter was suffering with a full-blown attack of falciparum malaria. One of his fellow test subjects, Ballou, shared his fate. Ballou has spent the past 20 years working on the development of a sub-unit vaccine and, unlike Hoffman, still believes that the original vaccine (RTS, S combined with AS02A and tentatively named Mosquirix) is the most promising vaccine candidate. He is critical of Hoffman's attenuated sporozoite approach, stating that it is impractical: large numbers of irradiated sporozoites must be repeatedly injected to provide solid protection; radiation attenuation is not precise and a proportion of the damaged parasites are ineffective; there are safety risks associated with the sterility and manufacture and use of such a vaccine; and there are huge challenges to produce it on a commercial scale including thousands of liters of infected blood, raising and infecting 200 million pathogen-free *Anopheles* and the tedious and time-consuming manual extraction of sporozoites from the salivary glands [36]. Further, Ballou argues that the dose and schedule will need to be worked out, there is no guarantee that vaccinated infants living in an endemic area will not be boosted, and there is a significant risk that a vaccine that completely prevents infection would shift the occurrence of severe disease to older age groups. Another critic of Hoffman's approach is Pierre Druilhe of the Pasteur Institute who says, "Even calling it a vaccine is a compliment. It has no chance of offering protection. It is like Captain Ahab trying to kill Moby Dick with a knife" [446]. Aside from the risk that underirradiated mosquitoes may cause a blood infection or that the irradiated mosquitoes may not reach the liver, Druilhe says that a frozen vaccine will never be practical in tropical endemic areas that are without proper facilities for refrigeration.

Not all agree with Ballou and Druilhe. Hoffman's proposal to develop a whole sporozoite vaccine has obtained $15 million in backing from the U.S. Army, a San Francisco non-profit pharmaceutical company called Institute for One World Health, and another $29 million from the Bill and Melinda Gates Foundation to allow the building of an assembly line to mass-produce the vaccine. This involves raising mosquitoes aseptically, supercharging them with far more parasites than nature does by membrane feeding the mosquitoes on blood containing *in vitro*-grown gametocytes, allowing two weeks for the sporozoites to mature in the mosquito salivary glands, irradiating the infected mosquitoes, and finally dissecting out the sporozoites from salivary glands. It has been claimed by Hoffman that four trained dissectors working in two biosafety hoods can aseptically isolate sporozoites from at least 75 mosquitoes per dissector; this he believes would be enough for 1,200 three-dose immunization regimens. With two shifts per day and 310 workdays per year, a small factory with 50 full-time dissectors per shift could produce 110 million doses of vaccine per year. These irradiated sporozoites would then be placed in suspended animation in liquid nitrogen. In September 2010, Reuters reported that, of 80 vaccinated volunteers, only a small number [5] were protected. Hoffman suggests that it may be necessary to administer the vaccine intravenously to increase its effectiveness. Undeterred, Sanaria continues to raise monies to produce irradiated sporozoites and intends to conduct further trials.

The PATH MVI and the Seattle Biomedical Research Institute are teaming up to build a new facility for testing the safety and efficacy of potential vaccines using paid volunteers. Christian Loucq, the PATH MVI Director, optimistically states [61]: "The target is that by 2025 there will be a vaccine with at least 80–85% efficacy. It is possible we will combine the Sanaria vaccine with protein-based vaccines…and we are going to be in a position to test it here in Seattle and later on in Africa." Even if the Sanaria vaccine is successful, questions remain. Can Sanaria actually produce enough irradiated sporozoites to deliver 200 million doses a year? What will the dosing regimen be? Will the vaccine create dangerous side-effects or will it induce a malaria infection? Will there be boosting when the immunized are exposed to natural infections by the bite of infected *Anopheles*?

Learning from the liver

The hunt for a vaccine against the liver stages of malaria began more than 75 years ago when Clay Huff (1900–1982), then a young

Assistant Professor in W.H. Taliaferro's Department of Microbiology at the University of Chicago, made an unexpected discovery: malaria parasites could develop in the tissues outside red blood cells (EE). In 1935, Huff and Bloom described EE multiplication of *P. elongatum* in the canary and in 1944, using the popular chicken malaria *P. gallinaceum*, it was possible for Huff and his graduate student Frederick Coulston to follow the entire development of the parasite from sporozoite to EE form to the stages growing and reproducing in the red blood cell [337]. In 1947, Huff left the University of Chicago and moved to the NMRI as the only parasitologist; he spent the remainder of his career at NMRI until retirement (in 1969). In 1964, Richard Beaudoin joined Huff and in 1974 Beaudoin, now as head of the NMRI malaria laboratory, came to realize that, although the bird malaria model for growing EE forms had been utilized for 30 years (thanks to the efforts of Huff and colleagues), little progress had been made with mammalian parasites. As a consequence, he initiated a program of research to grow EE forms in liver cells. The discovery that the human embryonic (WI38) cell line was susceptible to *P. berghei* led to the first successful culture of mammalian EE stages [324]. Following this, *in vitro* cultivation of various rodent species of *Plasmodium* was achieved in liver cells; however, the necessity for repeated isolation of these cells, which do not replicate or retain their differentiated functions, had restricted the value of such culture systems. This was soon overcome by the use of "immortalized" liver cells such as the human hepatoma line HepG2-A16 that supports liver development of *P. berghei* with about 8% of the sporozoites becoming EE forms [325].

In 1976, Beaudoin attended a WHO meeting in Geneva, and was asked whether cultivated EE parasites might be considered a source of antigen for malaria. He replied: "There is a recent demonstration that EE merozoites are immunogenic" and from these findings "it is reasonable to suspect that these stages may stimulate protective immunity" [54]. He went on to speculate that perhaps anti-EE immunity will be found to be related to the protection stimulated by irradiated sporozoites. Indeed, in his very first paper using irradiated sporozoites, he suggested: "The results of the present study do not imply that liver stages were unimportant in initiating the immune response" [56]. In effect, he questioned the prevailing view held since 1967 when using X-irradiated sporozoites in rodents and humans for protection: the dominant sporozoite antigen, CSP, was key to the induction of sterilizing immunity, and as such it alone could serve as the basis of a protective vaccine. The gauntlet had been thrown down and the research groups at NYU and NMRI as well as those at the Pasteur Institute in Paris soon became adversaries.

Using the mouse malaria model, *P. yoelii*, it was shown that, after mosquito inoculation, the irradiated sporozoites are deposited in the skin [236, 707], and then move into the regional lymph nodes where, after dendritic cell presentation, T cells were primed to recognize the parasite [498]. The activated T cells then moved on to the liver where sporozoite-infected liver cells were eliminated upon subsequent challenge. Only antigens expressed by both the sporozoite and the liver-stage parasites and not those expressed solely by the liver stages, it is claimed by Fidel Zavala's group at Johns Hopkins, are involved in protection. Further, they argue that CSP, expressed in both the liver and sporozoite stages, is the immunodominant target of the T cell response, and the anti-sporozoite CD8 T cell response originates not in the liver but in the lymphoid tissues at the site in the skin where the mosquito has bitten. Afterwards, when the CD8 T cells are activated, they travel to the liver where the sporozoite antigen is recognized by the liver-resident CD8 T cells [285]. This occurs rather quickly and, within a few hours, IFN-γ is produced within the liver, killing the parasites [123].

Zavala is convinced that studies of immunity using rodent malarias are important as they have the potential to identify important elements and to raise critical questions that can then be applied to the human disease. Indeed, he opines, almost everything involved in the RTS, S vaccine (see pp. 246–249) now in clinical trials, is based on the large amount of work done in rodents, and carried out by a legion of people including himself. He laments, "something not readily apparent when one reads papers on clinical trials with the sporozoite vaccine is that the seminal findings in rodents have been ignored and oftentimes it appears that basic knowledge sprang *de novo*. Alas, life is unfair..."

Not all malaria researchers who have studied pre-blood immunity place their entire faith in CSP as the key pre-blood-stage antigen. Indeed, CSP as a dominant antigen interferes with the development of T cell responses with antigens of different specificity and thus the result is an immunological response that is too narrow on specificity and much too regulated, as it does not seem to increase with time and repeated exposure. Clearly, antigens other than the CSP must exist, but it has been difficult to find them because of the dominance of CSP (Zavala, personal communication).

The suspicion that liver-stage antigens (LSAs) could act in the induction of immunity came to be raised by more and more observations, first in mouse and then with human malarias [315, 664]. With mouse malarias, it was found that the X-irradiated sporozoites had to be alive for some time in the liver: too low a dose of radiation led to a blood infection, whereas too high a dose resulted in an absence of protection. At the irradiation dose inducing protection, sporozoites were found to transform into young liver-stage parasites blocked in development and unable to divide, but still able to secrete antigen. Indeed, in the rodent malarias, young liver-stage forms could persist for up to a year after immunization with X-irradiated sporozoites, thus allowing for prolonged antigenic stimulation. Similarly, if live non-irradiated sporozoites were used to inoculate mice and then were treated with a drug that either blocks liver-stage parasites (α-difluoromethlornithine) or one that prevents blood stages (CQ), the animals were protected against a challenge with sporozoites but not with blood-stage parasites [58]. However, despite such evidence, for decades investigators were discouraged from studying LSAs in mammalian malarias. The presumption was that because the liver form was clinically silent it could have no immunologic importance.

There was another impediment to their study: EE forms are present in miniscule numbers within the very large liver. By way of example, if a human were to be bitten by an infected mosquito receiving, at a maximum, 100 sporozoites, this could result in 100 EE forms being present in a liver containing 100 billion cells. Although the liver of a mouse is much smaller than that of a human and contains only two billion liver cells, the chance of finding EE forms is still very slim. Further, because CSP was (and, to some malaria researchers, still is) considered to be the sole major antigen, the focus was almost entirely on it, and other antigenic proteins that might have been expressed by the sporozoite or liver stage were neglected. The disappointing results with CSP-based vaccines coupled with the early observations on irradiated sporozoites and drug treatments (see above) prompted some investigators to search for LSAs [126, 189].

Following clues provided by the studies of Verhave [711], Beaudoin et al. [55] made a seminal finding: if, during the course of immunization, the mice were treated with CQ to suppress the blood infection, and prior to sporozoite challenge were given a curative dose of primaquine to eliminate the liver-stage parasites, the mice resisted a challenge with sporozoites, but not with blood stages. They concluded that unaltered sporozoites were immunogenic and stimulated protection similar to that achieved by immunization with irradiated sporozoites. Vaccination studies by others (again using a mouse malaria, *P. yoelii*) with live sporozoites, followed by CQ treatment, showed that antibody was not implicated in protection. However, depletion of CD8 and CD4 T cells did result in a partial reversal against a challenge infection with sporozoites (and this was mirrored by elimination of liver-stage parasites). In this they agreed with Zavala *et al.* that the CD8 T cell-derived IFN-γ, as well as nitric oxide and other factors, affected liver-stage parasites. However, they went on: "The immunity induced by this procedure [i.e. live sporozoites with CQ treatment] is mainly directed at the liver stage parasites, which is also the case for irradiated sporozoites" and "the observations imply that a very short exposure to the immune system to liver stage parasites could result in efficient induction of protection...and only T cells are responsible" [58]. Maximal activation of CD8 T cells occurs within eight hours after immunization with irradiated sporozoites and then declines to imperceptible levels by 96 hours. In spite of this rather brief presentation of sporozoite antigen, it is sufficient to trigger T cell proliferation lasting several days [286]. Of some interest is the observation that the magnitude of the T cell response was not enhanced by repeated bites of infected mosquitoes to increase the amount of antigen presented. Zavala contends that these mouse studies with *P. yoelii* mimic the manner by which immunity to sporozoites is acquired under natural conditions in endemic areas and is similar to the sterile immunity to *P. falciparum* induced in human volunteers. It is unexplained, however, why in humans, under natural endemic conditions subjected to the bites of infected mosquitoes, sterile immunity does not occur.

It was suggested by some investigators that the antigens expressed in the maturing and mature liver-stage parasites were potent inducers of protective immunity. They also had the audacity to claim that the protection afforded by the sporozoite vaccine might in fact depend on LSAs. There are, however, skeptics who doubt the importance of LSAs in pre-blood immunity because liver cells do not prime T cells and because the liver is an organ that suppresses rather than induces T cell responses. The non-CSP antigens that may be effective in the induction of immunity are those expressed in both sporozoites (so that they can be induced after immunization) and liver stages (so that they can make the liver the target of activated T cells). Reinforcing the proposition that LSAs as well as CSP are involved in pre-blood-stage immunity is a recent report concerning Kenyan children with high levels of IgG with antibodies to LSA-1 as well as CSP and thrombospondin-related antigenic protein (TRAP). All correlated with a lowered risk of developing clinical malaria; however, the antibody to LSA-1 was the most strongly associated with protection [352].

By the late 1980s, Beaudoin's group had identified multiple non-CSP antigens in sporozoites that might be associated with liver stages [181, 182, 323]. However, after attempts to produce monoclonal antibodies for liver stages in mouse malarias failed, another approach was tried: screening of human sera to find those that reacted only with liver stages. This high-risk approach was pioneered by Pierre Druilhe of the Pasteur Institute. Rather than employing the more traditional methods of moving from bench experiments and then using the mouse immune system to screen for putative protective antigens, human IgG passive transfer experiments were used to identify blood-stage candidates [190, 191].

In vivo transfer of antibody from immune adult Africans reduced the severity of blood infections and, in laboratory studies, the antibodies were shown to act indirectly to kill the parasites in an antibody-dependent cellular inhibition (ADCI) involving monocytes. The ADCI mechanism involved two of the seven IgG subtypes (IgG_1 and IgG_3) and these were used to select protective blood-stage antigens, one of which was MSP-3. Antibodies raised in mice as well as in humans immunized with MSP-3 mediated killing of *P. falciparum in vitro* (ADCI assay) as well as *in vivo* (severe combined immunodeficiency [SCID] mouse model). In humans living in a malaria-endemic region in Senegal, a strong correlation was observed between protection against malaria attack and the levels of IgG_1 and IgG_3; the IgG_3 level against MSP-3 was predictive of clinical outcome; however, human vaccine trials are yet to be conducted.

Druilhe was able to successfully culture *P. yoeli, P. vivax* and *P. falciparum* liver stages *in vitro* and, using these, showed that anti-sporozoite antibodies affected not only invasion but also development within the liver cell. As pre-blood immunity depends on the successful transformation of irradiated sporozoites into liver stages, Druilhe, ever the iconoclast, has contended that what has been long considered as "anti-sporozoite immunity" is in fact "liver stage-dependent immunity."

The sera that Druilhe used to identify LSAs came from individuals who had been living in Africa for 15–26 years and who had taken daily doses of CQ to suppress the appearance of blood parasites [190]. A serum (designated as PM) from one of these individuals, a priest known as Père Mauvais, who had not suffered from a blood infection but who had repeatedly been bitten by infected mosquitoes, was used to screen a *P. falciparum* genomic DNA expression library (see p. 37). The priest's serum identified many new antigens and several were shown to react only with liver stages. Two of the antigens, LSA-1 and LSA-3, which were discovered then, have received special consideration as vaccine candidates because individuals living in malaria-endemic areas show a high prevalence of immune responses (antibody, CTL and helper T cells) to them.

LSA-1 is a very large protein consisting of a large central repeat (approximately 89 repeats of 17 amino acids that make up 77% of the protein), flanked by two non-repeats at the ends, and is exclusively found in liver-stage forms [313]. The gene for LSA-1 was cloned in the late 1980s [276]; however, it was only recently that scientists at WRAIR were able to develop an LSA-1-based recombinant protein (designated FMP011) [313]. When FMP011 was used to immunize human volunteers using the adjuvants AS02 and AS01, it was shown to be safe; however, both preparations failed to show protection against challenge with sporozoites and there was no delay in the appearance of parasites in the blood.

LSA-3 is an abundant large protein made up of non-repeated sequences flanking glutamic acid-rich repeated regions, and expressed in both sporozoite and liver stages. It is thought to be involved in the reproduction of the EE forms and their release from the liver cell. LSA-3 was able to confer strain-independent protection against *P. falciparum* sporozoite challenge in immunized chimpanzees and *Aotus l. griseimembra* monkeys [164, 165, 514]. Of nine chimps immunized with LSA-3 as either a lipopeptide or a polypeptide in saline or with the adjuvants, Montanide or SBAS2 (SmithKlineBeecham) resulted in full sterile protection in six on first challenge. In an *Aotus* trial, LSA-3 was adsorbed to polystyrene beads. Animals were vaccinated three times and challenged as late as five months after the last immunization by intravenous inoculation of 100,000 sporozoites, and sterile protection was achieved in three of five monkeys. Protection appeared to be associated with high levels of IFN-γ production. LSA-3 is now undergoing clinical trials in human volunteers for safety.

Vaccine approaches before sequencing the genome

Prior to the publication of the *P. falciparum* genome sequence, research on malaria vaccines moved along two parallel paths: those involving recombinant proteins formulated with novel adjuvants and the other involving vaccination with DNA plasmids, recombinant viral vectors and DNA–prime virus boost combinations. In the early 1990s, the main focus of the vaccine program at NMRI led by Stephen Hoffman was on DNA vaccines with special emphasis on pre-blood stages and immunity to these stages. Indeed, when the program was started, DNA vaccines seemed to overcome the impediments of more traditional recombinant protein sub-unit vaccines. DNA-based vaccines induce both CD8 T cell responses (to produce IL-12 and IFN-γ) and T helper cell responses (to release IL-4, IL-5, IL-6 and IL-10), are simple to design and modify, can be combined easily, are stable, do not require refrigeration and can be produced on a large scale. In 1993, a plasmid DNA encoding a portion of the CSP from *P. yoelii* was used to immunize mice [586]; as predicted from experience with other systems, the levels of CD8 T cells and IFN-γ increased. A year later, when mice immunized with a plasmid encoding another yoelii antigen were protected against sporozoite challenge, it encouraged the development of DNA-based vaccines for use in humans. Indeed, the combination of DNA plasmids was pursued by the NMRI in partnership with Vical Inc. in a program called MuStDo (MultiStage DNA Vaccine Operation) [180]. The initial clinical trial tested the efficacy of a five-plasmid mixture of pre-blood antigens (CSP, TRAP, LSA-1, LSA-3 and Exp-1) and, as with the mouse malaria, antigen-specific CD8 T cells and IFN-γ responses were found in the human volunteers. The vaccine was safe and well-tolerated when given at four-week intervals. Subsequent studies varied the effectiveness of the route of vaccination. "However, despite the induction of CTL and IFN-γ responses by each of three different routes of DNA administration...the frequency and magnitude of the responses was suboptimal and probably not adequate for protection...Antibodies were not detected in humans by any of the different routes of

DNA administration despite the same DNA vaccine having been shown to induce antibody responses in mice, rabbits and monkeys" [180]. Even in mice, strain differences in response were observed and, in those where there was protection, it did not withstand challenge doses and the protection was short-lived. Further, although the vaccine did provide evidence for priming in that the immune responses were boosted in the vaccinees after being exposed to sporozoite challenge, there was no protection [722].

Such disappointments using plasmid DNA vaccines [692] have led to another approach: virus vectors. Viruses, when modified, have the advantage of being able to serve as vaccine carriers (= vectors), being able to deliver DNA more efficiently than in the form of plasmids; however, they could have a drawback in that they may divert the immune response to irrelevant virus antigens and may neutralize pre-existing antibody responses to viral components. The NMRI vaccinologists elected to take advantage of both plasmid DNA and viruses. In a complex regimen, they used a priming dose of plasmid DNA followed by boosting with a recombinant virus. A cytokine-enhanced multi-antigen DNA plasmid encoding two sporozoite antigens (CSP and TRAP) plus two blood-stage antigens (AMA-1 and MSP-1_{42}) for priming, and a canary pox virus as a booster, was used with *P. knowlesi*; two out of 11 monkeys (18%) vaccinated showed sterile protection and seven out of nine (78%) vaccinated rhesus monkeys spontaneously recovered [555]. In a subsequent and more detailed study with priming by three doses of DNA plasmids containing a mix of the same four *P. knowlesi* antigens and then boosting with the pox virus, it was concluded that: (1) timing of vaccination and formulation were critical to protective efficacy; (2) protection could be achieved in some animals against both pre-blood- and blood-stage infections; (3) in a multi-stage vaccine the individual antigens may interfere with one another. A perplexing finding was that, although the vaccine was protective, the IFN-γ responses correlated with an earlier appearance of parasites in the blood [726]. The fowlpox virus (FP9) and a modified vaccinia Ankara (MVA) virus expressing peptides from TRAP and AMA-1 fused to a multi-epitope string of

T and B cell epitopes including copies of NANP, named FP9-MVA-TRAP, developed by the University of Oxford and shown to reduce by 90% the number of parasites emerging from the liver when tested in the Gambia, showed only 10.3% efficacy relative to controls. When PEV3A (prepared by Pevion Biotech in collaboration with the Swiss Tropical Institute), an influenza virus carrying the NANP B cell epitope and an AMA-1 peptide, was tested on its own or in combination with FP9-MVA-TRAP after sporozoite challenge, there was neither sterile protection nor clear evidence for a reduction in liver-stage parasites with the combination; however, with PEV3A on its own, there was evidence of reduced multiplication of asexual stages in the blood and abnormally-appearing parasites (called crisis forms; first described in monkey malaria by W. H. and Lucy Taliaferro in 1934 and later in mouse and human malarias). None of the volunteers was completely protected [668].

One of the more promising vaccine approaches has been to use adenovirus vectors, which induce both humoral and cell-mediated responses simultaneously, and which can protect after a single immunization. For clinical evaluation, the NMRI in collaboration with GenVac Inc. is testing a combination of two adenovirus vectors, one carrying CSP and another AMA-1, in humans in one- and two-dose regimens and subsequently in prime-boost combinations to determine their feasibility as a multi-stage vaccine [408]. Other groups, particularly one led by Adrian Hill at the Wellcome Trust Centre for Human Genetics, University of Oxford, are in the process of testing other virus vectors able to induce strong antigen-specific antibody responses [187, 545, 546, 641].

A priori, the obvious way to develop a protective vaccine for malaria is to employ multiple antigens against multiple stages. However, to date there is no multi-antigen, multi-stage vaccine tested in clinical trials that has been able to induce strong immune responses against each component simultaneously in more than a few volunteers; nor has it been possible to simultaneously induce both a humoral and cell-mediated immune response in humans. There is, however, no shortage of potential vaccine candidates — either composed of a single or several antigens — but, of all the vaccine candidates under clinical testing and listed in the 2010 WHO portfolio, most are based on just three antigens (CSP, MSP, and AMA) and these were cloned two decades ago. Because there is suggestive evidence that the repeat sequences in CSP and TRAP may act as a smoke screen allowing the parasite to evade the immune response, it remains imperative that efforts be directed at identifying other target antigens from among the 5,300 *P. falciparum* putative proteins; to date this approach to novel vaccine candidates has been limited.

Vaccine approaches after sequencing the genome

Since the late 1990s, malaria vaccinologists have been employing virus vectors to produce robust immune responses to malaria antigens. In 1998, it was demonstrated that one out of 35 sporozoite-challenged volunteers could be protected against malaria using the pox virus NYVAC with seven malaria antigens. In 2006, using TRAP and the MVA vector alone or in a heterologous prime-boost regimen, one out of eight volunteers were sterilely protected and, using a fowlpox strain (FP9) (see pp. 262–263), two out of 16 vaccinees were sterilely protected with an estimated reduction of 90% of parasites in the liver. However, field testing showed a drop in the potency of this vaccine when given to African children of one to six years of age, and 70% showed blood parasites and had to be treated with antimalarials. Because the evaluation of the T cell response indicated that, for sterile protection, a threshold 1,000 times higher than that achieved by that vaccine was required, it was suggested that more potent vectors would have to be used. Using four *P. knowlesi* antigens (CSP, SSP2, AMA-1 and MSP-1) with four platforms for prime or boost vaccinations, three out of five monkeys were sterilely protected with the DNA/attenuated pox virus [349]. No immune correlates of protection could be identified from this study; there was a CD4 T cell response but no CD 8 T cell response was found. None of the sterilely protected monkeys, when challenged four months later, was protected. This being so, the plan was to replace DNA priming with recombinant malaria proteins using novel adjuvants such as Vaxfectin, a cationic lipid formulation. When tested in mice, Vaxfectin enhanced the immunogenicity of a low-dose plasmid encoding five *P. falciparum* antigens (CSP, TRAP, LSA1, AMA-1, and MSP-1) [587]. As a consequence, it was suggested by this group that Vaxfectin-formulated malaria plasmid DNA vaccines be pursued.

In parallel studies, immunization of macaque monkeys using the simian adenovirus, AdCh63, and MVA with the LSA TRAP plus a multiple epitope (ME) string containing 14 CD8 T cell epitopes, three CD4 epitopes (from CSP, the tetanus toxin and BCG) and two B cell epitopes NANP x 4 (from CSP and TRAP) elicited strong CD8 and CD4 T cell responses, leading the investigators to suggest that its future use should be encouraged in clinical trials [106]. Using AMA-1 with AdCh63 or MVA, there was a strong antibody response as well as induction of CD4 and CD8 T cells. Boosting the AMA-1 protein with the adjuvants (Alhydrogel or CoVaccine) in a three-shot AdCh63–MVA regimen or a two-shot AdCh63–protein regimen induced memory B cells and high antibody titers, and T cell levels were maintained. These regimens have not been evaluated in human challenges.

It is evident from this survey that almost all of the vaccine trials (and there are now 30-plus in progress) using plasmids or adjuvants or prime-boost strategies have employed a limited number of malaria antigens, some of which were cloned more than 25 years ago and prior to the sequencing of the *P. falciparum* genome. Indeed, little has been done to evaluate the potential of the 5,000-plus *P. falciparum* proteins. An exception has been the attempts to use GAPs as a component of a protective vaccine. Indeed, the availability of genome sequences, the generation of stage-specific gene expression data and the ability to genetically manipulate the malaria parasite have enabled investigators to search for genes that have essential roles for parasite survival. The most successful use of GAPs has come from studies of sporozoite biology that were designed to identify virulence genes needed for the successful transmission of the parasite. Using suppression subtractive cDNA hybridization, 30 distinct genes in salivary gland sporozoites were found to be upregulated and, of these, 29 were not significantly expressed in blood-stage parasites. Using the rodent malaria *P. berghei* [436], it was possible to attenuate sporozoites by deletion of genes known as upregulated in infectious sporozoites (UISs). The genes for UIS3 and UIS4 encode proteins of the PVM found in liver-stage parasites and, when these were deleted, there was complete arrest of liver-stage development after sporozoite invasion. However, the UIS4 "knockout" showed occasional breakthrough infections when high numbers of infectious sporozoites were used. Deletion of two other sporozoite-expressed genes, P52 which encodes a GPI-anchored protein and P36 encoding a putative secretory protein, also aborted sporozoite development in the liver but these also yielded breakthrough infections. However, when both P52 and P36 were deleted, there was complete attenuation and no breakthrough. Immunization of mice with all four genes deleted in sporozoites induced long-lasting protection against challenge with infectious sporozoites [435, 710]. These GAPs induced protection mainly by CD8 T cells but antibody also contributed. Deletion of sporozoite asparagine-rich protein 1 (SAP1) also resulted in liver arrest and, when mice were immunized three times with 10,000 of these GAPs, they were protected against infectious sporozoite challenge for at least 210 days.

SAP1 is involved in the infectivity of sporozoites and it is for this reason that deletion of the gene encoding SAP1 confers long-lasting sterile protection against sporozoite challenge in mice [14]. More effective, however, are late-stage-arresting GAPs that involve the deletion of the FabB/F gene that encodes an essential enzyme of the apicoplast fatty acid (FAS II) biosynthesis pathway. When this gene was deleted, late-stage liver schizonts were arrested in development, and protection against challenge was elicited when sporozoites were administered sub-cutaneously or intradermally. Another advantage of this method is that vaccination with the FabB/f GAP engenders both a CD8 T cell response as well as protection against blood-stage infections [100].

GAPs have also been used with *P. falciparum*. When P52/P36 were deleted there was complete arrest of the liver stages *in vitro* and this line has been selected for a phase I/IIa clinical trial using human volunteers bitten by infected mosquitoes [701].

In the future, the most formidable obstacles for the live attenuated vaccine approach, whether it be by irradiation or GAPs, lies in the current need for production of sufficient numbers of sporozoites as well as for cryopreservation in liquid nitrogen.

■ Chapter 15 ■

New Medicines, Old Problems

Because there is no licensed malaria vaccine, today's most desperate need is to find new medicines for the control of malaria. Two advances linked to the Malaria Genome Project have enhanced drug discovery for malaria. "First, the sequencing of the whole genomes of *P. falciparum* and *P. vivax* has been important in identifying the full range of potential drug targets [and] provided the basis for comparisons of gene expression patterns at different life cycle stages and between different species. This data set provides the ability to search for target classes that have not been pursued before in drug discovery" [727]. "The second key development is high content screening: the ability to study the viability of the parasite in 384 and 1536-well formats. This means that large compound collections can be rapidly screened, with throughputs of 100,000 per month possible even in academic centers" [727].

Guiguemde *et al.* [277] used a chemically diverse library of 300,000 chemical compounds and found that more than 1,000 inhibited the *in vitro* growth of *P. falciparum* by 80% at a cut-off concentration of 7 μM. Of the 172 compounds that were scrutinized more closely, and with novel chemical structures, "at least 80% seemed to act on parasite targets distinct from those affected by the current drugs" [219]. Although "the genome sequence of *P. falciparum* offers a multitude of potential drug targets" [220] only a few drugs have actually been realized, and "target-based lead discovery has produced disappointing results, generally for lack of whole-cell activity" [247]. For example, in a test of the library of two million compounds of GSK, >80% inhibition of *in vitro* growth of *P. falciparum* was found in 135,333 compounds (at a cut-off concentration of 2 μM). The overall activity of 70% of the 413 refined targets was on G-protein-coupled receptors or protein kinases; however, none of these targets was used for further development, despite the fact that the *P. falciparum* genome encodes 85–99 kinases.

In the belief that the *P. falciparum* genome sequence might have the potential to identify new drug targets absent in humans, the Novartis Institute for Tropical Diseases began (in December 2007) an intensive study to improve the antimalarial activity of spiroindolones that had been identified earlier from a library of 12,000 natural compounds for their ability to kill the blood stages of *P. falciparum* and *P. vivax*. The primary screen identified 275 primary hits and those with toxicity against mammalian cells were discarded. Of the remainder (17 compounds), a synthetic compound related to the spiroazepineindole class with favorable pharmacologic properties was used for lead optimization and derivitization. Of the 200 derivatives, the best was NITP 609 (spirotetrahydro-β-carboline or spiroindolone). Spiroindolone killed blood parasites *in vitro*, was active against drug-resistant parasites, was not cyto-, cardio- or genotoxic and had properties that would allow it to be taken once daily as an oral drug. Using the genome sequence of *P. falciparum*, the target of spiroindolone was identified as the P-type cation transporter ATPase 4. The PfATPase4 was localized to the plasma membrane of the parasite [562]. To determine the basis of parasite resistance to the drug, genomic DNA was prepared and screened by microarray. There were 276 detectable changes among six mutants and seven mapped to the single gene encoding ATPase 4 with 11 being non-synonymous mutations. The molecular characteristics of this ATPase suggest that inhibitors could include compounds related to thapsigargin, cyclopiazonic acid, and lansoprazole, which are known inhibitors of ATPase transporters.

Malaria parasites possess an unusual vestigial remnant of an algal chloroplast, the apicoplast. The apicoplast has a 35 kb genome that has been sequenced. The apicoplast is essential for the survival of the blood stages and, as such, has been considered to be an Achilles' heel suitable for the development of novel antimalarials. A number of antibiotics, including azithromycin, doxycycline and clindamycin, kill the parasite by interfering with the products of the apicoplast genome; however, these antibiotics are slow acting and result in "delayed death" wherein the progeny rather than the treated parents die. Mupirocin treatment of blood-stage parasites results in delayed death by inhibiting an isoleucyl tRNA synthetase required in the apicoplast [343]. Using a high-throughput *in vitro* screen targeting the apicoplast, it was possible to identify known and novel antimalarials including kitasamycin, a macrolide, already in use for agricultural application; its IC_{50} was 50 nM and comparable to the drug azithromycin [203].

Another contribution of the Malaria Genome Project has been the identification of the mechanism of action of the bisthiazolium compound T4. When tested against *in vitro P. falciparum* there was no immediate detectable change in transcription; however, with increased incubation time there was cell cycle arrest and significant decreases in the level of choline ethanolamine phosphatidyl transferase (CEPT), a key enzyme involved in the final step of the synthesis of phosphatidylcholine, an essential phospholipid synthesized from choline and ethanolamine [400].

Based on studies with bacteria and yeast, it was anticipated that genome-wide expression microarrays of *P. falciparum* would reveal genes that could be the target of uncharacterized substances. It was assumed that, as with *E. coli* and *Saccharomyces* when subjected to a putative inhibitor, the parasites would quickly upregulate the transcription of genes encoding enzymes of the appropriate biosynthetic pathway and, in this way, avoid the lethal consequences of the inhibitor. However, a study of expression of the anti-folate compound WR99210 failed to show upregulation of transcription of the *Plasmodium* DHFR, the drug target, as well as other malaria parasite genes involved in the biosynthesis of pyrimidines [248]. It has been contended from these results that transcription by *P. falciparum* is hard-wired "possibly as a result of the evolution of the parasite in a stable environment" [177]. Indeed, unlike yeast where transcriptional changes occur within 15 minutes after the addition of a putative inhibitor, with malaria parasites exposure for several hours is needed before the changes become apparent. In a microarray investigation involving 20 chemical perturbants, extensive changes were observed in 59% of parasite genes; however, it was not possible to determine the mechanism of action from the data. Thus, the use of gene-expression data to determine the mechanism of action of a particular drug is not as straightforward as had been anticipated. Nevertheless, gene-expression signatures may provide for the capacity to sort compounds into different classes, such as those affecting a particular organelle or enzyme, and comparing this known signature with an unknown could allow for target prediction. Indeed, an examination of the *P. falciparum* sequence revealed a total of 131 putative ribosomal proteins of which 29 were probably targeted to the apicoplast, whereas 20 were putative mitochondrial ribosomal proteins. Whole genome expression during the development of the parasite within the red blood cell revealed that apicoplast-mitochondrial synthesis of ribosomal proteins occurred at a different time period from the cytoplasmic ribosomes [434]. Although a microarray study was unable to demonstrate upregulation of the target (a kinase), it was able to predict the time of activity of this enzyme inhibitor.

Microarray approaches are able to rapidly identify variable genomic regions and assess copy numbers at relatively low cost, thus obviating the need for conventional sequencing. Genomic microarrays have been used to discover "novel amplification events surrounding GTP cyclohydrolase I, the first enzyme in the folate synthesis pathway, and may be important for anti-folate resistance" [177]. (This amplification was missed in previous sequencing efforts.) Using high-density microarrays, Dharia *et al.* [177] showed that malaria parasites become resistant to fosmidomycin through the acquisition of extra copies of 1-deoxy-*D*-xylulose-5-phosphate reductoisomerase, which enables the parasite to overcome drug inhibition of the biosynthesis of isoprenoids in the apicoplast. The single large amplification event was mapped to a region on chromosome 14.

"Malaria drug discovery has indirectly benefited from genomics through the transfer of various data processing algorithms, from gene expression to chemical screening.... existing screening data may allow the mode of action of a compound to be predicted" [177]. For example, Plouffe *et al.* used a collection of 200 historical screens to demonstrate that compounds with similar targets tend to display correlated chemical screening patterns [521].

The old problem of resistance

Perhaps the earliest notion that tolerance to a drug could result from prolonged use was that of the Turkish King Mithiridates (119–63 BC), who "in order to protect himself from the effects of poison had the habit of taking such substances prophylactically with the result that, when ultimately defeated in battle by Pompey and in danger of capture by Rome, was unable to take his own life by poison and was obliged to perish by the sword" [63]. Although it was known since ancient times that narcotic drugs such as morphine gradually lost their effects by repeated usage, the recognition of the problem of drug resistance by microbes and insects first began with Paul Ehrlich's use of "magic bullets." Working with mice infected with trypanosomes (causing the cattle disease nagana), he gave the mice a red aniline dye (trypan red) that was curative, i.e. the parasites disappeared from the blood; however, shortly thereafter they reappeared. Further treatment of infected mice showed the dye to be ineffective, with mice dying rapidly. The dye-treated strain of trypanosomes, when inoculated into a batch of normal mice, produced infections that, even in the absence of the drug, were tolerant to the killing power of the dye. This suggested that a genetic change in the parasite resulted in their renewed strength to overcome the drug's lethal effects.

Today, we are acutely aware of the problems associated with bacterial strains that are no longer susceptible to sulfonamides, antibiotics such as penicillin and streptomycin, as well as the lack of effectiveness of the insect killing by DDT and dieldrin. This phenomenon, called resistance, is the failure of a drug (or an insecticide) to kill its intended target. Resistance is defined as the ability of a particular cell, such as a cancer cell or a virus, parasite or insect, to multiply and/or survive in the presence of concentrations of the drug (or insecticide) that normally either destroy or prevent the growth and multiplication of that cancer cell, parasite, virus or insect. Resistance to antimalarial drugs and insecticide pose the greatest challenge to reducing the mortality caused by *P. falciparum* and to effectively controling the disease.

What is the basis of resistance? Just as there is a non-genetic resilience in our species to adapt to new environmental challenges, in other species there is also a genetic resilience that enables them to survive the onslaught of a drug (or insecticide). Oftentimes, resilience lies in favorable mutations that permit an organism to survive an environmental threat. This capacity is then passed on to its offspring, i.e. survival of the fittest. The presence of the drug (or insecticide) acts as a selective agent — a sieve if you will — that culls those that are susceptible and allows only the resistant ones to pass through to the next generation. Drug resistance is the result of natural selection, i.e. those with a particular genetic make-up (genotype) that are the most able to survive and reproduce their kind pass on their genes to future generations and increase in frequency over time. Simply put, it might develop in this way: an average malaria patient may have a billion parasites in the blood and one in 10,000 of these carries a genetic change that allows that parasite to evade the lethal effects of a particular antimalarial drug. Once the patient is treated with that particular drug, only the mutant survives and these are the drug-resistant parasites. The result is that there are now ~10,000 parasites able to grow and reproduce in the presence of the drug, increase their numbers and, as a result, over time almost the entire population becomes resistant.

There are three basic mechanisms that allow malaria parasites to develop resistance: (1) they become impermeable to the drug or, if the drug does gain entry, they are able to pump out the drug from the cell so toxic levels are not reached within the cell; (2) they develop an altered enzyme, which has a lower affinity for the drug;

(3) they manufacture excessive amounts of enzyme, thus counteracting the drug, a phenomenon called gene amplification. In the case of insects, in addition to detoxification mechanisms, there may be behavioral resistance — the insect avoids coming into contact with the insecticide. It is important to note that drug resistance can develop without exposure to the drug, but once the drug is present then natural selection promotes the survival of the resistant individuals. Not only does drug resistance blunt our ability to eradicate a disease, it may also contribute to our inability to suppress virulence and transmission.

Chloroquine

Transporter genes have been attractive candidates for understanding the mechanisms of drug resistance. Although more than 100 genes encoding transporters were found in the *P. falciparum* genome, only two of the four known drug resistance genes in *P. falciparum* are well-characterized as transporters [432]. The best studied have been the genes involved in CQ resistance.

On 12 April 1946, an article in the *New York Times* declared, "Cure for malaria revealed after 4-year, $7,000,000 research. The curtain of secrecy behind which the multimillion dollar Government antimalaria program had been operating, in the most concentrated attack in history against this scourge, was completely lifted today for the first time, with the revelation of the most potent chemicals so far found." The next day, a *New York Times* editorial said: "When the scientific story of the war is written, we have here an epic that rivals that of the atomic bomb, the proximity fuse and radar." The drug receiving so much attention was CQ. In the US it was the result of the screening of some 13,000 compounds under the aegis of the National Research Council's Board for the Coordination of Malarial Studies during World War II.

The most important of these new compounds was SN-7618 (SN = Survey Number). SN-7618, renamed at a later date CQ, had been prepared by German chemists, discarded by the Germans, tested by the French, and rediscovered by the Americans [137, 659]. In

1934, chemists at Bayer's Elberfeld laboratories synthesized a color-bearing 4-aminoquinoline and made two salts, one of which, 2,4-dihydrobenzoic acid, was named Resochin (because by German terminology it was the RESOrcinate of a 4-aminoCHolin). When screened using a bird malaria, it was found to be as effective as the current therapeutic Atabrine. It was also effective against blood-induced *P. vivax* infections in paretics. However, based on tests in animals, it was considered too toxic for practical use in humans and so it was abandoned by Bayer. Undeterred, in 1936, Bayer chemists produced a methylated derivative of Resochin, called Sontochin. This was less toxic and as effective as Atabrine. Bayer patented both Resochin and Sontochin, and through the I. G. Farben cartel (of which Bayer was a member), the same patents were taken out in the US in 1941 by Winthrop Chemical Company (also a member of the cartel). To gain patent protection in France, where there was no protection for the production of medicines, Bayer sent samples of Sontochin to a French pharmaceutical company, which in turn sent it to Tunisia for testing in humans. In early 1942, Sontochin was found to be effective against *P. vivax* and without adverse reactions. In late 1942, when the Allies invaded North Africa, the French investigators offered the remaining supplies of Sontochin to the U.S. Army. This material "which had been captured from the enemy in North Africa" was then tested as SN-183 by the Division of Chemistry and Chemical Technology of the Board for the Coordination of Malaria Studies and found to be identical to the compound (Sontochin) that Winthrop had made under manufacturing directions received from I. G. Farben in 1939 [137]. Although the patent was not issued until 1941, it had already been tested and found to be effective in malaria-infected canaries by Maier and Coggeshall at the Rockefeller Foundation Laboratory in New York. Despite this, Winthrop, for some reason, put SN-183 on the shelf where it was forgotten. In 1943, when the data on SN-183 was reviewed by the Division of Chemistry and Chemical Technology, the Chairman (a pharmacologist) made the mistake of considering its structure to be that of an 8-aminoquinoline, and wrote: "There is no need to study additional 8-aminoquinolines." Thus, SN-183 was dropped from further testing. Later, when it was discovered that Winthrop's SN-183 was identical to Sontochin, SN-183 was dropped as a designation and a new SN number, SN-6911, was substituted. SN-6911 was found to be four to eight times as active as quinine in the bird malaria *P. lophurae*; however, in humans it was no more effective than Atabrine. But, by this time, American chemists had produced a more effective 4-aminoquinoline, a 7-chloro derivative (= SN-7618) [629].

In dogs, SN-7618 was nine times as toxic as SN-6911 (Sontochin); however, in monkeys it was one-fourth as toxic. When in 1944 it was tested for toxicity on conscientious objectors at the Massachusetts General Hospital in Boston, no adverse toxic symptoms were observed. Further trials in malaria therapy were conducted at the Boston Psychiatric Hospital. SN-7618 persisted longer in the blood so that it could be used prophylactically, and a single weekly dose of 0.3 g was able to maintain a blood plasma level for complete suppression of malaria. Harry Most at the NYU Medical Center showed that 1.5 g given for three days was effective against acute attacks of *P. vivax* as well as *P. falciparum*. Clearly, SN-7618 was the drug of choice for the management of malaria and it was given the name CQ. Late in 1944, the Board was informed that SN-7618 was not patentable as a new antimalarial as its structure was identical to Resochin, and already covered by two patents owned by Winthrop Chemical Company.

The U.S. Army expressed interest in moving ahead with a large-scale field study of the suppressive capability of CQ as soon as positive results in prisoner volunteers became available. In Australia, CQ tests were conducted in soldier volunteers [215, 659]. The Australian Army received 2,500 CQ tablets to begin its trial and, over time, the program grew larger so that eventually 500 pounds were delivered. The Australian clinical work was definitive for CQ treatment and prophylaxis. With a daily dose of 100 mg (after a build-up of 200 mg twice per day for four days) neither the *P. falciparum* nor the *P. vivax* group showed any evidence of malaria. In recipients receiving 200 ml of blood from donors who had stopped taking the drug three weeks previously, none developed malaria except in the case of vivax where there were breakthroughs between nine and 33 days after ceasing to take the drug. "The findings suggest that with malignant tertian malaria (*P. falciparum*) the drug does not act as a causal prophylactic but destroys the young asexual parasites... as they emerge... into the circulating blood from the seventh day onward. Further, neither did the drug destroy the pre-blood tissue stages in *P. vivax* infections" [215].

The story of the discovery of CQ makes clear that the convenience and ease of using bird malarias can be misleading unless the proper host–parasite combination is used. For example, the course of infection by *P. relictum* in canaries was not significantly altered by sulfonamides, whereas another bird malaria, *P. lophurae* in ducklings, was particularly useful to workers at the Squibb Institute for Medical Research. The Johns Hopkins biochemist William Mansfield Clark (Chairman of the Committee on Chemotherapy = Division of Chemistry and Chemical Technology of the Board for Coordination of Malaria Studies) observed that it would seem wise to pay less attention to the toxicity of a drug in the bird and more attention to the toxicity of the drug in mammals. Indeed, although bird malarias were essential to the wartime effort aimed at drug development for use in humans, it became apparent that there were limitations to their utility and there was a need for more suitable models [441, 629].

Respected malaria researcher Robert Desowitz, in his book *The Malaria Capers*, wrote: "Chloroquine spread like a therapeutic ripple throughout the tropical world. In stately homes, the bottle of chloroquine would be a fixture, along with the condiments on the family table. In military cantonments the troops would assemble for a 'chloroquine parade'. In hospitals and rural health centers the malarious of all skin colors were treated with chloroquine..." [176]. In 1955, the WHO, confident in the power of CQ to kill malaria parasites and DDT to kill the malaria-carrying *Anopheles* mosquitoes, launched the Global Malaria Eradication Campaign. As a result, malaria was eliminated from Europe, Australia, and other developed countries, and large areas of tropical Asia and Latin America were freed from the risk of infection. However, the campaign had excluded sub-Saharan Africa. The WHO estimated that the campaign saved 500 million human lives that would otherwise have been lost due to malaria, but the vast majority of the 500 million lives saved were not Africans. In 1972, however, when it became clear that resistance to CQ was widespread in malaria-endemic areas so that the drug was rendered ineffective in control, the Global Eradication Program was formally declared dead.

In its time, the 4-aminoquinoline, CQ, was efficacious, of low toxicity, and affordable (less than $0.02 for a three-day adult course of treatment). In the early 1960s, the mode of action of CQ was generally believed to be through binding to DNA, and the evidence, particularly from Fred Hahn's group at WRAIR, was thought to be conclusive. However, subsequent studies by David Warhurst and colleagues demonstrated that this mechanism was incorrect. In 1980, Coy Fitch's group discovered that ferriprotoporphyrin IX (FP), a product of hemoglobin breakdown, had the specificity and affinity characteristics of the CQ receptor, as had been suspected earlier. His group also found that FP and its complex with CQ are toxic for erythrocytes and malaria parasites [45, 223]. These observations led Fitch to propose that FP mediates the chemotherapeutic action of CQ and to ask how malaria parasites escape FP toxicity while ingesting and metabolizing hemoglobin.

Believing the lack of toxicity to result from FP sequestration in hemozoin, Fitch *et al.* decided to better characterize the pigment of hemozoin. This pigment appeared to be an aggregate of FP that was similar, perhaps identical, to β-hematin [224]. Several years later, β-hematin was found to be aggregated dimers of FP. As FP is rendered non-toxic by converting it to β-hematin, it was now logical to study this process. In 1992, Chou and Fitch [134] discovered that FP dimerization could be measured easily and that CQ treatment reduces FP dimerization *in vivo*. At first, Chou and Fitch and others assumed that an enzyme catalyzed the reaction; however, that assumption was wrong. Instead, it was found that the reaction is catalyzed in malaria parasites by unsaturated fatty acids, probably predominantly linoleic acid. Furthermore, Fitch and co-workers obtained evidence that CQ treatment causes this catalyst to be masked (i.e. unavailable to catalyze FP dimerization) in CQ-susceptible malaria parasites, thus explaining the accumulation of toxic, undimerized FP in response to CQ treatment.

The worldwide spread of CQ resistance has led to a significant resurgence of malarial morbidity and mortality concomitant with interest in the manner by which resistance develops. Investigations into the mechanisms of CQ resistance have generated several different models, including reduced influx of CQ, increased efflux of CQ, pH effects on drug accumulation and/or receptor availability, glutathione degradation of FP or formation of CQ–FP complexes. Today, it is generally accepted that CQ enters the acidic food vacuole by passive diffusion as an uncharged molecule and becomes trapped in the vacuole in its positively charged, membrane-impermeable form; the charged CQ is retained in the food vacuole as an FP–CQ complex. Resistance involves restricted access of CQ to FP such that there are reduced drug levels in the digestive vacuole. This might be achieved by: (1) efflux of CQ from the digestive vacuole via an energy-coupled mechanism; (2) leakage of CQ out of the digestive vacuole down its concentration gradient, with energy driving the vacuole proton pump such that a concentration gradient of charged CQ is maintained (rather than energy being coupled to drug movement *per se*); (3) a pH-dependent reduction in CQ accumulation; (4) passive outward movement of positively charged CQ [484]. It is generally thought that differences in the digestive vacuole pH are not primarily responsible for CQ resistance; therefore, other models have received much more attention and all probably involve the *P. falciparum* CQ resistance transporter, named PfCRT [698].

The PfCRT gene was identified through the analysis of a genetic cross between a CQ-resistant and a CQ-sensitive clone by David Fidock and Thomas Wellems at the NIH. The transporter protein, made up of 424 amino acids, has been localized to the parasite's food vacuole membrane and it shows extraordinary amino acid sequence diversity among geographic isolates, involving as many as 15 amino acids and in a single resistant line there can be four to eight individual mutations. Indeed, when the K76T (lysine to threonine) mutation is removed, resistant parasites become sensitive to CQ. A recent study [128] using experimental data combined with genomic sequence and structure analysis strongly suggests that PfCRT is a carrier belonging to the drug/metabolite transporter superfamily and the transported species is the protonated form of CQ. The outward movement of CQ from the food vacuole reduces the binding of CQ to FP, decreasing the CQ concentration within the food vacuole in resistant parasites [658].

Studies of the distinct sets of genetic polymorphisms in PfCRT show that CQ-resistant strains originated from at least six geographic areas and spread across Southeast Asia, Latin America and the Pacific. Almost all resistant strains in Africa have their origin in a single foundation event, a strain apparently brought over from Southeast Asia [575].

Resistance to anti-folate drugs

All of the genes for the folate pathway enzymes in *Plasmodium* have been cloned and characterized [598]. Elucidation of their gene structure has provided the impetus for developing new therapeutics as well as a better understanding of the genetic basis for anti-folate drug resistance. The potency of different DHFR inhibitors against *P. falciparum* parasites varies widely. WR99210 (4,6-diamino-1,2-dihydro-2,2-dimethyl-1-(2,4,5-trichlorophenoxy)propyloxy)-1,3,5 triazine hydrobromide) is the most potent, whereas cycloguanil (a metabolite of proguanil) and chlorcycloguanil (the active metabolite of chlorproguanil) are more potent than pyrimethamine [272]. Because parasite resistance to anti-folates was found shortly after their introduction, attempts were made to increase their parasite-killing activity and to reduce the emergence of resistance. One of the first combinations was pyrimethamine plus sulfadoxine (Fansidar, Roche). Presently, other attempts to delay resistance involved the use of synergistic combinations such as chlorproguanil–dapsone (Lap-Dap) and proguanil–atovaquone

(Malarone). It is most likely that this synergism is due to the fact that these drugs act on different enzymes in a common biosynthetic pathway [154]. To further counteract the development of resistance, a triple combination (chlorproguanil–dapsone–artesunate) is being developed. However, a major disadvantage of these combinations is their high cost.

Sulfadoxine, the most commonly used sulfa drug, is a structural analog of para-aminobenzoic acid (pABA) and acts as a competitive inhibitor of malarial parasite dihydropteroate synthase (DHPS). Resistance to sulfadoxine in *P. falciparum* involves amino acid substitutions in DHPS that alter the enzyme's function. Resistance to sulfamethaxozole and sulfathiazole, as well as sulphones, also involves amino acid substitutions in DHPS — especially in residues Serine→Alanine 436, Alanine→Glycine 437, Lysine→Glutamic Acid 540, Alanine→Glycine 581 and Alanine→Serine 613—and, as such, there is cross-resistance to these drugs. The Alanine→Glycine 437 and Lysine→Glutamic Acid 540 mutations appear to be the initial and most important for resistance and higher levels require multiple mutations in addition to these [689, 721].

In *P. falciparum*, pyrimethamine resistance is due to decreased affinity in the binding of the drug to DHFR, due to mutations in the gene sequence specifying amino acid residues: Serine→Asparagine 108, Asparagine→Isoleucine 51, Cysteine→Arginine 59 and Isoleucine→Leucine 164; and resistance to cycloguanil was found to be linked to a pair of mutations in DHFR: Alanine→Valine 16 and Serine→Threonine 108. Evidence that DHFR mutations gave rise to resistance was strengthened by the fact that a recombinant mutant enzyme showed decreased drug binding, and transfection of wildtype parasites with DNA constructs bearing mutant forms gave rise to resistant phenotypes. Interestingly, the mutations for resistance appear to have arisen in Asia and then spread to Africa [559]. Although the mechanism for folate resistance usually involves mutations that alter drug binding in some strains, increased expression of DHFR as well as the ability to salvage folates bypassing the *de novo* pathway can also contribute to the level of drug resistance. When a gene knockout was carried out in the laboratory, it was found that complete disablement of the DHPS gene product could not be tolerated and the parasites died. Thus, it appears that the biosynthetic capability of folate is essential in spite of the parasite's ability to salvage folate [720].

Genetic polymorphisms include microsatellites (consisting of repeats of a short nucleotide sequence), single nucleotide polymorphisms (SNPs) or small insertions or deletions in the DNA. An examination of the number of SNPs in the genome of *P. falciparum* identified one SNP every 400–800 base pairs. The data revealed a region on chromosome 7 that encompassed PfCRT, the gene responsible for CQ resistance. The number of polymorphisms varies for different regions of the chromosome and for different gene classes [202].

Analysis of microsatellites surrounding the DHFR gene showed that the most common form found in Africa was characterized by three point mutations and was associated with related types that originated in Southeast Asia. The potential for rapidly identifying resistance genes from sampling clinical isolates stands in marked contrast to the classical genetic approach of crossing a resistant and a sensitive clone that culminated in the identification of the first gene for resistance — PfCRT [202].

It has been possible to crystallize DHFR and DHPS of *P. falciparum* and this has allowed a better appreciation of the action of folate inhibitors. With DHFR, most inhibitors fit between amino acid residues 108 and 54 within the active site [755]. WR99210 most closely resembles the flexibility seen in the natural substrate, dihydrofolate (DHF), and perhaps this is the reason for its greater potency and reduced susceptibility to point mutations. Modeling of the falciparum DHPS shows that the amino acid residues 436, 437, and 540 all line the channel of the active site where both substrate and inhibitor binding occurs; 581 and 613 are only one to three positions away from the channel, suggesting that these mutations may be compensatory in function, and explaining why the two residues are not seen in isolation but only in association with mutations at 436, 437, or 540 [272].

Differences in cross-reaction patterns and point mutations have raised the hope that new anti-folates can be developed against resistant strains, providing that they are able to retain binding affinity to the mutant forms [753, 754]. Also, there may be another possibility for retarding the emergence of resistance. It has been found that point mutations in the dhfr gene also result in resistance to pyrimethamine in *P. vivax*; however, the mutations that confer resistance to pyrimethamine also render the DHFR of *P. vivax* exquisitely sensitive to WR99210. Hastings and Sibley [296] suggest that pyrimethamine and WR99210 could "exert opposing selective forces on the *P. vivax* population, and if used in combination... could force the selection by some mechanism other than simple point mutations and greatly slow the selection of parasites resistant to both drugs." Indeed, it is possible that by combining them with other drugs having an entirely different mechanism of action, the selection of resistance could be greatly retarded [296, 298].

Mefloquine and multidrug resistance

In the 1960s, it was clear that the honeymoon for CQ was over. Although the Vietnam War began in 1959, and the first American combat troops did not actually arrive in South Vietnam until early 1965, by 1962 it was already apparent to the U.S. military that the malaria situation in the country was serious. Unlike malaria during the Korean War (1950–1953), the prevalent parasite in Vietnam was not *P. vivax* but the deadly *P. falciparum*, and worse still it was drug-resistant [73]. By 1963, the incidence of malaria in U.S. troops in Vietnam had risen to 3.74 cases per 1,000 compared with a worldwide figure for military personnel of 0.13 per 1,000 [104]. It was estimated that 1% of American soldiers were acquiring malaria for every day that they spent in combat in South Vietnam despite receiving CQ [517]. During 1964–1966, the treatment failure rate with CQ in troops from Vietnam with *P. falciparum* was 80–89%. Quinine, used as a single drug treatment, was only a little more effective, with a failure rate of 50–74%. In desperation, triple combinations such as quinine, pyrimethamine, and dapsone or CQ, pyrimethamine, and sulfadiazine were used and, although more effective as a cure (~98%), the relapse rates were high. Further, there were multiple side-effects from these combinations [105]. By 1969, there were 12,000 cases of malaria in the troops in Vietnam with a loss of 250,000 man days and direct medical costs of $11 million.

In 1963, Colonel William Tigertt, the Director of WRAIR, set into motion the machinery for the U.S. Army Research Program on Malaria in what was to prove to be the largest screening program ever undertaken, with a mission to identify and develop antimalarial agents effective against the emerging drug-resistant strains of *P. falciparum* [516]. Most of the Program was devoted to a search for new "magic bullets." Under the U.S. Army Medical Research and Development Command, the WRAIR conducted a program of drug development similar to that during World War II [743] by enlisting the collaboration of academic institutions, pharmaceutical companies and individuals in the US and elsewhere to provide compounds for screening and evaluation. Between 1963 and 1974, ~ 400 4-quinoline methanol derivatives, i.e. drugs with a quinoline nucleus and an amino alcohol side chain, were submitted to WRAIR. These putative drugs were tested on the asexual stages of malaria at the Rane Laboratory in Florida. In the Rane test, mice were infected with a million *P. berghei*-infected red cells — a dose designed to kill the mice within six days. On the first day, the test drug sample was given sub-cutaneously at a dose range of 600, 300, 150 and 75 mg/kg and survival times were compared with controls. An extension of survival to 12 or more days was considered significant [537].

In malaria-infected mice, seven of the new quinoline methanols were found to be effective against parasites resistant to CQ; at an effective dose of 10 mg/kg, some were eight times better than CQ and 64 times better than quinine [563]. At 40 mg/kg, one amino alcohol in particular, WR142,490, was 100% curative. From 1970, hundreds of quinoline-methanols were examined for antimalarial activity in *Aotus* monkeys infected with *P. falciparum*. Twelve specially selected 4-quinoline-methanols were tested with CQ-resistant and CQ-sensitive *P. falciparum* and, from this screening, five derivatives were as active or more active than CQ against strains susceptible to this drug, and equally effective against strains that were resistant to CQ, quinine, and pyrimethamine. WR142,490, examined in the *P. falciparum* owl monkey model, was the most active and it emerged as the flagship response to CQ-resistant *P. falciparum* [585, 660].

Administration of a single dose was as effective as the same amount of drug divided into three or seven doses over as many days. Intravenous administration of the drug in monkeys was also feasible. In human volunteers, when taken orally, the drug was safe and effective at 1,500 mg for one day or at 500 mg weekly for 52 weeks. The peak concentration of the drug in the blood was at 12–36 hours and the mean half-life was 14 days. Non-immune volunteers developed no *P. falciparum* infection after a single dose and, in those infected when treated, there was rapid clearance of fever and blood parasites in the CQ-sensitive and CQ-resistant strains. WR142,490 produced 100% cure against CQ-resistant *P. falciparum* compared with 89% with Fansidar, the best commercially available drug against CQ-resistant *P. falciparum* [687].

Under a contract with the U.S. Army, high antimalarial activity was found in animal models [2, 585] for SN-10,275 (α-(2-piperidyl)-6,8-dichloro-2-phenyl-quinoline methanol) and when tested in human prisoner volunteers infected with the Chesson strain of *P. vivax* [533]. Although toxic effects on vision prevented the general use of this compound, it was found to be effective against asexual blood stages and had a remarkably long duration of activity. An analog of SN-10,275 with the number WR142,490 (α-(2-piperidyl)-2,8-bis-(trifluoromethyl)-4-quinolinemethanol) was prepared and, in clinical studies, it had the longer duration of SN-10,275, showed no evidence of vision toxicity, and was more effective than WR30,090. This 4-quinoline methanol, named mefloquine, was given as a single oral dose to 47 volunteers infected with *P. falciparum* and resulted in rapid clearance of fever and parasites in the blood and there was no relapse in CQ-sensitive infections. More significantly, at 1 g it cured ten of 12 patients, and cured all patients who received 1.5 g with CQ-resistant strains.

Mefloquine (marketed as Lariam) is structurally related to quinine, is selective in killing asexual-stage parasites, and does not have activity against gametocytes or liver stages [501, 660]. Mefloquine and quinine are both lipophilic ("lipid loving") and bind tightly to serum and high-density lipoproteins and this may facilitate entry into the parasite. Mefloquine also binds to the membranes of the red blood cell and accumulates in the parasite food vacuoles, as does CQ. However, it is not clear that both drugs share the same mechanisms of action.

Multidrug resistance (MDR) in tumor cells (i.e. reduced drug accumulation and reversal by verapamil) has some features in common with drug resistance by malarial parasites. A parasite homolog of the mammalian MDR gene, pfmdr, was cloned, sequenced and characterized [227]. The gene is located on chromosome 5 and its product, a P-glycoprotein, PfPgh-1, with an Mr of 162 kDa, is a member of the ATP-binding cassette transporters, is located on the membrane of the food vacuole, and is expressed during blood-stage development [558]. There is a strong association between pfmdr amplification and mefloquine treatment failure and *in vitro* resistance, as well with resistance *in vitro* to halofantrine and quinine. When mefloquine resistance was induced *in vitro*, the amplification of gene copy number for pfmdr was one to two copies every 10^8 parasites and amplification of two to three copies occurred once every 10^3 parasites [528]. This work suggests that amplification of the pfmdr gene is a frequent event and is directly associated with mefloquine insensitivity (at least *in vitro*). However, in the absence of mefloquine, the presence of multiple copies of the pfmdr gene is disadvantageous. Decreasing the copy number of pfmdr heightens susceptibility to lumefantrine, halofantrine, quinine, artemisinin, and mefloquine [618].

The physiological function of PfMDR1 is unknown and one can only extrapolate findings from homologs in plants and mammals in which it transports ions, lipids, and glucosides. Whatever its function, PfMDR1 is essential for the survival of *Plasmodium* inasmuch as the gene cannot be knocked out [575].

The fall of mefloquine

In an intensive and fruitful collaboration, the WHO sponsored more than 12 clinical trials of mefloquine in Latin America, Zambia, and Thailand and, in 1979, Hoffmann-La Roche launched the drug by itself (as a monotherapy) under the trade name Lariam (or Mephaquine); it was licensed in the US in 1988. Lariam became the drug of choice for travelers and visitors to areas where CQ-resistant malaria was present.

In the early 1980s, the first reports of resistance to mefloquine appeared [78] and this prompted the WHO to issue a publication in 1984 expressing reservations concerning the widespread use of the drug and suggesting that it be used in combination with another antimalarial; this, the WHO said, might preserve the effectiveness of mefloquine by delaying the emergence of resistance as well as potentiating its activity. Hoffmann-La Roche had produced such a formulation by combining mefloquine with Fansidar (marketed as Fansimef) [218]. Clinical trials with Fansimef (mefloquine combined with sulfadoxine/pyrimethamine [250/500/25 mg, respectively]) carried out between 1982 and 1986 in Africa showed similar effectiveness against falciparum malaria as mefloquine on its own. However, in Southeast Asia, particularly on the Cambodian border with Thailand and Vietnam, some clinicians questioned the wisdom of the use of mefloquine in fixed combination with Fansidar in countries where there was already resistance to sulfadoxine/pyrimethamine.

There were further complications with the prophylactic use of mefloquine: neuropsychiatric episodes, including insomnia, strange or vivid dreams, paranoia, dizziness or vertigo, depression and attempts at suicide. The overall risk varies with ethnicity (highest in Caucasians and Africans rather than Asians) as well as with differences in health, cultural, and geographical background so it is not certain what the actual reasons are for the differential adverse reactions. Due to concerns over the safety of mefloquine prophylaxis in Western countries, the packet insert has been revised (July 2002) stating: "Use of Lariam (mefloquine Roche) is contraindicated in patients with known hypersensitivity to mefloquine or related compounds such as quinine and quinidine. Lariam should not be prescribed for prophylaxis in patients with active depression, generalized anxiety disorder, psychosis or other major psychiatric disorders, or with a history of convulsions" (http://www.lariam.com). These reasons, as well as the development of mefloquine resistance in endemic areas in parts of Southeast Asia and the loss of efficacy in such areas, exemplify the danger of introducing a long-acting antimalarial where there is high transmission and where other control strategies are blocked [592]. Therefore, "it is unlikely that mefloquine alone will ever regain its clinical efficacy in such areas" [744].

Atovaquone

When Winston Gutteridge, Dilip Dave and W. H. G. (Harry) Richards, working on a collaborative project between the University of Kent at Canterbury, UK, and the Wellcome Research Laboratories, discovered that the conversion of DHO to orotate was mediated by DHODH, a key reaction in the *de novo* synthesis of pyrimidines, they postulated that DHODH was intimately connected to the electron transport chain of *Plasmodium*. Although the enzyme was similar to the mammalian enzyme, it was predicted to be a useful drug target [598].

In many-celled animals, such as ourselves, the electron transport chain consists of four integral membrane complexes localized to the inner mitochondrial membrane: complex I (NADH–ubiquinoine oxidoreductase), complex II (succinate–ubiquinone oxidoreductase), complex III (ubiquinol–cytochrome c oxidoreductase), and complex IV (cytochrome c oxidase), plus coenzyme Q (ubiquinone) and cytochrome c. In malaria parasites, however, it is different. Fry and Beesley [238], working in the Wellcome Research Laboratories, were able to isolate functional mitochondria from *P. yoelii* and

P. falciparum. They found a reduction of cytochrome c by NADH, α-glycerophosphate and succinate, as well as low rates of oxidation of proline, DHO, and glutamate. Although the oxidation of either α-glycerophosphate or succinate was inhibited by standard mitochondrial electron transport inhibitors, including a number of complex III inhibitors, the mitochondria were unable to oxidize NAD-linked substrates and there was insensitivity of NADH oxidation by rotenone in the presence of fumarate with or without cytochrome c, suggesting the absence of complex I of the respiratory chain. Thus, Fry and Beesley showed the plasmodial electron transport chain to differ from other animals in lacking complex I; however, a single sub-unit NADH dehydrogenase was present and homologous to that found in plants, bacteria, and yeast but not in animals [384, 693, 694, 700].

Electron flow in the mitochondrion of the malaria parasite is as follows: the substrates NADH, succinate, malate, DHO, and glycerol-3-phosphate located within the intermembrane space or the matrix are acted upon by the NADH dehydrogenase (complex I), succinate dehydrogenase (complex II), malate–quinone oxidoreductase, DHODH, and glycerol-3-phosphate dehydrogenase present within the inner membrane. The electron acceptor of all of these dehydrogenases is ubiquinone, oxidized by complex III (cytochrome bc_1 complex), the complex that transfers electrons to cytochrome c. The reduced cytochrome c is oxidized by complex IV (cytochrome c oxidase) that transfers electrons to oxygen. It is the oxidation of cytochrome c by molecular oxygen that explains why malaria parasites require only small amounts of oxygen, and why the parasites growing in Trager–Jensen culture system require only 5% oxygen.

But, you might ask, what does all this discussion of the malaria parasite's mitochondrion have to do with the development of antimalarial drugs? It turns out quite a bit. Shortages of quinine during World War II provoked an interest in the US in developing new antimalarials. Under the aegis of the Board for the Coordination of Malaria Studies, several hundred hydroxynaphthoquinones were synthesized and some, when administered orally in ducklings infected with *P. lophurae*, had greater activity than quinine. During this same period, Wendel [728], after testing 76 naphthoquinones, found that 69 had anti-respiratory activity in *P. lophurae* that roughly corresponded to their antimalarial activity. A year later when Ball *et al.* [34] found that 2-hydroxy-3-alkyl-naphthoquinones inhibited the respiration of beef heart succinoxidase, they concluded that the inhibition was due to "some step in the chain of electron transport below cytochrome c and above cytochrome b."

Sponsored by the Board's Malaria Research Committee, Fieser *et al.* [221] synthesized an array of naphthoquinones and found that menoctone (3-(8-cyclohexyl)-octyl-2-hydroxy-1,4-naphthoquinone) was a potent inhibitor of NADH- and succinate-cytochrome c reductases of yeast and mitochondrial particles [222]. However, little further work was done with naphthoquinones in humans because, when administered by mouth, there was poor absorption and rapid degradation, hence there was little antimalarial activity. "Heroic attempts to solve these problems both then and again in the 1960s...an orally effective quinone proved elusive until one compound, lapinone, when given intravenously did have clinical activity towards *P. vivax*." Because poor absorption after oral administration required high doses and intravenous administration, lapinone was not developed any further [29, 217].

In the late 1960s, the ubiquinones of malaria parasites were found to differ from those of the host: *P. lophurae*, *P. cynomolgi*, and *P. berghei* synthesized (from p-hydroxybenzoic acid) ubiquinones 8 and 9, whereas the mitochondria of mammals and birds contained ubiquinone 10. Based on these differences, it was suggested that structural analogs of ubiquinone could act as antimalarials by serving as inhibitors of the electron transfer mechanisms in malaria parasites.

In the 1970s, a project at Wellcome Research Laboratories in Beckenham, UK, was to study hydroxynaphthoquinones and electron transport [598]. The use of naphthoquinones as chemotherapeutics had languished for decades until it was discovered that menoctone was effective against the protozoan causing East Coast fever (*Theileria parva*) [443]. Structural studies of analogs of menoctone using *in vitro* cultures of *T. parva* resulted in the identification of a more suitable and effective agent, 2-cyclohexyl-3-hydroxy-1,4-naphthoquinone (parvaquone) [444]. At the Wellcome Research Laboratories, this finding, together with the discovery that menoctone blocked electron transport/pyrimidine biosynthesis in *Plasmodium* [278], led to renewed interest in the possibility of using naphthoquinones for other infections including malaria, and so several analogs of parvaquone were synthesized by Alan Hudson, the senior chemist for the parasitology area at Wellcome. One, 2-(4-t-butylcyclohexyl)-3-hydroxy-1,4-naphthoquinone, coded as BW58C, had exceptional activity against several mouse malarias [336]. However, BW58C was dropped from further consideration when it was tested in humans and found to be metabolized by the liver to an inactive red-colored compound [414]. The key to developing a successful drug from the naphthoquinone series was to understand the metabolism of the compound by the host liver (microsomes), and to prevent degradation to inactive metabolites. Analogs of the 2-cyclohexyl-3-hydroxy-1,4-naphthoquinone were synthesized and the metabolically labile 4' position in the cyclohexyl ring was replaced by a variety of substituents to prevent metabolism in an *in vitro* liver microsome preparation. (The metabolism of BW58C to produce the inactive red-colored metabolite was replicated *in vitro* using human liver microsomes.) A chlorophenyl substitution of the cyclohexyl ring, coded as BW566C, was metabolically stable. At the same time, with *P. falciparum* available both *in vitro* and *in vivo* (*Aotus* monkeys), it became possible to evaluate the structure and clinical activity of BW566C against human malaria. It was more active than the standard antimalarials using *in vitro* cultures of *P. falciparum*. It was curative in the mouse malarias *P. berghei* and *P. yoelii* and when administered orally in *Aotus* monkeys infected with *P. falciparum*. Compound BW566C was named atovaquone [335]. In human volunteers it was well tolerated and had a 70 hour plasma half-life of elimination. Sufficient antimalarial activity in plasma was achieved with a single oral dose of 225 mg.

Studies with BW58C (and BW566C) showed that the inhibition of respiration of isolated mitochondria from malaria parasites was

1,000-fold more sensitive than with mammal and bird mitochondria. Subsequently, it was shown that BW566C acts on the cytochrome bc_1 complex (in which electrons are transferred from ubiquinol to cytochrome c) to block either ubiquinol oxidation or ubiquinone reduction.

Atovaquone is also effective as a monotherapy against *Pneumocystis carinii* infections in patients with AIDS and was registered by Wellcome as Mepron for this indication; however, when it was used on its own for treatment of *P. falciparum* infections in humans there was a 30% failure rate, with atovaquone-resistant parasites emerging 28 days after treatment. To counter the problems associated with the treatment of *P. falciparum* by atovaquone alone, it has been combined with proguanil hydrochloride (marketed as Malarone) [414] and the combination has been found to be more effective than either component alone or mefloquine, CQ, or pyrimethamine/sulfadoxine in areas where parasites are resistant to these drugs. The high production costs of the drug precluded its use in developing countries, but its exquisite safety and potent activity means that it is the benchmark antimalarial drug (or so Wellcome stated) for the traveler market.

The most common mutation observed in atovaquone-resistant isolates of *P. falciparum* associated with Malarone failure in infected patients appeared to involve miss-sense mutations in a region of cytochrome b, especially near a highly conserved amino acid (proline, glutamic acid, tryptophan, tyrosine = PEWY) sequence, important for the way that ubiquinone (or an inhibitor) fits into the binding site of the malaria cytochrome bc_1 complex [48].

Artemisia **to artemisinins**

The story of the discovery of artemisinin had its basis in the political climate in China. During the Vietnam War (1959–1975), the Chinese government supported the Vietnamese Communists against the US, which was becoming increasingly involved in the fighting. Malaria was rampant in Vietnam, causing casualties on both sides, and Ho Chi Minh asked Mao Zedong for new antimalarial drugs to assist the Vietnamese troops in their fight against the imperialists [407]. In May 1969, the China Institute of Sciences set up Office 523 with the express purpose of drug discovery and it was there that hundreds of scientists were committed to the search [310, 374, 409, 688]. Zeuxing Wei, Professor of Traditional Medicine at the Research Institute of Shandon Province, claims to have rediscovered qinghaosu. Indeed, the drug first appears in a Chinese recipe book (*Recipes for 52 Kinds of Diseases*) in 168BC as a treatment for hemorrhoids, and there is a reference to it in the *Handbook of Prescriptions for Emergency Treatment* written by Ge Hang in 340AD. Ge Hang gives instructions to those suffering from intermittent fevers to soak a handful of the leaves of the sweet wormwood plant, *Artemisia annua*, in about a liter of water, squeeze out the juice and drink the remaining liquid. In 1596, Li Shizen, the great Chinese herbalist, wrote in his *Compendium of Materia Medica* that the chills and fever of malaria can be combated with preparations of qinghaosu (literally the active principle of qinghao or *A. annua*).

Professor Wei claimed that, after several failed attempts to isolate the fever-reducing principle in 1970, he succeeded in obtaining 30 mg of pure crystalline product that was not toxic (by testing it on himself!). In another account, it was Professor Tu Youyou and her research group, also working in Office 523 from July to December 1971, after screening more than 100 drugs and Chinese herbals, who isolated the active principle from dried *A. annua* using low heat and ethyl ether extraction after they discarded those parts of the herbal extract that made it sour. In May 1972, the first official report stated that, when the preparation was fed to mice (50 mg/kg daily for three days) infected with *P. berghei*, it killed asexual-stage parasites and was 95–100% effective as a cure. Monkeys infected with sporozoites of *P. cynomolgi*, after the appearance of parasites in the blood, were given 200 mg/kg of qinghaosu daily for three days via gastric tube and the blood was cleared of infected red blood cells in two to three days although there were relapses. Qinghaosu had no effect on liver-stage parasites in *P. cynomolgi* infection or in chickens infected with *P. gallinaceum*. Seven months later, there was another report of the successful treatment in Beijing of 21 patients with malaria and over 90% of them were said to have recovered from infections with either *P. falciparum* or *P. vivax*.

In 1973, when Tu Youyou and her colleagues synthesized the compound dihydroartemisinin (DHA) to prove that the active material had a ketone group, they were unaware that DHA would be more effective than the natural substance (qinghaosu). Only later would it be shown that DHA is the substance produced in the body after ingestion of artemisinin, acting to clear the malarial fever. In 1975, Professor X. T. Liang found that qinghaosu with the empirical formula $C_{15}H_{22}O_5$ contained a stable peroxide in a complicated compound, a sesquiterpene lactone bearing an endoperoxide located at one side of the molecule in almost a semi-spiral form.

At this time, immediately after the Vietnam War and with the Cultural Revolution still influencing attitudes in the People's Republic of China, the Chinese were skeptical about sharing information on the techniques for extraction and crystallization for fear this would be used by pharmaceutical companies in the West for monetary gain [407]. As a consequence, in 1982 Daniel Klayman (1929–1992), Chief of Medicinal Chemistry at WRAIR's Division of Experimental Therapeutics, decided to take a closer look at the Chinese drug [374, 375]. He recruited botanists in the Washington, DC, area to determine whether *A. annua* existed in North America. To his amazement, it was found growing on the banks of the Potomac River just down the road from WRAIR's headquarters. The plants were harvested with the help of a Boy Scout troop and then, through a difficult process of trial and error, in 1984 Klayman was finally able to duplicate the Chinese method of extraction using dried leaves or flowers. Artemisinin is a volatile, aromatic oil that is poorly soluble in water. It is, as Klayman and Liang found, an endoperoxide, consisting of two atoms of oxygen in a form never seen before. Klayman believed that the two oxygen atoms were like a bomb waiting to explode at the right time and all it took to set the "bomb" off was iron. He hypothesized that, when artemisinin encounters the iron-rich heme in malaria pigment (hemozoin), the molecule falls apart, triggering the rapid release of free radicals toxic to the malaria parasite [756].

Various derivatives of artemisinin [310, 407, 688, 756]—artemether (AM) and arteether (AE)—were prepared. Arteether was selected by WRAIR and WHO for further development as a sesame oil preparation to be used as an intramuscular injectable for the emergency treatment of severe malaria. However, toxicology studies with rats and dogs (but not humans) showed both AE and AM to result in neurological defects such as loss of posture, loss of pain response reflexes, and loss of brain stem reflexes leading to cardiorespiratory collapse and frequently death. As a result, by 1994 WRAIR had abandoned further work on AE and AM.

In China, oil suspensions suitable for injection or in tablet form or suppositories were produced and the Qinghaosu Antimalarial Coordinating Research group proceeded to treat and document the therapeutic response in 1,511 cases of *P. vivax* and *P. falciparum* in humans. Artemisinin was effective in CQ-resistant cases and produced a rapid recovery in 141 cases of cerebral malaria. There was no evidence of serious toxicity but there was a high relapse rate.

Artemether was studied at the Shanghai Institute of Materia Medica, and artemisinin and artesunate (AS) were developed by the Institute of Chinese Materia Medica, the Academia of Chinese Traditional Medicine, and the Guilin pharmaceutical plant. Artemether (formulated in oil for oral use, but mainly in tablet form) and AS (a water-soluble hemisuccinate derivative in an injectable formulation) were approved as new antimalarial drugs by the Chinese authorities in 1986 and 1987 and DHA was approved in 1991. A. J. Lin at WRAIR was also able to produce water-soluble artemisinins that preserved the active endoperoxide bridge. Although relatively unstable compared with the fat-soluble forms they were not as toxic as AE and AM. The most active form was AS, which had originally been isolated by the Chinese. It is manufactured in China and Vietnam for oral and intravenous use. WRAIR elected to develop their own derivatives: one was a semisynthetic artemisinin derivative, artelinic acid, and another was artemotil (= arteether). In March 2000, the Dutch company

ARTECEF registered artemotil. Although artelinic acid is much more stable than AS, the U.S. Army favored the continuing development of AS over artelinic acid and, within two years, the preclinical work had been completed. In 2004, an Investigational New Drug application was filed with the FDA. The drug was found to be safe in normal healthy volunteers and was granted Orphan Drug Status. In August 2005, the South East Asian Quinine Artesunate Malaria Trial Group reported that, in 1,500 patients with severe malaria, AS was superior to quinine in that it prevented mortality. A partnership with the Italian subsidiary in the US, Sigma-Tau Pharmaceuticals, was sealed in 2007. Artesunate is available from the CDC for compassionate use in the US. (Parenthetically, the problem encountered with water-soluble AS formulations is their rapid hydrolysis to DHA, and within minutes it crystallizes out in the syringes. Regulatory bodies don't like that!) The artemisinin derivatives work quickly and are eliminated quickly. After oral or intravenous administration, they are converted to DHA in the body. The DHA is eliminated with a half-life of ~ 1 hour.

How do artemisinin and its derivatives work to kill the malaria parasite? It has been hypothesized [88, 450–452] that, in the parasite, heme (in the form of hemozoin) catalyzes the breakdown of artemisinin to produce free radicals that act to inflict damage on the parasite by irreversibly denaturing (as would boiling an egg) the mitochondrial and food vacuole membrane proteins so they cannot function, and growth is inhibited. Alternatively, the free radical may bind to heme and block the formation of hemozoin from hemoglobin. These heme-based hypotheses may explain the drug's effectiveness against the hemozoin-containing asexual feeding and dividing forms. This mode of action seems, at first glance, to be paradoxical to artemisinin's killing of the youngest blood forms, the rings, without apparent malaria pigment by light microscopy; however, a recent study using the electron microscope clearly shows that even the ring stages digest hemoglobin and produce hemozoin. Indeed, by artemisinin removing ring stages, the number of parasites that mature and reproduce is severely reduced and this is the basis for the rapid clinical response and life-saving benefit of these drugs in severe malaria when compared with quinine. Another hypothesis for the mechanism of action of the drug posits that it is transported into the parasite from the red blood cell by means of parasite-derived vesicles, and within the parasite the artemisinin is activated by free iron or heme and then it specifically and selectively inhibits the Ca^{2+}-ATPase 6 (adenosine triphosphatase) in the sarco/endoplasmic reticulum to reduce the levels of Ca^{2+} below those necessary for mitochondrial function and to slow parasite growth [301].

Expression studies of PfATPase6 in *Xenopus* oocytes clearly showed that artemisinins inhibited the enzyme. Resistance has been demonstrated to be due to polymorphisms in the gene for PfATPase6. In French Guiana, isolates carrying the serine to asparagine mutation at position 769 have increased resistance to AM. However, the same mutation is not associated with resistance to DHA when assayed in African isolates; here a different mutation (histidine to tyrosine in position 243) may modulate sensitivity.

The worsening problem of drug resistance in many parts of the world, coupled with the limited number of antimalarials available, has provoked a search for new drugs, as well as the development of guidelines to protect the effectiveness of available drugs by limiting the emergence of resistance. Observations since the 1960s have shown that where there was increased mortality from malaria it was directly related to the continued use of increasingly ineffective antimalarials such as CQ and Fansidar (sulfadoxine–pyrimethamine). Thus, there has been considerable concern that resistance in the field would emerge to the artemisinins as it has to almost every other class of antimalarials. Although it has been difficult to induce a stable and high level of resistance to artemisinins in the laboratory, it has been possible to select strains of rodent malarias that are five to ten times as insusceptible, and by *in vitro* selection the same level of resistance has been accomplished with *P. falciparum*. After 2001, the use of artemisinins as a single drug for treatment decreased significantly to prevent the emergence of resistance [488]. More importantly, resistance to artemisinin began to appear in isolated areas where it had been used on its own. In 2006, the WHO requested discontinuance of the manufacturing and marketing of all artemisinin monotherapies except for the treatment of severe malaria [490].

The WHO recommends four artemisinin-based combination therapies (ACTs): AM–lumefantrine (benflumetol), AS–mefloquine (marketed as Artequin), AS–amodiaquine, and AS–Fansidar. In 1992, oral AM–lumefantrine was the first registered, fixed, ACT treatment. The first ACT to be evaluated systematically was AS–mefloquine — deployed in 1994 on the Northwest border of Thailand where there was mefloquine-resistant *P. falciparum* — and shown to be efficacious over a 14-year period. However, recently, the efficacy of this combination has suffered, apparently as a result of further significant reduction of the sensitivity of *P. falciparum* to mefloquine and artemisinin in the Thailand/Myanmar border area. In 1994, Novartis formed a collaborative agreement with the Chinese Academy of Military Sciences, Kunming Pharmaceutical Factory and the China International Trust and Investment Corporation to further develop this combination and eventually it was registered as Coartem and Riamet. Coartem constitutes about 70% of all current clinically used ACTs. Coartem (Novartis), consisting of AM/lumefantrine, 20 mg/120 mg, in tablet form, was approved by the FDA in April 2009. It is highly effective when given for three days (six doses) and with a small amount of fat to ensure adequate absorption and thus enhance efficacy.

A single daily dose of artemisinin is sufficient for maximal killing of artemisinin-sensitive parasites. In an ACT, the artemisinin component acts quickly over a three-day course of treatment and provides antimalarial activity for two asexual parasite cycles, resulting in a staggering reduction of the billion blood parasites within an infected patient. Although some parasites may remain, the partner drug should remove these (and this residual killing of parasites would be variably assisted by an immune response). Therefore, "the artemisinin component of the ACT reduces the probability that a mutant resistant to the partner drug will arise from the primary infection and, if effective, the partner should kill any artemisinin-resistant parasites that arose" [735].

"The technological advances in genomics research are beginning to afford unprecedented genome-wide discovery of mutations, thus facilitating high-throughput approaches for discovering genes that are involved in drug resistance" [178]. The method of using high-tiling microarrays allows for the accurate detection of SNPs in the parasite and can be applied to screening populations of *P. falciparum* and for tracking the spread of drug resistance, to discover the genetic basis of other phenotypic changes, and to determine how the parasite is evolving worldwide both in the field and in the laboratory [178].

Chapter 16

Prospice: Looking to the Future

A celebration of the completion of the human genome sequence was held at the White House on 26 June 2000. The President at the time, Bill Clinton, said, "it will have a real impact on all our lives — and even more, on the lives of our children. It will revolutionize the diagnosis, prevention and treatment, of most, if not all, human diseases." Those who, like Clinton, expected dramatic results overnight, would be disappointed in the years since the announcement. Among these would be Craig Venter who stood next to President Clinton and was involved in the race to sequence the human genome. In a Spiegel Online interview on 29 July 2010 he reflected on the Project: "We have learned nothing from the genome and the medical benefits so far have been close to zero." In that same year Francis Collins, who shared with Venter in the completion of the human genome sequence, and was also alongside Clinton at the celebration, put it more gently: "It is fair to say that the Human Genome Project has not directly affected the healthcare of most individuals" [147].

Two years after the announcement of the sequencing of the human genome, the genomes of *P. falciparum* and *A. gambiae* were decoded and they too were celebrated. The promise was that "it will greatly speed the development of much needed drugs...and help vaccine designers" [717]. The impact for clinical medicine of the human genome sequence as well as that of the malaria and mosquito genomes has thus far been quite modest. Why? Sequencing of genomes obeys the First Law of Technology: we invariably overestimate the short-term impacts of new technologies and underestimate their long-term effects [147]. This being so, what can be expected now that we are witness to the dashed hopes for the post-genomic era?

It is reasonable to expect that, in the next decade, some of the significant gains from the Malaria Genome Projects will significantly reduce the gap between basic research and clinical application. With a continuing appreciation of the complex biology of the parasite and its vector there will be a trickle of therapies, not a flood; there will be improved diagnostics; we will better understand how to deal with drug resistance; and progress will be made in identifying novel vaccine candidates. There will be better control but eradication will remain an elusive goal. Novel drugs will be based not only on known gene targets but on the hypothetical genes of *Plasmodium*.

We will see:

1. Drugs that target the apicoplast will become potential antimalarials. *Plasmodium falciparum* has 545 proteins — approximately 70% with unknown function — predicted from the genome sequence to be apicoplast targeted. The apicoplast is essential to the survival of *P. falciparum* because it supplies essential metabolites for the synthesis of isoprenoids, fatty acids, and heme. It is expected that the import of apicoplast-targeted proteins will be the antimalarials of the future [411]. The challenge, however, will be to not only understand how the permeome of the apicoplast works but to discover drugs that are able to reach, without harm to the host, their *in vivo* plasmodial target.
2. There will be new malaria vaccine candidates but no protective vaccine will emerge in the next decade. Identifying the thousands of genes expressed by the sporozoite and the liver stages that contribute to immunity will remain a challenge, as will the optimization of vaccine delivery systems. High-throughput immune assays will be developed that will aid in the identification of pre-erythrocytic antigens targeted by the immune system [723]. An immune correlate to protection — currently the

greatest impediment to the development of a blood-stage vaccine — will be found. This will enable the screening of many more blood-stage antigens that have already been identified through genome sequencing. Putting these antigens, singly and in combination, into a priority pipeline for clinical evaluation will remain an obstacle before there can be any thought of mass vaccination [152].

3. Understanding the genetic basis of malarial drug resistance will be used to track the spread of resistance in a population and to map the distribution and spread of resistance mutations, thereby leading to better treatment policies as well as suggesting ways in which drugs might be modified to restore efficacy [18]. Knowledge of the genetic basis of resistance will provide the tools needed to better understand disease evolution and allow for new strategies to curtail drug resistance by *Plasmodium*.

4. By mining the genomes of *P. falciparum* and *P. vivax*, new diagnostic targets will be identified. Indeed, a recent report shows that, by using bioinformatics, it is possible to identify novel genetic targets and to apply these through a single-round, unnested PCR reaction [173]. However, for these novel targets to be used in the field for rapid diagnosis of active malaria infections will require that they become cheaper, less labor intensive, thermostable, accurate, and highly sensitive. Once these impediments have been overcome then there can be better control of disease through treatment.

5. In the search for funding for the Malaria Genome Project, the proposers suggested that "the main immediate impact of the genome project is likely to be the discovery of new targets for drugs" [99]. As of today no such novel drug is on the market. Using high-density microarrays it will be possible to identify the site of genetic change that is the target of an antimalarial drug [177, 178]. This will allow for improvements in drug design. Target identification has already been shown with known compounds that have the potential to interfere with the utilization of isoleucine — an amino acid critical to the growth of blood-stage parasites [343]. The bottlenecks in drug research

will be to develop suitable, readily available, and affordable animal models, as well as the necessity to convince the pharmaceutical industry that investments in antimalarial drugs should receive a high priority.
6. Progress will be made to develop genetically engineered malaria-resistant mosquitoes and measures to spread a genetic modification from laboratory to field populations. Selfish genetic elements will be exploited for this purpose [740]. Blocking of malaria transmission through genetic modification of mosquitoes will continue to be tested in small-scale field trials. Genetically modified mosquitoes (and TBVs), however, will have little impact on the rates of transmission in endemic areas and there will be continued reliance on insecticide-treated bednets.

Finally, a significant barrier to the full and effective use of the findings from the Malaria Genome Projects will be to maintain a continuing interest and commitment by funding agencies over the 10–15 year timeline that is required for the translation from bench to practical use [1].

References

1. 2011. A research agenda for malaria eradication: vaccines. *PLoS Med.* **8**:e1000398.
2. Abdulla, S., R. Oberholzer, O. Juma, *et al.* 2008. Safety and immunogenicity of RTS,S/AS02D malaria vaccine in infants. *N. Engl. J. Med.* **359**:2533–2544.
3. Abu Bakar, N., N. Klonis, E. Hanssen, *et al.* 2010. Digestive-vacuole genesis and endocytic processes in the early intraerythrocytic stages of Plasmodium falciparum. *J. Cell Sci.* **123**:441–450.
4. Acosta, C. J., C. M. Galindo, D. Schellenberg, *et al.* 1999. Evaluation of the SPf66 vaccine for malaria control when delivered through the EPI scheme in Tanzania. *Trop. Med. Int. Health* **4**:368–376.
5. Adachi, J. and M. Hasegawa. 1996. Tempo and mode of synonymous substitutions in mitochondrial DNA of primates. *Mol. Biol. Evol.* **13**:200–208.
6. Adams, J. H., P. L. Blair, O. Kaneko, and D. S. Peterson. 2001. An expanding ebl family of Plasmodium falciparum. *Trends Parasitol.* **17**:297–299.
7. Adams, J. H., D. E. Hudson, M. Torii, *et al.* 1990. The Duffy receptor family of Plasmodium knowlesi is located within the micronemes of invasive malaria merozoites. *Cell* **63**:141–153.
8. Aikawa, M., L. H. Miller, J. Johnson, and J. Rabbege. 1978. Erythrocyte entry by malarial parasites. A moving junction between erythrocyte and parasite. *J. Cell Biol.* **77**:72–82.
9. Aikawa, M., J. R. Rabbege, I. Udeinya, and L. H. Miller. 1983. Electron microscopy of knobs in Plasmodium falciparum-infected erythrocytes. *J. Parasitol.* **69**:435–437.
10. Aldritt, S. M., J. T. Joseph, and D. F. Wirth. 1989. Sequence identification of cytochrome b in Plasmodium gallinaceum. *Mol. Cell Biol.* **9**:3614–3620.
11. Alkhalil, A., J. V. Cohn, M. A. Wagner, *et al.* 2004. Plasmodium falciparum likely encodes the principal anion channel on infected human erythrocytes. *Blood* **104**:4279–4286.

12. Alonso, P. L., T. Smith, J. R. Schellenberg, *et al.* 1994. Randomised trial of efficacy of SPf66 vaccine against Plasmodium falciparum malaria in children in southern Tanzania. *Lancet* **344**:1175–1181.
13. Altman, L. K. 1987. *Who Goes First? The Story of Self-Experimentation in Medicine*, 1st edition. Random House, New York.
14. Aly, A. S., S. E. Lindner, D. C. MacKellar, *et al.* 2011. SAP1 is a critical posttranscriptional regulator of infectivity in malaria parasite sporozoite stages. *Mol. Microbiol.* **79**:929–939.
15. Aly, A. S. and K. Matuschewski. 2005. A malarial cysteine protease is necessary for Plasmodium sporozoite egress from oocysts. *J. Exp. Med.* **202**:225–230.
16. Ambrose, K., D. Henry, and A. Weiss. 2002. *Washington Weather. The Weather Sourcebook for the DC Area*. Historical Enterprises, Fairfax.
17. Ancelin, M. L., M. Parant, M. J. Thuet, *et al.* 1991. Increased permeability to choline in simian erythrocytes after Plasmodium knowlesi infection. *Biochem. J.* **273 (Pt 3)**:701–709.
18. Anderson, T., S. Nkhoma, A. Ecker, and D. Fidock. 2011. How can we identify parasite genes that underlie antimalarial drug resistance? *Pharmacogenomics* **12**:59–85.
19. Andrews, K. T., T. N. Tran, N. C. Wheatley, and D. P. Fairlie. 2009. Targeting histone deacetylase inhibitors for anti-malarial therapy. *Curr. Top. Med. Chem.* **9**:292–308.
20. Andrews, K. T., A. Walduck, M. J. Kelso, *et al.* 2000. Antimalarial effect of histone deacetylation inhibitors and mammalian tumour cytodifferentiating agents. *Int. J. Parasitol.* **30**:761–768.
21. Anonymous. 1989. AID's Malaria Vaccine Research Activities, pp. 1–55. General Accounting Office, Washington, DC.
22. Anonymous. 2004. Daniel Carucci, Director, Grand Challenges in Global Health, National Institutes of Health. *Nature* **430**:384.
23. Anonymous. 1975. Editorial. Malaria vaccine on the horizon. *BMJ* **1**:231–232.
24. Aponte, J. J., P. Aide, M. Renom, *et al.* 2007. Safety of the RTS,S/AS02D candidate malaria vaccine in infants living in a highly endemic area of Mozambique: a double blind randomised controlled phase I/IIb trial. *Lancet* **370**:1543–1551.
25. Arastu-Kapur, S., E. L. Ponder, U. P. Fonovic, *et al.* 2008. Identification of proteases that regulate erythrocyte rupture by the malaria parasite Plasmodium falciparum. *Nat. Chem. Biol.* **4**:203–213.

26. Ardeshir, F., J. E. Flint, and R. T. Reese. 1985. Expression of Plasmodium falciparum surface antigens in Escherichia coli. *Proc. Natl. Acad. Sci. U.S.A.* **82**:2518–2522.
27. Ardeshir, F., J. E. Flint, S. J. Richman, and R. T. Reese. 1987. A 75 kd merozoite surface protein of Plasmodium falciparum which is related to the 70 kd heat-shock proteins. *EMBO J.* **6**:493–499.
28. Atkinson, C. T. and M. Aikawa. 1990. Ultrastructure of malaria-infected erythrocytes. *Blood Cells* **16**:351–368.
29. Aviado, D. M. and D. H. Will. 1969. Pharmacology of naphthoquinones, with special reference to the antimalarial activity of Lapinone (WR 26, 041). *Am. J. Trop. Med. Hyg.* **18**:188–198.
30. Babon, J. J., W. D. Morgan, G. Kelly, et al. 2007. Structural studies on Plasmodium vivax merozoite surface protein-1. *Mol. Biochem. Parasitol.* **153**:31–40.
31. Bai, T., M. Becker, A. Gupta, et al. 2005. Structure of AMA1 from Plasmodium falciparum reveals a clustering of polymorphisms that surround a conserved hydrophobic pocket. *Proc. Natl. Acad. Sci. U.S.A.* **102**: 12736–12741.
32. Baker, R. P., R. Wijetilaka, and S. Urban. 2006. Two Plasmodium rhomboid proteases preferentially cleave different adhesins implicated in all invasive stages of malaria. *PLoS Path.* **2**:e113.
33. Baldwin, S. A., G. A. McConkey, C. E. Cass, and J. D. Young. 2007. Nucleoside transport as a potential target for chemotherapy in malaria. *Curr. Pharm. Des.* **13**:569–580.
34. Ball, E., C. B. Anfinsen, and O. Cooper. 1947. The inhibitory action of naphthoquinones on respiratory processes. *J. Biol. Chem.* **168**:257–270.
35. Ball, E. G., R. W. McKee, C. B. Anfinsen, et al. 1948. Studies on malarial parasites. IX. Chemical and metabolic changes during growth and multiplication *in vivo* and *in vitro*. *J. Biol. Chem.* **168**:547–571.
36. Ballou, W. R. 2007. Obstacles to the development of a safe and effective attenuated pre-erythrocytic stage malaria vaccine. *Microbes Infect./Inst. Past.* **9**:761–766.
37. Ballou, W. R. and C. P. Cahill. 2007. Two decades of commitment to malaria vaccine development: GlaxoSmithKline Biologicals. *Am. J. Trop. Med. Hyg.* **77**:289–295.
38. Ballou, W. R., S. L. Hoffman, J. A. Sherwood, et al. 1987. Safety and efficacy of a recombinant DNA Plasmodium falciparum sporozoite vaccine. *Lancet* **1**:1277–1281.

39. Bannister, L., A. Margos, and J. Hopkins. 2005. Making a home for Plasmodium post-genomics: ultrastructural organization of the blood stages, pp. 24–49. In I. W. Sherman (ed.), *Molecular Approaches to Malaria*. ASM Press, Washington, DC.
40. Bannister, L. H., G. A. Butcher, E. D. Dennis, and G. H. Mitchell. 1975. Structure and invasive behaviour of Plasmodium knowlesi merozoites *in vitro*. *Parasitology* **71**:483–491.
41. Bannister, L. H., G. A. Butcher, E. D. Dennis, and G. H. Mitchell. 1975. Studies on the structure and invasive behaviour of merozoites of Plasmodium knowlesi. *Trans. R. Soc. Trop. Med. Hyg.* **69**:5.
42. Bannister, L. H., J. M. Hopkins, A. R. Dluzewski, *et al.* 2003. Plasmodium falciparum apical membrane antigen 1 (PfAMA-1) is translocated within micronemes along subpellicular microtubules during merozoite development. *J. Cell Sci.* **116**:3825–3834.
43. Bannister, L. H. and G. H. Mitchell. 2009. The malaria merozoite, forty years on. *Parasitology* **136**:1435–1444.
44. Bannister, L. H. and I. W. Sherman. 2009. Plasmodium, pp. 1–12. In *Encyclopedia of the Life Sciences*. John Wiley and Sons, Chichester.
45. Banyal, H. S. and C. D. Fitch. 1982. Ferriprotoporphyrin IX binding substances and the mode of action of chloroquine against malaria. *Life Sci.* **31**:1141–1144.
46. Bardes, C. L. 2008. *Pale Faces. The Masks of Anemia*. Bellevue, New York.
47. Barnwell, J. W., M. E. Nichols, and P. Rubinstein. 1989. *In vitro* evaluation of the role of the Duffy blood group in erythrocyte invasion by Plasmodium vivax. *J. Exp. Med.* **169**:1795–1802.
48. Barton, V., N. Fisher, G. A. Biagini, *et al.* 2010. Inhibiting Plasmodium cytochrome bc1: a complex issue. *Curr. Opin. Chem. Biol.* **14**:440–446.
49. Baruch, D. I., B. L. Pasloske, H. B. Singh, *et al.* 1995. Cloning the P. falciparum gene encoding PfEMP1, a malarial variant antigen and adherence receptor on the surface of parasitized human erythrocytes. *Cell* **82**:77–87.
50. Baumeister, S., M. Winterberg, C. Duranton, *et al.* 2006. Evidence for the involvement of Plasmodium falciparum proteins in the formation of new permeability pathways in the erythrocyte membrane. *Mol. Microbiol.* **60**:493–504.
51. Baumeister, S., M. Winterberg, J. M. Przyborski, and K. Lingelbach. 2010. The malaria parasite Plasmodium falciparum: cell biological peculiarities and nutritional consequences. *Protoplasma* **240**:3–12.
52. Beach, D. H., I. W. Sherman, and G. G. Holz, Jr. 1977. Lipids of Plasmodium lophurae, and of erythrocytes and plasma of normal and P. lophurae-infected Pekin ducklings. *J. Parasitol.* **63**:62–75.

53. Beaty, B. J., D. J. Prager, A. A. James, *et al.* 2009. From Tucson to genomics and transgenics: the vector biology network and the emergence of modern vector biology. *PLoS Neglected Trop. Dis.* **3**:e343.
54. Beaudoin, R. L. 1977. Should cultivated exoerythrocytic parasites be considered as a source of antigen for a malaria vaccine? *Bull. World Health Organ.* **55**:373–376.
55. Beaudoin, R. L., C. P. Strome, F. Mitchell, and T. A. Tubergen. 1977. Plasmodium berghei: immunization of mice against the ANKA strain using the unaltered sporozoite as an antigen. *Exp. Parasitol.* **42**:1–5.
56. Beaudoin, R. L., C. P. Strome, T. A. Tubergen, and F. Mitchell. 1976. Plasmodium berghei berghei: irradiated sporozoites of the ANKA strain as immunizing antigens in mice. *Exp. Parasitol.* **39**:438–443.
57. Becker, K., L. Tilley, J. L. Vennerstrom, *et al..* 2004. Oxidative stress in malaria parasite-infected erythrocytes: host-parasite interactions. *Int. J. Parasitol.* **34**:163–189.
58. Belnoue, E., F. T. Costa, T. Frankenberg, *et al.* 2004. Protective T cell immunity against malaria liver stage after vaccination with live sporozoites under chloroquine treatment. *J. Immunol.* **172**:2487–2495.
59. Bhattacharjee, S., C. van Ooij, B. Balu, *et al.* 2008. Maurer's clefts of Plasmodium falciparum are secretory organelles that concentrate virulence protein reporters for delivery to the host erythrocyte. *Blood* **111**: 2418–2426.
60. Biagini, G. A., E. M. Pasini, R. Hughes, *et al.* 2004. Characterization of the choline carrier of Plasmodium falciparum: a route for the selective delivery of novel antimalarial drugs. *Blood* **104**:3372–3377.
61. Biello, D. 2008. Lending a helping arm: volunteers risk malaria to test vaccine. *Sci. Am. Online.* Available at: http://www.scientificamerican.com/article.cfm?id=lending-a-helping-arm-to-test-malaria-vaccine [Accessed 20 November 2011].
62. Bignami, A. and G. Bastianelli. 1890. Observazioni suile febbri malariche estive-autunnali. *La Rif. Med.* **232**:1334–1335.
63. Bishop, A. 1958. Drug resistance in protozoa. *Biol. Rev.* **34**:445–500.
64. Blackman, M. J. 2008. Malarial proteases and host cell egress: an 'emerging' cascade. *Cell. Microbiol.* **10**:1925–1934.
65. Blackman, M. J. 2000. Proteases involved in erythrocyte invasion by the malaria parasite: function and potential as chemotherapeutic targets. *Curr. Drug Targets* **1**:59–83.
66. Blackman, M. J., T. J. Scott-Finnigan, S. Shai, and A. A. Holder. 1994. Antibodies inhibit the protease-mediated processing of a malaria merozoite surface protein. *J. Exp. Med.* **180**:389–393.

67. Blandin, S. and E. A. Levashina. 2004. Thioester-containing proteins and insect immunity. *Mol. Immunol.* **40**:903–908.
68. Blandin, S., S. H. Shiao, L. F. Moita, *et al.* 2004. Complement-like protein TEP1 is a determinant of vectorial capacity in the malaria vector Anopheles gambiae. *Cell* **116**:661–670.
69. Blandin, S. A., E. Marois, and E. A. Levashina. 2008. Antimalarial responses in Anopheles gambiae: from a complement-like protein to a complement-like pathway. *Cell Host Microbe* **3**:364–374.
70. Blandin, S. A., R. Wang-Sattler, M. Lamacchia, *et al.* 2009. Dissecting the genetic basis of resistance to malaria parasites in Anopheles gambiae. *Science* **326**:147–150.
71. Blisnick, T., L. Vincensini, J. C. Barale, *et al.* 2005. LANCL1, an erythrocyte protein recruited to the Maurer's clefts during Plasmodium falciparum development. *Mol. Biochem. Parasitol.* **141**:39–47.
72. Blisnick, T., L. Vincensini, G. Fall, and C. Braun-Breton. 2006. Protein phosphatase 1, a Plasmodium falciparum essential enzyme, is exported to the host cell and implicated in the release of infectious merozoites. *Cell. Microbiol.* **8**:591–601.
73. Blount, R. 1967. Management of chloroquine resistant falciparum malaria. *Trans. Am. Clin. Climatol. Ass.* **78**:196–206.
74. Blume, M., M. Hliscs, D. Rodriguez-Contreras, *et al.* 2010. A constitutive pan-hexose permease for the Plasmodium life cycle and transgenic models for screening of antimalarial sugar analogs. *Faseb J.* **25**:1–12.
75. Boffey, P. M. 1984. Malaria vaccine is near, U.S. health officials say. *New York Times*, 3 August.
76. Boffey, P. M. 1988. U.S. Malaria Program, roiled by harassment dispute, faces new security. *New York Times*, 17 July.
77. Bojang, K. A., S. K. Obaro, U. D'Alessandro, *et al.* 1998. An efficacy trial of the malaria vaccine SPf66 in Gambian infants — second year of follow-up. *Vaccine* **16**:62–67.
78. Boudreau, E. F., H. K. Webster, K. Pavanand, and L. Thosingha. 1982. Type II mefloquine resistance in Thailand. *Lancet* **2**:1335.
79. Bouyer, G., S. Egee, and S. L. Thomas. 2006. Three types of spontaneously active anionic channels in malaria-infected human red blood cells. *Blood Cells Mol. Dis.* **36**:248–254.
80. Bovarnick, M., A. Lindsay, and L. Hellerman. 1946. Metabolism of the malarial parasite, with reference particularly to the action of antimalarial agents. II. Atabrine (quinacrine) inhibition of glucose oxidation in parasites initially depleted of substrate. Reversal by adenylic acid. *J. Biol. Chem.* **163**:535–551.

81. Bowman, I. B., P. T. Grant, W. O. Kermack, and D. Ogston. 1961. The metabolism of Plasmodium berghei, the malaria parasite of rodents. 2. An effect of mepacrine on the metabolism of glucose by the parasite separated from its host cell. *Biochem. J.* **78**:472–478.
82. Bowman, S., D. Lawson, D. Basham, *et al.* 1999. The complete nucleotide sequence of chromosome 3 of Plasmodium falciparum. *Nature* **400**:532–538.
83. Bozdech, Z., M. Llinas, B. L. Pulliam, *et al.* 2003. The transcriptome of the intraerythrocytic developmental cycle of Plasmodium falciparum. *PLoS Biol.* **1**:85–100.
84. Bozdech, Z., S. Mok, G. Hu, *et al.* 2008. The transcriptome of Plasmodium vivax reveals divergence and diversity of transcriptional regulation in malaria parasites. *Proc. Natl. Acad. Sci. U.S.A.* **105**:16290–16295.
85. Brackett, S., E. Waletzky, and M. Baker. 1946. The relation between pantothenic acid and Plasmodium gallinaceum infections in the chicken and the antimalarial activity of analogues of pantothenic acid. *J. Parasitol.* **32**:453–462.
86. Braun-Breton, C. and L. H. Pereira da Silva. 1993. Malaria proteases and red blood cell invasion. *Parasitol. Today* **9**:92–96.
87. Braun-Breton, C., T. Blisnick, M. E. Morales-Betoulle, *et al.* 1994. Malaria parasites: enzymes involved in red blood cell invasion. *Braz. J. Med. Biol. Res.* **27**:363–367.
88. Bray, P. G., S. A. Ward, and P. M. O'Neill. 2005. Quinolines and artemisinin: chemistry, biology and history. *Curr. Top. Microbiol. Immunol.* **295**:3–38.
89. Bromham, L. and D. Penny. 2003. The modern molecular clock. *Nat. Rev. Genet.* **4**:216–224.
90. Brooke, J. 1993. Colombian physician challenges malaria on the home front. *New York Times*, 23 March.
91. Brown, K. N. and I. N. Brown. 1965. Immunity to malaria: antigenic variation in chronic infections of Plasmodium knowlesi. *Nature* **208**:1286–1288.
92. Brown, P. 1991. Trials and tribulations of a malaria vaccine: One of the most hopeful candidates for a malaria vaccine emerged from an area where the disease is endemic. Its results have raised expectations and scientific controversy. *New Scientist* **129**:18–20.
93. Brown, W. M., C. A. Yowell, A. Hoard, *et al.* 2004. Comparative structural analysis and kinetic properties of lactate dehydrogenases from the four species of human malarial parasites. *Biochemistry* **43**:6219–6229.
94. Bueding, E. 1962. Comparative biochemistry of parasitic helminths. *Comp. Biochem. Physiol.* **4**:343–351.
95. Bueding, E. 1959. Mechanisms of action of schistosomicidal agents. *J. Pharm. Pharmacol.* **11**:385–392.

96. Bueding, E. and J. A. Mackinnon. 1955. Studies of the phosphoglucose isomerase of Schistosoma mansoni. *J. Biol. Chem.* **215**:507–513.
97. Bueding, E. and J. M. Mansour. 1957. The relationship between inhibition of phosphofructokinase activity and the mode of action of trivalent organic antimonials on Schistosoma mansoni. *Br. J. Pharmacol. Chemother.* **12**:159–165.
98. Butcher, G. A., G. H. Mitchell, and S. Cohen. 1973. Letter: Mechanism of host specificity in malarial infection. *Nature* **244**:40–41.
99. Butler, D. 1997. Funding assured for international sequencing project. *Nature* **388**:701.
100. Butler, N. S., N. W. Schmidt, A. M. Vaughan, *et al.* 2011. Superior antimalarial immunity after vaccination with late liver stage-arresting genetically attenuated parasites. *Cell Host Microbe* **9**:451–462.
101. Cabantchik, Z. I. and A. Rothstein. 1972. The nature of the membrane sites controlling anion permeability of human red blood cells as determined by studies with disulfonic stilbene derivatives. *J. Membr. Biol.* **10**:311–330.
102. Cameron, A., J. Read, R. Tranter, *et al.* 2004. Identification and activity of a series of azole-based compounds with lactate dehydrogenase-directed antimalarial activity. *J. Biol. Chem.* **279**:31429–31439.
103. Camus, D. and T. J. Hadley. 1985. A Plasmodium falciparum antigen that binds to host erythrocytes and merozoites. *Science* **230**:553–556.
104. Canfield, C. J. 1972. Malaria in USA military personnel 1965–1971. *Proc. Helm. Soc. Washington DC* **39**:15–18.
105. Canfield, C. J., A. P. Hall, B. S. MacDonald, *et al.* 1973. Treatment of falciparum malaria from Vietnam with a phenanthrene methanol (WR 33063) and a quinoline methanol (WR 30090). *Antimicrob. Agents Chemother.* **3**:224–227.
106. Capone, S. R. S. A., M. Naddeo, L. Siani, *et al.* 2010. Immune responses against a liver-stage malaria antigen induced by simian adenoviral vector AdCh63 and MVA prime-boost immunisation in non-human primates. *Vaccine* **29**:256–265.
107. Carlton, J. 2003. The Plasmodium vivax genome sequencing project. *Trends Parasitol.* **19**:227–231.
108. Carlton, J., J. Silva, and N. Hall. 2005. The genome of model malaria parasites, and comparative genomics. *Curr. Issues Mol. Biol.* **7**:23–37.
109. Carlton, J. M., S. V. Angiuoli, B. B. Suh, *et al.* 2002. Genome sequence and comparative analysis of the model rodent malaria parasite Plasmodium yoelii yoelii. *Nature* **419**:512–519.
110. Carlton, J. M., A. A. Escalante, D. Neafsey, and S. K. Volkman. 2008. Comparative evolutionary genomics of human malaria parasites. *Trends Parasitol.* **24**:545–550.

111. Carter, N. S., C. Ben Mamoun, W. Liu, *et al.* 2000. Isolation and functional characterization of the PfNT1 nucleoside transporter gene from Plasmodium falciparum. *J. Biol. Chem.* **275**:10683–10691.
112. Carter, R. 2002. Spatial simulation of malaria transmission and its control by malaria transmission blocking vaccination. *Int. J. Parasitol.* **32**:1617–1624.
113. Carter, R. 2003. Speculations on the origins of Plasmodium vivax malaria. *Trends Parasitol.* **19**:214–219.
114. Carter, R. 2001. Transmission blocking malaria vaccines. *Vaccine* **19**:2309–2314.
115. Carter, R. and D. H. Chen. 1976. Malaria transmission blocked by immunisation with gametes of the malaria parasite. *Nature* **263**:57–60.
116. Carter, R. and D. C. Kaushal. 1984. Characterization of antigens on mosquito midgut stages of Plasmodium gallinaceum. III. Changes in zygote surface proteins during transformation to mature ookinete. *Mol. Biochem. Parasitol.* **13**:235–241.
117. Carter, R. and K. N. Mendis. 2002. Evolutionary and historical aspects of the burden of malaria. *Clin. Microbiol. Rev.* **15**:564–594.
118. Carter, V., A. M. Nacer, A. Underhill, *et al.* 2007. Minimum requirements for ookinete to oocyst transformation in Plasmodium. *Int. J. Parasitol.* **37**:1221–1232.
119. Carucci, D. J., M. J. Gardner, H. Tettelin, *et al.* 1998. Sequencing the genome of Plasmodium falciparum. *Curr. Opin. Infect. Dis.* **11**:531–534.
120. Catteruccia, F. 2007. Malaria vector control in the third millennium: progress and perspectives of molecular approaches. *Pest Manage. Sci.* **63**:634–640.
121. Cavasini, C. E., L. C. Mattos, A. A. Couto, *et al.* 2007. Plasmodium vivax infection among Duffy antigen-negative individuals from the Brazilian Amazon region: an exception? *Trans. R. Soc. Trop. Med. Hyg.* **101**:1042–1044.
122. Chaikuad, A., V. Fairweather, R. Conners, *et al.* 2005. Structure of lactate dehydrogenase from Plasmodium vivax: complexes with NADH and APADH. *Biochemistry* **44**:16221–16228.
123. Chakravarty, S., I. A. Cockburn, S. Kuk, *et al.* 2007. CD8+ T lymphocytes protective against malaria liver stages are primed in skin-draining lymph nodes. *Nat. Med.* **13**:1035–1041.
124. Chandramohanadas, R., P. H. Davis, D. P. Beiting, *et al.* 2009. Apicomplexan parasites co-opt host calpains to facilitate their escape from infected cells. *Science* **324**:794–797.
125. Chargaff, E. 1978. Heraclitean Fire. *Sketches from Life before Nature.* The Rockefeller University Press, New York.

126. Chatterjee, S., J. L. Perignon, E. Van Marck, and P. Druilhe. 2006. How reliable are models for malaria vaccine development? Lessons from irradiated sporozoite immunizations. *J. Postgrad. Med.* **52**:321–324.
127. Chen, X. G., G. Mathur, and A. A. James. 2008. Gene expression studies in mosquitoes. *Adv. Gen.* **64**:19–50.
128. Chinappi, M., A. Via, P. Marcatili, and A. Tramontano. 2010. On the mechanism of chloroquine resistance in Plasmodium falciparum. *PloS One* **5**:e14064.
129. Chiodini, P. L., K. Bowers, P. Jorgensen, *et al.* 2007. The heat stability of Plasmodium lactate dehydrogenase-based and histidine-rich protein 2-based malaria rapid diagnostic tests. *Trans. R. Soc. Trop. Med. Hyg.* **101**:331–337.
130. Chitnis, C. E. 2001. Molecular insights into receptors used by malaria parasites for erythrocyte invasion. *Curr. Opin. Hematol.* **8**:85–91.
131. Chitnis, C. E., A. Chaudhuri, R. Horuk, *et al.* 1996. The domain on the Duffy blood group antigen for binding Plasmodium vivax and P. knowlesi malarial parasites to erythrocytes. *J. Exp. Med.* **184**:1531–1536.
132. Chitnis, C. E. and L. H. Miller. 1994. Identification of the erythrocyte binding domains of Plasmodium vivax and Plasmodium knowlesi proteins involved in erythrocyte invasion. *J. Exp. Med.* **180**:497–506.
133. Choe, H., M. J. Moore, C. M. Owens, *et al.* 2005. Sulphated tyrosines mediate association of chemokines and Plasmodium vivax Duffy binding protein with the Duffy antigen/receptor for chemokines (DARC). *Mol. Microbiol.* **55**:1413–1422.
134. Chou, A. C. and C. D. Fitch. 1992. Heme polymerase: modulation by chloroquine treatment of a rodent malaria. *Life Sci.* **51**:2073–2078.
135. Clyde, D. F. 1990. Immunity to falciparum and vivax malaria induced by irradiated sporozoites: a review of the University of Maryland studies 1971–75. *Bull. World Health Organ.* **68 (Suppl 1)**:9–12.
136. Clyde, D. F. 1975. Immunization of man against falciparum and vivax malaria by use of attenuated sporozoites. *Am. J. Trop. Med. Hyg.* **24**: 397–401.
137. Coatney, G. R. 1963. Pitfalls in a discovery: the chronicle of quinine. *Am. J. Trop. Med. Hyg.* **12**:122–137.
138. Cobbold, S. A., R. E. Martin, and K. Kirk. 2011. Methionine transport in the malaria parasite Plasmodium falciparum. *Int. J. Parasitol.* **41**:125–135.
139. Cochrane, A. H., R. Nussenzweig, and E. Nardin. 1980. Immunization against sporozoites, pp. 163–202. In J. P. Kreier (ed.), *Malaria*, Vol. 3. Academic Press, San Diego.

140. Coggeshall, L. T., J. Maier, and C. A. Best. 1941. The effectiveness of two new types of chemotherapeutic agents in malaria. *J. Am. Med. Assoc.* **117**:1077–1081.
141. Coggeshall, L. T. and M. Eaton. 1938. Complement fixation reaction in monkey malaria. *J. Exp. Med.* **66**:177–190.
142. Coggeshall, L. T. and H. Kum. 1937. Demonstration of passive immunity in experimental monkey malaria. *J. Exp. Med.* **66**:177–190.
143. Coggeshall, L. T. and H. Kum. 1938. Effect of repeated superinfection upon the patency of immune serum of monkeys harboring chronic infections of Plasmodium knowlesi. *J. Exp. Med.* **68**:17–27.
144. Cohen, S. 1963. Gamma-globulin metabolism. *BMJ* **19**:202–206.
145. Cohen, S., G. A. Butcher, and R. B. Crandall. 1969. Action of malarial antibody *in vitro*. *Nature* **223**:368–371.
146. Cohen, S., I. McGregor, and S. Carrington. 1961. Gamma-globulin and acquired immunity to human malaria. *Nature* **192**:733–737.
147. Collins, F. 2010. Has the revolution arrived? *Nature* **464**:674–675.
148. Collins, F. H., R. K. Sakai, K. D. Vernick, *et al.* 1986. Genetic selection of a Plasmodium-refractory strain of the malaria vector Anopheles gambiae. *Science* **234**:607–610.
149. Collins, W. E. and J. W. Barnwell. 2009. Plasmodium knowlesi: finally being recognized. *J. Infect. Dis.* **199**:1107–1108.
150. Cooke, B. M., D. W. Buckingham, F. K. Glenister, *et al.* 2006. A Maurer's cleft-associated protein is essential for expression of the major malaria virulence antigen on the surface of infected red blood cells. *J. Cell Biol.* **172**:899–908.
151. Cooper, R. A., J. Papakrivos, K. D. Lane, *et al.*. 2005. PfCG2, a Plasmodium falciparum protein peripherally associated with the parasitophorous vacuolar membrane, is expressed in the period of maximum hemoglobin uptake and digestion by trophozoites. *Mol. Biochem. Parasitol.* **144**:167–176.
152. Coppel, R. L. 2009. Vaccinating with the genome: a Sisyphean task? *Trends Parasitol.* **25**:205–212.
153. Cornejo, O. E. and A. A. Escalante. 2006. The origin and age of Plasmodium vivax. *Trends Parasitol.* **22**:558–563.
154. Cowman, A. F. 1998. The molecular basis of resistance to the sulfones, sulfonamides, and dihydrofolate reductase inhibitors, pp. 317–330. In I. W. Sherman (ed.), *Malaria. Parasite Biology. Pathogenesis and Protection*. ASM Press, Washington, DC.
155. Cowman, A. F., D. L. Baldi, J. Healer, *et al.* 2000. Functional analysis of proteins involved in Plasmodium falciparum merozoite invasion of red blood cells. *FEBS Lett.* **476**:84–88.

156. Cowman, A. F. and B. S. Crabb. 2006. Invasion of red blood cells by malaria parasites. *Cell* **124**:755–766.
157. Cowman, A. F. and C. J. Tonkin. 2011. Microbiology. A tail of division. *Science* **331**:409–410.
158. Cox, F. E. G. 2010. History of the discovery of the malaria parasites and their vectors. *Parasites Vectors* **3**:1–9.
159. Crewther, P. E., J. G. Culvenor, A. Silva, *et al.* 1990. Plasmodium falciparum: two antigens of similar size are located in different compartments of the rhoptry. *Exp. Parasitol.* **70**:193–206.
160. Cunningham, D. A., W. Jarra, S. Koernig, *et al.* 2005. Host immunity modulates transcriptional changes in a multigene family (yir) of rodent malaria. *Mol. Microbiol.* **58**:636–647.
161. Curnoe, D. and A. Thorne. 2003. Number of ancestral human species: a molecular perspective. *Homo* **53**:201–224.
162. Dahm, R. 2008. Discovering DNA: Friedrich Miescher and the early years of nucleic acid research. *Hum. Genet.* **122**:565–581.
163. Dame, J. B., J. L. Williams, T. F. McCutchan, *et al.* 1984. Structure of the gene encoding the immunodominant surface antigen on the sporozoite of the human malaria parasite Plasmodium falciparum. *Science* **225**:593–599.
164. Daubersies, P., B. Ollomo, J. P. Sauzet, *et al.* 2008. Genetic immunisation by liver stage antigen 3 protects chimpanzees against malaria despite low immune responses. *PloS One* **3**:e2659.
165. Daubersies, P., A. W. Thomas, P. Millet, *et al.* 2000. Protection against Plasmodium falciparum malaria in chimpanzees by immunization with the conserved pre-erythrocytic liver-stage antigen 3. *Nat. Med.* **6**:1258–1263.
166. de Koning-Ward, T. F., P. R. Gilson, J. A. Boddey, *et al.* 2009. A newly discovered protein export machine in malaria parasites. *Nature* **459**:945–949.
167. de Kruif, P. 1926. *Microbe Hunters*. Harcourt Brace, San Diego, CA.
168. Deans, A. M., S. Nery, D. J. Conway, *et al.* 2007. Invasion pathways and malaria severity in Kenyan Plasmodium falciparum clinical isolates. *Infect. Immun.* **75**:3014–3020.
169. Deans, J. A. 1984. Protective antigens of blood stage Plasmodium knowelsi parasites. *Phil. Trans. R. Soc. Lond. B* **307**:159–169.
170. Deans, J. A., T. Alderson, A. W. Thomas, *et al.* 1982. Rat monoclonal antibodies which inhibit the *in vitro* multiplication of Plasmodium knowlesi. *Clin. Exp. Immunol.* **49**:297–309.
171. Deans, J. A., A. M. Knight, W. C. Jean, A. P. Waters, S. Cohen, and G. H. Mitchell. 1988. Vaccination trials in rhesus monkeys with a minor, invariant, Plasmodium knowlesi 66 kD merozoite antigen. *Parasite Immunol.* **10**:535–552.

172. Decherf, G., S. Egee, H. M. Staines, *et al.* 2004. Anionic channels in malaria-infected human red blood cells. *Blood Cells Mol. Dis.* **32**:366–371.
173. Demas, A., J. Oberstaller, J. Debarry, *et al.* 2011. Applied genomics: Data mining reveals species-specific malaria diagnostic targets more sensitive than 18S rRNA. *J. Clin. Microbiol.* **49**: 2411–2418.
174. DeRisi, J. L., V. R. Iyer, and P. O. Brown. 1997. Exploring the metabolic and genetic control of gene expression on a genomic scale. *Science* **278**:680–686.
175. Desai, S. A., S. M. Bezrukov, and J. Zimmerberg. 2000. A voltage-dependent channel involved in nutrient uptake by red blood cells infected with the malaria parasite. *Nature* **406**:1001–1005.
176. Desowitz, R. 1991. *The Malaria Capers. More Tales of Parasites and People, Research and Reality.* Norton, New York.
177. Dharia, N. V., A. Chatterjee, and E. A. Winzeler. 2010. Genomics and systems biology in malaria drug discovery. *Curr. Opin. Invest. Drugs* **11**:131–138.
178. Dharia, N. V., A. B. Sidhu, M. B. Cassera, *et al.* 2009. Use of high-density tiling microarrays to identify mutations globally and elucidate mechanisms of drug resistance in Plasmodium falciparum. *Genome Biol.* **10**:R21.
179. Doolan, D. L., J. Dobano, and K. Baird. 2008. Acquired immunity to malaria. *Clin. Microbiol. Rev.* **22**:13–36.
180. Doolan, D. L. and S. L. Hoffman. 2002. Nucleic acid vaccines against malaria, pp. 308–321. In P. Perlmann and M. Troye-Blomberg (eds.), *Malaria Immunology*, Vol. 80. Karger, Basle.
181. Doolan, D. L. and S. L. Hoffman. 2000. The complexity of protective immunity against liver-stage malaria. *J. Immunol.* **165**:1453–1462.
182. Doolan, D. L. and N. Martinez-Alier. 2006. Immune response to pre-erythrocytic stages of malaria parasites. *Curr. Mol. Med.* **6**:169–185.
183. dos Santos, S. C., S. Tenreiro, M. Palma, *et al.* 2009. Transcriptomic profiling of the Saccharomyces cerevisiae response to quinine reveals a glucose limitation response attributable to drug-induced inhibition of glucose uptake. *Antimicrob. Agents Chemother.* **53**:5213–5223.
184. Downie, M. J., K. El Bissati, A. M. Bobenchik, *et al.* 2010. PfNT2, a permease of the equilibrative nucleoside transporter family in the endoplasmic reticulum of Plasmodium falciparum. *J. Biol. Chem.* **285**:20827–20833.
185. Downie, M. J., K. J. Saliba, S. Broer, *et al.* 2008. Purine nucleobase transport in the intraerythrocytic malaria parasite. *Int. J. Parasitol.* **38**:203–209.
186. Downie, M. J., K. J. Saliba, S. M. Howitt, *et al.* 2006. Transport of nucleosides across the Plasmodium falciparum parasite plasma membrane has characteristics of PfENT1. *Mol. Microbiol.* **60**:738–748.
187. Draper, S. J., A. C. Moore, A. L. Goodman, *et al.* 2008. Effective induction of high-titer antibodies by viral vector vaccines. *Nat. Med.* **14**:819–821.

188. Drew, D. R., R. A. O'Donnell, B. J. Smith, and B. S. Crabb. 2004. A common cross-species function for the double epidermal growth factor-like modules of the highly divergent plasmodium surface proteins MSP-1 and MSP-8. *J. Biol. Chem.* **279**:20147–20153.
189. Druilhe, P. and J. W. Barnwell. 2007. Pre-erythrocytic stage malaria vaccines: time for a change in path. *Curr. Opin. Microbiol.* **10**:371–378.
190. Druilhe, P. and C. Marchand. 1989. From sporozoite to liver stages: the saga of the irradiated sporozoite vaccine, pp. 39–50. In K. P. W. McAdams (ed.), *Frontiers of Infectious Diseases. New Strategies in Parasitology.* Churchill Livingstone, London.
191. Druilhe, P., L. Renia, and D. Fidock. 1998. Immunity to liver stages, pp. 513–544. In I. W. Sherman (ed.), *Malaria. Parsite Biology, Pathogenesis and Protection.* ASM Press, Washington, DC.
192. Dunn, M. J. 1969. Alterations of red blood cell sodium transport during malarial infection. *J. Clin. Invest.* **48**:674–684.
193. Duraisingh, M. T., A. G. Maier, T. Triglia, and A. F. Cowman. 2003. Erythrocyte-binding antigen 175 mediates invasion in Plasmodium falciparum utilizing sialic acid-dependent and -independent pathways. *Proc. Natl. Acad. Sci. U.S.A.* **100**:4796–4801.
194. Duranton, C., S. M. Huber, V. Tanneur, et al. 2004. Organic osmolyte permeabilities of the malaria-induced anion conductances in human erythrocytes. *J. Gen. Physiol.* **123**:417–426.
195. Duval, L., M. Fourment, E. Nerrienet, et al. 2010. African apes as reservoirs of Plasmodium falciparum and the origin and diversification of the Laverania subgenus. *Proc. Natl. Acad. Sci. U.S.A.* **107**:10561–10566.
196. Dvorak, J. A., L. H. Miller, W. C. Whitehouse, and T. Shiroishi. 1975. Invasion of erythrocytes by malaria merozoites. *Science* **187**:748–750.
197. Dvorin, J. D., D. C. Martyn, S. D. Patel, et al. 2010. A plant-like kinase in Plasmodium falciparum regulates parasite egress from erythrocytes. *Science* **328**:910–912.
198. Dzikowski, R. and K. W. Deitsch. 2009. Genetics of antigenic variation in Plasmodium falciparum. *Curr. Genet.* **55**:103–110.
199. Eaton, M. D. and L. T. Coggeshall. 1939. Production in monkeys of complement fixing antibodies without active immunity by injection of killed Plasmodium knowlesi. *J. Exp. Med.* **70**:141–146.
200. Egee, S., F. Lapaix, G. Decherf, et al. 2002. A stretch-activated anion channel is up-regulated by the malaria parasite Plasmodium falciparum. *J. Physiol.* **542**:795–801.
201. Ekland, E. H., M. H. Akabas, and D. A. Fidock. 2011. Taking charge: feeding malaria via anion channels. *Cell* **145**:645–647.

202. Ekland, E. H. and D. A. Fidock. 2007. Advances in understanding the genetic basis of antimalarial drug resistance. *Curr. Opin. Microbiol.* **10**:363–370.
203. Ekland, E. H., J. Schneider, and D. A. Fidock. 2011. Identifying apicoplast-targeting antimalarials using high-throughput compatible approaches. *Faseb J.* **25**:3583–3593.
204. El Bissati, K., R. Zufferey, W. H. Witola, *et al.* 2006. The plasma membrane permease PfNT1 is essential for purine salvage in the human malaria parasite Plasmodium falciparum. *Proc. Natl. Acad. Sci. U.S.A.* **103**:9286–9291.
205. Elford, B. C., J. D. Haynes, J. D. Chulay, and R. J. Wilson. 1985. Selective stage-specific changes in the permeability to small hydrophilic solutes of human erythrocytes infected with Plasmodium falciparum. *Mol. Biochem. Parasitol.* **16**:43–60.
206. Ellis, J., L. S. Ozaki, R. W. Gwadz, *et al.* 1983. Cloning and expression in E. coli of the malarial sporozoite surface antigen gene from Plasmodium knowlesi. *Nature* **302**:536–538.
207. Enea, V., D. Arnot, E. C. Schmidt, *et al.* 1984. Circumsporozoite gene of Plasmodium cynomolgi (Gombak): cDNA cloning and expression of the repetitive circumsporozoite epitope. *Proc. Natl. Acad. Sci. U.S.A.* **81**:7520–7524.
208. Enserink, M. and E. Pennisi. 2002. Reserachers crack malaria genome. *Science* **295**:1207.
209. Epstein, J. E., S. Rao, F. Williams, *et al.* 2007. Safety and clinical outcome of experimental challenge of human volunteers with Plasmodium falciparum-infected mosquitoes: an update. *J. Infect. Dis.* **196**:145–154.
210. Erdman, L. K. and K. C. Kain. 2008. Molecular diagnostic and surveillance tools for global malaria control. *Trav. Med. Infect. Dis.* **6**:82–99.
211. Escalante, A. A. and F. J. Ayala. 1994. Phylogeny of the malarial genus Plasmodium, derived from rRNA gene sequences. *Proc. Natl. Acad. Sci. U.S.A.* **91**:11373–11377.
212. Escalante, A. A., O. E. Cornejo, D. E. Freeland, *et al.* 2005. A monkey's tale: the origin of Plasmodium vivax as a human malaria parasite. *Proc. Natl. Acad. Sci. U.S.A.* **102**:1980–1985.
213. Eyles, D. 1952. Studies on Plasmodium gallinaceum. I. Factors associated with the malaria infection in the vertebrate host which influence the degree of infection in the mosquito. *Am. J. Hyg.* **55**:386–391.
214. Eyles, D. 1952. Studies on Plasmodium gallinaceum. II. Factors in the blood of the vertebrate host influencing mosquito infection. *Am. J. Hyg.* **55**:276–290.

215. Fairley, N. H. 1957. Experiments with antimalarial drugs in man. VI. The value of chloroquine diphosphate as a suppressive drug in volunteers exposed to experimental mosquito transmitted malaria. *Trans. R. Soc. Trop. Med. Hyg.* **51**:493–501.
216. Fang, X. D., D. C. Kaslow, J. H. Adams, and L. H. Miller. 1991. Cloning of the Plasmodium vivax Duffy receptor. *Mol. Biochem. Parasitol.* **44**: 125–132.
217. Fawaz, G. and F. S. Haddad. 1951. The effect of lapinone (M-2350) on P. vivax infection in man. *Am. J. Trop. Med. Hyg.* **31**:569–571.
218. Fernex, M., C. Jaquet, M. Mittelholzer, *et al.* 1991. Neuer medikamente fur die behandlung der malaria tropica. *Schweiz. Rund. Med.* **80**:67–71.
219. Fidock, D. A. 2010. Drug discovery: Priming the antimalarial pipeline. *Nature* **465**:297–298.
220. Fidock, D. A., P. J. Rosenthal, S. L. Croft, *et al.* 2004. Antimalarial drug discovery: efficacy models for compound screening. *Nat. Rev. Drug Discov.* **3**:509–520.
221. Fieser, L. F., E. Berlinger, F. Bondhus, *et al.* 1948. Naphthoquinone antimalarials. *JACS* **70**:3151–3156.
222. Fieser, L. F., J. P. Schirmer, S. Archer, *et al.* 1967. Naphthoquinone antimalarial. XXIX. 2-hydroxy-3-w-cyclohexylalkyl-1,4-naphthoquinoines. *J. Med. Chem.* **10**:513–522.
223. Fitch, C. D., R. Chevli, H. S. Banyal, *et al.* 1982. Lysis of Plasmodium falciparum by ferriprotoporphyrin IX and a chloroquine-ferriprotoporphyrin IX complex. *Antimicrob. Agents Chemother.* **21**:819–822.
224. Fitch, C. D. and P. Kanjananggulpan. 1987. The state of ferriprotoporphyrin IX in malaria pigment. *J. Biol. Chem.* **262**:15552–15555.
225. Fleischmann, R. D., M. D. Adams, O. White, *et al.* 1995. Whole-genome random sequencing and assembly of Haemophilus influenzae Rd. *Science* **269**:496–512.
226. Flueck, C., R. Bartfai, J. Volz, *et al.* 2009. Plasmodium falciparum heterochromatin protein 1 marks genomic loci linked to phenotypic variation of exported virulence factors. *PLoS Path.* **5**:e1000569.
227. Foote, S. J., J. K. Thompson, A. F. Cowman, and D. J. Kemp. 1989. Amplification of the multidrug resistance gene in some chloroquine-resistant isolates of P. falciparum. *Cell* **57**:921–930.
228. Fraiture, M., R. H. Baxter, S. Steinert, *et al.* 2009. Two mosquito LRR proteins function as complement control factors in the TEP1-mediated killing of Plasmodium. *Cell Host Microbe* **5**:273–284.

229. Frankland, S., A. Adisa, P. Horrocks, et al. 2006. Delivery of the malaria virulence protein PfEMP1 to the erythrocyte surface requires cholesterol-rich domains. *Eukaryot. Cell* **5**:849–860.
230. Fraser, C. M., S. Casjens, W. M. Huang, et al. 1997. Genomic sequence of a Lyme disease spirochaete, Borrelia burgdorferi. *Nature* **390**:580–586.
231. Fraser, C. M., J. D. Gocayne, O. White, et al. 1995. The minimal gene complement of Mycoplasma genitalium. *Science* **270**:397–403.
232. Freeman, R. R., A. J. Trejdosiewicz, and G. A. Cross. 1980. Protective monoclonal antibodies recognising stage-specific merozoite antigens of a rodent malaria parasite. *Nature* **284**:366–368.
233. Freund, J., K. J. Thomson, H. E. Sommer, et al. 1945. Immunization of Rhesus monkeys against malarial infection (P. knowlesi) with killed parasites and adjuvants. *Science* **102**:202–204.
234. Freund, J., K. J. Thomson, H. E. Sommer, et al. 1948. Immunization of monkeys against malaria by means of killed parasites with adjuvants. *Am. J. Trop. Med.* **28**:1–22.
235. Freund, J., A. W. Walter and H. E. Sommer. 1945. Immunization against malaria: vaccination of ducks with killed parasites incorporated with adjuvants. *Science* **102**:200–202.
236. Frevert, U. and E. Nardin. 2008. Cellular effector mechanisms against Plasmodium liver stages. *Cell. Microbiol.* **10**:1956–1967.
237. Frolet, C., M. Thoma, S. Blandin, et al. 2006. Boosting NF-kappaB-dependent basal immunity of Anopheles gambiae aborts development of Plasmodium berghei. *Immunity* **25**:677–685.
238. Fry, M. and J. E. Beesley. 1991. Mitochondria of mammalian Plasmodium spp. *Parasitology* **102**:17–26.
239. Fulton, J. D. 1939. Experiments on the utilization of sugars by malarial parasites (Plasmodium knowlesi). *Ann. Trop. Med. Parasitol.* **32**:217–227.
240. Fulton, J. D. 1951. The metabolism of malaria parasites. *Br. Med. Bull.* **8**:22–27.
241. Fulton, J. D. and D. F. Spooner. 1956. The *in vitro* respiratory metabolism of erythrocytic forms of Plasmodium berghei. *Exp. Parasitol.* **5**:59–78.
242. Galinski, M. R. and J. W. Barnwell. 2008. Plasmodium vivax: who cares? *Malaria J.* **7 (Suppl 1)**:S9.
243. Galinski, M. R., A. R. Dluzewski, and J. W. Barnwell. 2005. A mechanistic approach to merozoite invasion of red blood cells: merozoite biogenesis, rupture, and invasion of erythrocytes, pp. 113–168. In I. W. Sherman (ed.), *Molecular Approaches to Malaria*. ASM Press, Washington, DC.

244. Galinski, M. R., C. C. Medina, P. Ingravallo, and J. W. Barnwell. 1992. A reticulocyte-binding protein complex of Plasmodium vivax merozoites. *Cell* 69:1213–1226.
245. Gama, B. E., F. Do E. Silva-Pires, M. N. Lopes, *et al.* 2007. Real-time PCR versus conventional PCR for malaria parasite detection in low-grade parasitemia. *Exp. Parasitol.* **116**:427–432.
246. Gamage-Mendis, A. C., J. Rajakaruna, R. Carter, and K. N. Mendis. 1992. Transmission blocking immunity to human Plasmodium vivax malaria in an endemic population in Kataragama, Sri Lanka. *Parasite Immunol.* **14**:385–396.
247. Gamo, F. J., L. M. Sanz, J. Vidal, *et al.* 2010. Thousands of chemical starting points for antimalarial lead identification. *Nature* **465**:305–310.
248. Ganesan, K., N. Ponmee, L. Jiang, *et al.* 2008. A genetically hard-wired metabolic transcriptome in Plasmodium falciparum fails to mount protective responses to lethal antifolates. *PLoS Path.* **4**:e1000214.
249. Gardner, M. J., N. Hall, E. Fung, *et al.* 2002. Genome sequence of the human malaria parasite Plasmodium falciparum. *Nature* **419**:498–511.
250. Gardner, M. J., S. J. Shallom, J. M. Carlton, *et al.* 2002. Sequence of Plasmodium falciparum chromosomes 2, 10, 11 and 14. *Nature* **419**:531–534.
251. Gardner, M. J., H. Tettelin, D. J. Carucci, *et al.* 1998. Chromosome 2 sequence of the human malaria parasite Plasmodium falciparum. *Science* **282**:1126–1132.
252. Garnham, P. C. C. 1947. Exo-erythrocytic schizogony in Plasmodium kochi: a preliminary note. *Trans. R. Soc. Trop. Med. Hyg.* **40**:719–722.
253. Garnham, P. C. C. 1966. *Malaria Parasites and other Haemosporidia*. Blackwell, Oxford.
254. Garnham, P. C. C. 1967. Presidential address: reflections on Laveran, Marchiafava, Golgi, Koch and Danilewsky after sixty years. *Trans. R. Soc. Trop. Med. Hyg.* **61**:753–764.
255. Gaur, D., D. C. Mayer, and L. H. Miller. 2004. Parasite ligand-host receptor interactions during invasion of erythrocytes by Plasmodium merozoites. *Int. J. Parasitol.* **34**:1413–1429.
256. Gaur, D., S. Singh, S. Singh, *et al.* 2007. Recombinant Plasmodium falciparum reticulocyte homology protein 4 binds to erythrocytes and blocks invasion. *Proc. Natl. Acad. Sci. U.S.A.* **104**:17789–17794.
257. Gaur, D., J. R. Storry, M. E. Reid, *et al.* 2003. Plasmodium falciparum is able to invade erythrocytes through a trypsin-resistant pathway independent of glycophorin B. *Infect. Immun.* **71**:6742–6746.

258. Gazarini, M. L., A. P. Thomas, T. Pozzan, and C. R. Garcia. 2003. Calcium signaling in a low calcium environment: how the intracellular malaria parasite solves the problem. *J. Cell Biol.* **161**:103–110.
259. Geiman, Q. M. and M. J. Meagher. 1967. Susceptibility of a New World monkey to Plasmodium falciparum from man. *Nature* **215**:437–439.
260. Gilberger, T. W., J. K. Thompson, T. Triglia, *et al.* 2003. A novel erythrocyte binding antigen-175 paralogue from Plasmodium falciparum defines a new trypsin-resistant receptor on human erythrocytes. *J. Biol. Chem.* **278**:14480–14486.
261. Ginsburg, H. and K. Kirk. 1998. Membrane transport in the malaria-infected erythrocyte, pp. 219–232. In I. W. Sherman (ed.), *Malaria: Parasite Biology, Pathogenesis and Protection*. ASM Press, Washington, DC.
262. Ginsburg, H., M. Krugliak, O. Eidelman, and Z. I. Cabantchik. 1983. New permeability pathways induced in membranes of Plasmodium falciparum infected erythrocytes. *Mol. Biochem. Parasitol.* **8**:177–190.
263. Ginsburg, H., S. Kutner, M. Krugliak, and Z. I. Cabantchik. 1985. Characterization of permeation pathways appearing in the host membrane of Plasmodium falciparum-infected red blood cells. *Mol. Biochem. Parasitol.* **14**:313–322.
264. Ginsburg, H. and W. D. Stein. 2004. The new permeability pathways induced by the malaria parasite in the membrane of the infected erythrocyte: comparison of results using different experimental techniques. *J. Membr. Biol.* **197**:113–134.
265. Glenister, F. K., R. L. Coppel, A. F. Cowman, *et al.* 2002. Contribution of parasite proteins to altered mechanical properties of malaria-infected red blood cells. *Blood* **99**:1060–1063.
266. Glushakova, S., G. Humphrey, E. Leikina, *et al.* 2011. New stages in the program of malaria parasite egress imaged in normal and sickle erythrocytes. *Curr. Biol.* **20**:1117–1121.
267. Glushakova, S., J. Mazar, M. F. Hohmann-Marriott, *et al.* 2009. Irreversible effect of cysteine protease inhibitors on the release of malaria parasites from infected erythrocytes. *Cell. Microbiol.* **11**:95–105.
268. Glushakova, S., D. Yin, T. Li, and J. Zimmerberg. 2005. Membrane transformation during malaria parasite release from human red blood cells. *Curr. Biol.* **15**:1645–1650.
269. Graves, P., H. Gelband, and P. Garner. 1998. The SPf66 malaria vaccine: what is the evidence for efficacy? *Parasitol. Today* **14**:218–220.
270. Graves, P. M., R. Carter, T. R. Burkot, *et al.* 1988. Antibodies to Plasmodium falciparum gamete surface antigens in Papua New Guinea sera. *Parasite Immunol.* **10**:209–218.

271. Greenwood, B. 2011. Immunological correlates of protection for the RTS,S candidate malaria vaccine. *Lancet Infect. Dis.* **11**:75–76.
272. Gregson, A. and C. V. Plowe. 2005. Mechanisms of resistance of malaria parasites to antifolates. *Pharmacol. Rev.* **57**:117–145.
273. Grotendorst, C. A., N. Kumar, R. Carter, and D. C. Kaushal. 1984. A surface protein expressed during the transformation of zygotes of Plasmodium gallinaceum is a target of transmission-blocking antibodies. *Infect. Immun.* **45**:775–777.
274. Gruenberg, J., D. R. Allred, and I. W. Sherman. 1983. Scanning electron microscope-analysis of the protrusions (knobs) present on the surface of Plasmodium falciparum-infected erythrocytes. *J. Cell Biol.* **97**:795–802.
275. Gruner, A. C., M. Mauduit, R. Tewari, *et al.* 2007. Sterile protection against malaria is independent of immune responses to the circumsporozoite protein. *PloS One* **2**:e1371.
276. Guerin-Marchand, C., P. Druilhe, B. Galey, *et al.* 1987. A liver-stage-specific antigen of Plasmodium falciparum characterized by gene cloning. *Nature* **329**:164–167.
277. Guiguemde, W. A., A. A. Shelat, D. Bouck, *et al.* 2010. Chemical genetics of Plasmodium falciparum. *Nature* **465**:311–315.
278. Gutteridge, W. E., D. Dave, and W. H. Richards. 1979. Conversion of dihydroorotate to orotate in parasitic protozoa. *Biochim. Biophys. Acta* **582**: 390–401.
279. Gwadz, R. W. 1976. Successful immunization against the sexual stages of Plasmodium gallinaceum. *Science* **193**:1150–1151.
280. Haase, S. and T. F. de Koning-Ward. 2010. New insights into protein export in malaria parasites. *Cell. Microbiol.* **12**:580–587.
281. Hadley, T., M. Aikawa, and L. H. Miller. 1983. Plasmodium knowlesi: studies on invasion of rhesus erythrocytes by merozoites in the presence of protease inhibitors. *Exp. Parasitol.* **55**:306–311.
282. Hadley, T. J., F. W. Klotz, and L. H. Miller. 1986. Invasion of erythrocytes by malaria parasites: a cellular and molecular overview. *Ann. Rev. Microbiol.* **40**:451–477.
283. Hadley, T. J. and L. H. Miller. 1988. Invasion of erythrocytes by malaria parasites: erythrocyte ligands and parasite receptors. *Prog. Allergy* **41**:49–71.
284. Haeggstrom, M., A. von Euler, F. Kironde, *et al.* 2007. Characterization of Maurer's clefts in Plasmodium falciparum-infected erythrocytes. *Am. J. Trop. Med. Hyg.* **76**:27–32.

285. Hafalla, J. C., A. Morrot, G. Sano, et al. 2003. Early self-regulatory mechanisms control the magnitude of CD8+ T cell responses against liver stages of murine malaria. *J. Immunol.* **171**:964–970.
286. Hafalla, J. C., G. Sano, L. H. Carvalho, et al. 2002. Short-term antigen presentation and single clonal burst limit the magnitude of the CD8(+) T cell responses to malaria liver stages. *Proc. Natl. Acad. Sci. U.S.A.* **99**:11819–11824.
287. Haldar, K. and N. Mohandas. 2007. Erythrocyte remodeling by malaria parasites. *Curr. Opin. Hematol.* **14**:203–209.
288. Haldar, K., N. Mohandas, S. Bhattacharjee, et al. 2005. Trafficking and the tubulovesicular membrane network, pp. 253–271. In I. W. Sherman (ed.), *Molecular Approaches to Malaria.* ASM Press, Washington, DC.
289. Hamblin, M. and A. Di Rienzo. 2000. Detecting the signature of natural selection in humans: evidence from the Duffy blood group locus. *Am. J. Hum. Gen.* **66**:1669–1679.
290. Hans, D., P. Pattnaik, A. Bhattacharyya, et al. 2005. Mapping binding residues in the Plasmodium vivax domain that binds Duffy antigen during red cell invasion. *Mol. Microbiol.* **55**:1423–1434.
291. Hanssen, E., P. Carlton, S. Deed, et al. 2010. Whole cell imaging reveals novel modular features of the exomembrane system of the malaria parasite, Plasmodium falciparum. *Int. J. Parasitol.* **40**:123–134.
292. Hanssen, E., P. J. McMillan, and L. Tilley. 2010. Cellular architecture of Plasmodium falciparum-infected erythrocytes. *Int. J. Parasitol.* **40**:1127–1135.
293. Harris, P. K., S. Yeoh, A. R. Dluzewski, et al. 2005. Molecular identification of a malaria merozoite surface sheddase. *PLoS Path.* **1**:241–251.
294. Harrison, G. 1978. *Mosquitoes, Malaria and Man: A History of Hostilities Since 1880.* Dutton, New York.
295. Hasegawa, M., A. Di Rienzo, T. D. Kocher, and A. C. Wilson. 1993. Toward a more accurate time scale for the human mitochondrial DNA tree. *J. Mol. Evol.* **37**:347–354.
296. Hastings, M. D. and C. H. Sibley. 2002. Pyrimethamine and WR99210 exert opposing selection on dihydrofolate reductase from Plasmodium vivax. *Proc. Natl. Acad. Sci. U.S.A.* **99**:13137–13141.
297. Hawkes, M. and K. C. Kain. 2007. Advances in malaria diagnosis. *Exp. Rev. Anti-Infect. Ther.* **5**:485–495.
298. Hawkins, V. N., H. Joshi, K. Rungsihirunrat, et al. 2007. Antifolates can have a role in the treatment of Plasmodium vivax. *Trends Parasitol.* **23**:213–222.

299. Hay, S. I., C. A. Guerra, P. W. Gething, et al. 2009. A world malaria map: Plasmodium falciparum endemicity in 2007. *PLoS Med.* **6**:e1000048.
300. Haynes, J. D., J. P. Dalton, F. W. Klotz, et al. 1988. Receptor-like specificity of a Plasmodium knowlesi malarial protein that binds to Duffy antigen ligands on erythrocytes. *J. Exp. Med.* **167**:1873–1881.
301. Haynes, R. K. and S. Krishna. 2004. Artemisinins: activities and actions. *Microbes Infect./Instit. Pasteur* **6**:1339–1346.
302. Heidelberger, M., W. A. Coates, and M. Mayer. 1946. Studies in human malaria. II Attempts to influence relapsing vivax malaria by treatment of patients with vaccine (P. vivax). *J. Immunol.* **52**:101–107.
303. Heidelberger, M., M. Mayer, and C. Demarest. 1946. Studies in human malaria. I. The preparation of vaccines and suspensions containing plasmodia. *J. Immunol.* **52**:325–330.
304. Heidelberger, M., C. Prout, J. Hindle, and A. Rose. 1946. Studies in human malaria. III. An attempt at vaccination of paretics against blood-borne infection with Plasmodium vivax. *J. Immunol.* **52**:109–118.
305. Henion, W. F., T. E. Mansour, and E. Bueding. 1955. The immunological specificity of lactic dehydrogenase of Schistosoma mansoni. *Exp. Parasitol.* **4**:40–44.
306. Heppner, D. G., D. M. Gordon, M. Gross, et al. 1996. Safety, immunogenicity, and efficacy of Plasmodium falciparum repeatless circumsporozoite protein vaccine encapsulated in liposomes. *J. Infect. Dis.* **174**:361–366.
307. Heppner, D. G., Jr., K. E. Kester, C. F. Ockenhouse, et al. 2005. Towards an RTS,S-based, multi-stage, multi-antigen vaccine against falciparum malaria: progress at the Walter Reed Army Institute of Research. *Vaccine* **23**:2243–2250.
308. Herman, Y. F., R. A. Ward, and R. H. Herman. 1966. Stimulation of the utilization of 1-14C-glucose in chicken red blood cells infected with Plasmodium gallinaceum. *Am. J. Trop. Med. Hyg.* **15**:276–280.
309. Heussler, V., A. Rennenberg, and R. Stanway. 2010. Host cell death induced by the egress of intracellular Plasmodium parasites. *Apoptosis* **15**:376–385.
310. Hien, T. T. and N. J. White. 1983. Qinghaosu. *Lancet* **341**:603–608.
311. Hill, D. A., A. D. Pillai, F. Nawaz, et al. 2007. A blasticidin S-resistant Plasmodium falciparum mutant with a defective plasmodial surface anion channel. *Proc. Natl. Acad. Sci. U.S.A.* **104**:1063–1068.
312. Hiller, N. L., S. Bhattacharjee, C. van Ooij, et al. 2004. A host-targeting signal in virulence proteins reveals a secretome in malarial infection. *Science* **306**:1934–1937.
313. Hillier, C. J., L. A. Ware, A. Barbosa, et al. 2005. Process development and analysis of liver-stage antigen 1, a preerythrocyte-stage protein-based vaccine for Plasmodium falciparum. *Infect. Immun.* **73**:2109–2115.

314. Hoffman, S. L., W. H. Bancroft, M. Gottlieb, *et al.* 1997. Funding for malaria genome sequencing. *Nature* **387**:647.
315. Hoffman, S. L., E. D. Franke, M. R. Hollingdale, and P. Druilhe. 1996. Attacking the infected hepatocyte, pp. 35–75. In S. L. Hoffman (ed.), *Malaria Vaccine Development: A Multi-Immune Response Approach.* ASM Press, Washington, DC.
316. Hoffman, S. L., L. M. Goh, T. C. Luke, *et al.* 2002. Protection of humans against malaria by immunization with radiation-attenuated Plasmodium falciparum sporozoites. *J. Infect. Dis.* **185**:1155–1164.
317. Holder, A. A. 1994. Proteins on the surface of the malaria parasite and cell invasion. *Parasitology* **108 (Suppl 1)**:S5–18.
318. Holder, A. A. and M. J. Blackman. 1994. What is the function of MSP-I on the malaria merozoite? *Parasitol. Today* **10**:182–184.
319. Holder, A. A. and R. R. Freeman. 1981. Immunization against blood-stage rodent malaria using purified parasite antigens. *Nature* **294**:361–364.
320. Holder, A. A. and R. R. Freeman. 1982. Biosynthesis and processing of a Plasmodium falciparum schizont antigen recognized by immune serum and a monoclonal antibody. *J. Exp. Med.* **156**:1528–1538.
321. Holder, A. A., R. R. Freeman, and S. C. Nicholls. 1988. Immunization against Plasmodium falciparum with recombinant polypeptides produced in Escherichia coli. *Parasite Immunol.* **10**:607–617.
322. Holder, A. A., J. A. Guevara Patino, C. Uthaipibull, *et al.* 1999. Merozoite surface protein 1, immune evasion, and vaccines against asexual blood stage malaria. *Parassitologia* **41**:409–414.
323. Hollingdale, M. R. and U. Krzych. 2002. Immune responses to liver-stage parasites: implications for vaccine development. *Chem. Immunol.* **80**:97–124.
324. Hollingdale, M. R., J. L. Leef, M. McCullough, and R. L. Beaudoin. 1981. *In vitro* cultivation of the exoerythrocytic stage of Plasmodium berghei from sporozoites. *Science* **213**:1021–1022.
325. Hollingdale, M. R., P. Leland, and A. L. Schwartz. 1983. *In vitro* cultivation of the exoerythrocytic stage of Plasmodium berghei in a hepatoma cell line. *Am. J. Trop. Med. Hyg.* **32**:682–684.
326. Holt, R. A., G. M. Subramanian, A. Halpern, *et al.* 2002. The genome sequence of the malaria mosquito Anopheles gambiae. *Science* **298**:129–149.
327. Homewood, C. A. and K. D. Neame. 1974. Malaria and the permeability of the host erythrocyte. *Nature* **252**:718–719.
328. Howard, R. F. and R. T. Reese. 1990. Plasmodium falciparum: hetero-oligomeric complexes of rhoptry polypeptides. *Exp. Parasitol.* **71**:330–342.

329. Howard, R. F., H. A. Stanley, and R. T. Reese. 1988. Characterization of a high-molecular-weight phosphoprotein synthesized by the human malarial parasite Plasmodium falciparum. *Gene* **64**:65–75.
330. Howard, R. F., A. Varki, and R. T. Reese. 1984. Merozoite proteins synthesized in P. falciparum schizonts. *Prog. Clin. Biol. Res.* **155**:45–61.
331. Howard, R. J., J. D. Haynes, M. H. McGinniss, and L. H. Miller. 1982. Studies on the role of red blood cell glycoproteins as receptors for invasion by Plasmodium falciparum merozoites. *Mol. Biochem. Parasitol.* **6**:303–315.
332. Howell, S. A., I. Well, S. L. Fleck, *et al.* 2003. A single malaria merozoite serine protease mediates shedding of multiple surface proteins by juxtamembrane cleavage. *J. Biol. Chem.* **278**:23890–23898.
333. Howitt, C., D. Wilinski, M. Llinas, *et al.* 2009. Clonally variant gene families in Plasmodium falciparum share a common acivation factor. *Mol. Microbiol.* **73**:1171–1185.
334. Huber, S. M., A. C. Uhlemann, N. L. Gamper, *et al.* 2002. Plasmodium falciparum activates endogenous Cl(-) channels of human erythrocytes by membrane oxidation. *EMBO J.* **21**:22–30.
335. Hudson, A. T. 1993. Atovaquone — a novel broad-spectrum anti-infective drug. *Parasitol. Today* **9**:66–68.
336. Hudson, A. T., A. W. Randall, M. Fry, *et al.* 1985. Novel anti-malarial hydroxynaphthoquinones with potent broad spectrum anti-protozoal activity. *Parasitology* **90 (Pt 1)**:45–55.
337. Huff, C. G. and F. Coulston. 1944. The development of Plasmodium gallinaceum from sporozoite to erythrocytic trophozoite. *J.Infect. Dis.* **75**:231–249.
338. Hughes, A. L. and F. Verra. 2010. Malaria parasite sequences from chimpanzee support the co-speciation hypothesis for the origin of virulent human malaria (Plasmodium falciparum). *Mol. Phylogenet. Evol.* **57**:135–143.
339. Hui, G. S. and W. A. Siddiqui. 1987. Serum from Pf195 protected Aotus monkeys inhibit Plasmodium falciparum growth *in vitro*. *Exp. Parasitol.* **64**:519–522.
340. Hui, G. S., L. Q. Tam, S. P. Chang, *et al.* 1991. Synthetic low-toxicity muramyl dipeptide and monophosphoryl lipid A replace Freund complete adjuvant in inducing growth-inhibitory antibodies to the Plasmodium falciparum major merozoite surface protein, gp195. *Infect. Immun.* **59**:1585–1591.
341. Ishino, T., B. Boisson, Y. Orito, *et al.* 2009. LISP1 is important for the egress of Plasmodium berghei parasites from liver cells. *Cell. Microbiol.* **11**:1329–1339.
342. Isikoff, M. 1989. Mismanagement in Malaria Program. *Washington Post*, 18 October.

343. Istvan, E. S., N. V. Dharia, S. E. Bopp, I et al. 2011. Validation of isoleucine utilization targets in Plasmodium falciparum. *Proc. Natl. Acad. Sci. U.S.A.* **108**:1627–1632.
344. Iyer, J., A. C. Gruner, L. Renia, et al. 2007. Invasion of host cells by malaria parasites: a tale of two protein families. *Mol. Microbiol.* **65**:231–249.
345. Jacobs, H. 1943. Immunization against malaria. *Am. J. Trop. Med.* **23**:597–606.
346. Jalah, R., R. Sarin, N. Sud, et al. 2005. Identification, expression, localization and serological characterization of a tryptophan-rich antigen from the human malaria parasite Plasmodium vivax. *Mol. Biochem. Parasitol.* **142**:158–169.
347. Jarcho, S. 1984. Laveran's discovery in the retrospect of a century. *Bull. Hist. Med.* **58**:215–224.
348. Jensen, J. B. 2005. Reflections on the continuous cultivation of Plasmodium falciparum. *J. Parasitol.* **91**:487–491.
349. Jiang, G., M. Shi, S. Conteh, et al. 2009. Sterile protection against Plasmodium knowlesi in rhesus monkeys from a malaria vaccine: comparison of heterologous prime boost strategies. *PloS One* **4**:e6559.
350. Joet, T., K. Chotivanich, K. Silamut, et al. 2004. Analysis of Plasmodium vivax hexose transporters and effects of a parasitocidal inhibitor. *Biochem. J.* **381**:905–909.
351. Joet, T., U. Eckstein-Ludwig, C. Morin, and S. Krishna. 2003. Validation of the hexose transporter of Plasmodium falciparum as a novel drug target. *Proc. Natl. Acad. Sci. U.S.A.* **100**:7476–7479.
352. John, C. C., A. J. Tande, A. M. Moormann, et al. 2008. Antibodies to pre-erythrocytic Plasmodium falciparum antigens and risk of clinical malaria in Kenyan children. *J. Infect. Dis.* **197**:519–526.
353. Joy, D. A., X. Feng, J. Mu, et al. 2003. Early origin and recent expansion of Plasmodium falciparum. *Science* **300**:318–321.
354. Jungery, M., G. Pasvol, C. I. Newbold, and D. J. Weatherall. 1983. A lectin-like receptor is involved in invasion of erythrocytes by Plasmodium falciparum. *Proc. Natl. Acad. Sci. U.S.A.* **80**:1018–1022.
355. Kalanon, M. and G. McFadden. 2010. Malaria, Plasmodium falciparum and its apicoplast. *Biochem. Soc. Trans.* **38**:775–782.
356. Kan, S. C., K. M. Yamaga, K. J. Kramer, et al. 1984. Plasmodium falciparum: protein antigens identified by analysis of serum samples from vaccinated Aotus monkeys. *Infect. Immun.* **43**:276–282.
357. Kaneko, O. 2007. Erythrocyte invasion: vocabulary and grammar of the Plasmodium rhoptry. *Parasitol. Int.* **56**:255–262.

358. Kaneko, O., B. Y. Yim Lim, H. Iriko, et al. 2005. Apical expression of three RhopH1/Clag proteins as components of the Plasmodium falciparum RhopH complex. *Mol. Biochem. Parasitol.* **143**:20–28.
359. Kang, M., G. Lisk, S. Hollingworth, et al. 2005. Malaria parasites are rapidly killed by dantrolene derivatives specific for the plasmodial surface anion channel. *Mol. Pharmacol.* **68**:34–40.
360. Kapuscinsk, R. 2002. *Shadow of the Sun.* Vantage, New York.
361. Kaslow, D. C. 2002. Transmission blocking vaccines, pp. 287–307. In P. Perlmann and M. Troye-Blomberg (eds.), *Chemical Immunology*, vol. 80. ASM Press, Washington, DC.
362. Kean, B., K. E. Mott, and A. J. Russell. 1978. *Tropical Medicine and Parasitology: Classic Investigations*, pp. 23–34, vol. 1. Cornell University Press, Ithaca.
363. Keeley, A. and D. Soldati. 2004. The glideosome: a molecular machine powering motility and host-cell invasion by Apicomplexa. *Trends Cell Biol.* **14**:528–532.
364. Kemp, D. J., R. L. Coppel, A. F. Cowman, et al. 1983. Expression of Plasmodium falciparum blood-stage antigens in Escherichia coli: detection with antibodies from immune humans. *Proc. Natl. Acad. Sci. U.S.A.* **80**:3787–3791.
365. Kester, K. E., J. F. Cummings, C. F. Ockenhouse, et al. 2008. Phase 2a trial of 0, 1, and 3 month and 0, 7, and 28 day immunization schedules of malaria vaccine RTS,S/AS02 in malaria-naive adults at the Walter Reed Army Institute of Research. *Vaccine* **26**:2191–2202.
366. Kilejian, A. 1978. Histidine-rich protein as a model malaria vaccine. *Science* **201**:922–924.
367. Kilejian, A., S. Chen, and A. Sloma. 1985. The biosynthesis of the histidine-rich protein of Plasmodium lophurae and the cloning of its gene in Escherichia coli. *Mol. Biochem. Parasitol.* **14**:1–10.
368. King, C. L., P. Michon, A. R. Shakri, et al. 2008. Naturally acquired Duffy-binding protein-specific binding inhibitory antibodies confer protection from blood-stage Plasmodium vivax infection. *Proc. Natl. Acad. Sci. U.S.A.* **105**:8363–8368.
369. Kirk, K. 2001. Membrane transport in the malaria-infected erythrocyte. *Physiol. Rev.* **81**:495–537.
370. Kirk, K. 2004. Channels and transporters as drug targets in the Plasmodium-infected erythrocyte. *Acta Trop.* **89**:285–298.
371. Kirk, K., H. A. Horner, B. C. Elford, et al. 1994. Transport of diverse substrates into malaria-infected erythrocytes *via* a pathway showing functional characteristics of a chloride channel. *J. Biol. Chem.* **269**:3339–3347.

372. Kirk, K., R. E. Martin, S. Broer, et al. 2005. Plasmodium permeomics: membrane transport proteins in the malaria parasite. *Curr. Top. Microbiol. Immunol.* **295**:325–356.
373. Kirk, K. and K. J. Saliba. 2007. Targeting nutrient uptake mechanisms in Plasmodium. *Curr. Drug Targ.* **8**:75–88.
374. Klayman, D. 1985. Qinghaosu (artemisinin): an antimalarial drug from China. *Science* **228**:1049–1055.
375. Klayman, D. 1989. Weeding out malaria. *Nat. Hist.* **10**:18–24.
376. Knuepfer, E., M. Rug, and A. F. Cowman. 2005. Function of the plasmodium export element can be blocked by green fluorescent protein. *Mol. Biochem. Parasitol.* **142**:258–262.
377. Kokoza, V., A. Ahmed, S. Woon Shin, et al. 2010. Blocking of Plasmodium transmission by cooperative action of Cecropin A and Defensin A in transgenic Aedes aegypti mosquitoes. *Proc. Natl. Acad. Sci. U.S.A.* **107**:8111–8116.
378. Koussis, K., C. Withers-Martinez, S. Yeoh, et al. 2009. A multifunctional serine protease primes the malaria parasite for red blood cell invasion. *EMBO J.* **28**:725–735.
379. Kremer, M. and R. Glennerster. 2004. Strong Medicine. *Creating Incentives for Phamaceutical Research on Neglected Diseases*. Princeton University Press, Princeton, NJ.
380. Krief, S., A. A. Escalante, M. A. Pacheco, et al. 2010. On the diversity of malaria parasites in African apes and the origin of Plasmodium falciparum from Bonobos. *PLoS Pathog.* **6**:e1000765.
381. Kriek, N., L. Tilley, P. Horrocks, et al. 2003. Characterization of the pathway for transport of the cytoadherence-mediating protein, PfEMP1, to the host cell surface in malaria parasite-infected erythrocytes. *Mol. Microbiol.* **50**:1215–1227.
382. Krotoski, W. A., D. M. Krotoski, P. C. Garnham, et al. 1980. Relapses in primate malaria: discovery of two populations of exoerythrocytic stages. Preliminary note. *BMJ* **280**:153–154.
383. Krugliak, M. and H. Ginsburg. 2006. The evolution of the new permeability pathways in Plasmodium falciparum — infected erythrocytes — a kinetic analysis. *Exp. Parasitol.* **114**:253–258.
384. Krungkrai, J. 2004. The multiple roles of the mitochondrion of the malarial parasite. *Parasitology* **129**:511–524.
385. Kuehn, A. and G. Pradel. 2010. The coming-out of malaria gametocytes. *J. Biomed. Biotechnol.* **2010**:ID number 976827.
386. Kumar, K. A., G. Sano, S. Boscardin, et al. 2006. The circumsporozoite protein is an immunodominant protective antigen in irradiated sporozoites. *Nature* **444**:937–940.

387. Kumar, N. and R. Carter. 1985. Biosynthesis of two stage-specific membrane proteins during transformation of Plasmodium gallinaceum zygotes into ookinetes. *Mol. Biochem. Parasitol.* **14**:127–139.
388. Kun, J. F. and E. G. de Carvalho. 2009. Novel therapeutic targets in Plasmodium falciparum: aquaglyceroporins. *Exp. Opin. Therapeut. Targ.* **13**:385–394.
389. Kutner, S., D. Baruch, H. Ginsburg, and Z. I. Cabantchik. 1982. Alterations in membrane permeability of malaria-infected human erythrocytes are related to the growth stage of the parasite. *Biochim. Biophys. Acta* **687**:113–117.
390. Kutner, S., H. Ginsburg, and Z. I. Cabantchik. 1983. Permselectivity changes in malaria (Plasmodium falciparum) infected human red blood cell membranes. *J. Cell Physiol.* **114**:245–251.
391. Kyes, S. A., S. M. Kraemer, and J. D. Smith. 2007. Antigenic variation in Plasmodium falciparum: gene organization and regulation of the var multigene family. *Eukaryot. Cell* **6**:1511–1520.
392. Lagerkvist, U. 1998. *DNA Pioneers and their Legacy.* Yale University Press, New Haven.
393. Lal, A. A., V. F. de la Cruz, G. H. Campbell, *et al.* 1988. Structure of the circumsporozoite gene of Plasmodium malariae. *Mol. Biochem. Parasitol.* **30**:291–294.
394. Lang-Unnasch, N. 1992. Purification and properties of Plasmodium falciparum malate dehydrogenase. *Mol. Biochem. Parasitol.* **50**:17–25.
395. Lang-Unnasch, N. 1995. Plasmodium falciparum: antiserum to malate dehydrogenase. *Exp. Parasitol.* **80**:357–359.
396. Langreth, S. G., J. B. Jensen, R. T. Reese, and W. Trager. 1978. Fine structure of human malaria *in vitro*. *J. Protozool.* **25**:443–452.
397. Langreth, S. G., and R. T. Reese. 1979. Antigenicity of the infected-erythrocyte and merozoite surfaces in Falciparum malaria. *J. Exp. Med.* **150**:1241–1254.
398. Lanzer, M., H. Wickert, G. Krohne, *et al.* 2006. Maurer's clefts: a novel multifunctional organelle in the cytoplasm of Plasmodium falciparum-infected erythrocytes. *Int. J. Parasitol.* **36**:23–36.
399. Lauer, S. A., P. K. Rathod, N. Ghori, and K. Haldar. 1997. A membrane network for nutrient import in red cells infected with the malaria parasite. *Science* **276**:1122–1125.
400. Le Roch, K. G., J. R. Johnson, H. Ahiboh, *et al.* 2008. A systematic approach to understand the mechanism of action of the bisthiazolium compound T4 on the human malaria parasite, Plasmodium falciparum. *BMC Genom.* **9**:513.

401. Le Roch, K. G., Y. Zhou, P. L. Blair, *et al.* 2003. Discovery of gene function by expression profiling of the malaria parasite life cycle. *Science* **301**:1503–1508.
402. Leclerc, M. C., P. Durand, C. Gauthier, *et al.* 2004. Meager genetic variability of the human malaria agent Plasmodium vivax. *Proc. Natl. Acad. Sci. U.S.A.* **101**:14455–14460.
403. Lee, P., Z. Ye, K. Van Dyke, and R. G. Kirk. 1988. X-ray microanalysis of Plasmodium falciparum and infected red blood cells: effects of qinghaosu and chloroquine on potassium, sodium, and phosphorus composition. *Am. J. Trop. Med. Hyg.* **39**:157–165.
404. Leech, J. H., J. W. Barnwell, L. H. Miller, and R. J. Howard. 1984. Identification of a strain-specific malarial antigen exposed on the surface of Plasmodium falciparum-infected erythrocytes. *J. Exp. Med.* **159**:1567–1575.
405. Lehane, A. M., K. J. Saliba, R. J. Allen, and K. Kirk. 2004. Choline uptake into the malaria parasite is energized by the membrane potential. *Biochem. Biophys. Res. Commun.* **320**:311–317.
406. Li, J., W. E. Collins, R. A. Wirtz, *et al.* 2001. Geographic subdivision of the range of the malaria parasite Plasmodium vivax. *Emerg. Infect. Dis.* **7**:35–42.
407. Li, Q., W. K. Milhous, and P. J. Weina. 2007. *Artemisinin in Malaria Therapy.* Nova Biomedical Books, New York.
408. Li, S., E. Locke, J. Bruder, *et al.* 2007. Viral vectors for malaria vaccine development. *Vaccine* **25**:2567–2574.
409. Li, Y. and W.-L. Wu. 2003. An over four millenium story behind qinghaosu (artemisinin) — a fantastic antimalarial drug from a traditional Chinese herb. *Curr. Med. Chem.* **10**:2197–2230.
410. Lim, L. and G. I. McFadden. 2010. The evolution, metabolism and functions of the apicoplast. *Phil. Trans. R. Soc. Biol. Sci.* **365**:749–763.
411. Lim, L. and G. I. McFadden. 2010. The evolution, metabolism and functions of the apicoplast. *Phil. Trans. R. Soc. Biol. Sci.* **365**:749–763.
412. Lisk, G., M. Kang, J. V. Cohn, and S. A. Desai. 2006. Specific inhibition of the plasmodial surface anion channel by dantrolene. *Eukaryot. Cell* **5**:1882–1893.
413. Lobo, C. A., M. Rodriguez, M. Reid, and S. Lustigman. 2003. Glycophorin C is the receptor for the Plasmodium falciparum erythrocyte binding ligand PfEBP-2 (baebl). *Blood* **101**:4628–4631.
414. Looareesuwan, S., C. Viravan, H. K. Webster, *et al.* 1996. Clinical studies of atovaquone, alone or in combination with other antimalarial drugs, for treatment of acute uncomplicated malaria in Thailand. *Am. J. Trop. Med. Hyg.* **54**:62–66.

415. Lopez-Rubio, J. J., L. Mancio-Silva, and A. Scherf. 2009. Genome-wide analysis of heterochromatin associates clonally variant gene regulation with perinuclear repressive centers in malaria parasites. *Cell Host Microbe* **5**:179–190.
416. Luke, T. C. and S. L. Hoffman. 2003. Rationale and plans for developing a non-replicating, metabolically active, radiation-attenuated Plasmodium falciparum sporozoite vaccine. *J. Exp. Biol.* **206**:3803–3808.
417. Lyon, J. A., J. D. Haynes, C. L. Diggs, et al. 1986. Plasmodium falciparum antigens synthesized by schizonts and stabilized at the merozoite surface by antibodies when schizonts mature in the presence of growth inhibitory immune serum. *J. Immunol.* **136**:2252–2258.
418. MacCallum, W. G. 1898. On the hematozoan infections of birds. *J. Exp. Med.* **3**:117–135.
419. Maguire, P. A. and I. W. Sherman. 1990. Phospholipid composition, cholesterol content and cholesterol exchange in Plasmodium falciparum-infected red cells. *Mol. Biochem. Parasitol.* **38**:105–112.
420. Maier, A. G., M. T. Duraisingh, J. C. Reeder, et al. 2003. Plasmodium falciparum erythrocyte invasion through glycophorin C and selection for Gerbich negativity in human populations. *Nat. Med.* **9**:87–92.
421. Maier, A. G., M. Rug, M. T. O'Neill, et al. 2008. Exported proteins required for virulence and rigidity of Plasmodium falciparum-infected human erythrocytes. *Cell* **134**:48–61.
422. Makler, M. T., R. C. Piper, and W. K. Milhous. 1998. Lactate dehydrogenase and the diagnosis of malaria. *Parasitol. Today* **14**:376–377.
423. Malkin, E. M., D. J. Diemert, J. H. McArthur, et al. 2005. Phase 1 clinical trial of apical membrane antigen 1: an asexual blood-stage vaccine for Plasmodium falciparum malaria. *Infect. Immun.* **73**:3677–3685.
424. Manson-Bahr, P. 1962. *Patrick Manson*. Thomas Nelson and Sons, London.
425. Manson-Bahr, P. 1963. The story of malaria: the drama and actors. *Int. Rev. Trop. Med.* **2**:329–390.
426. Markert, C. L. and F. Møller. 1959. Multiple forms of enzymes: tissue, ontogenetic, and species specific patterns. *Proc. Natl. Acad. Sci. U.S.A.* **45**:753–763.
427. Marshall, E. 1989. Malaria reseracher indicted. *Science* **245**:1326.
428. Marshall, E. 1996. Serious setback for Patarroyo. *Science* **273**:1652.
429. Marshall, J. M. and C. E. Taylor. 2009. Malaria control with transgenic mosquitoes. *PLoS Med.* **6**:e20.
430. Marti, M., R. T. Good, M. Rug, et al. 2004. Targeting malaria virulence and remodeling proteins to the host erythrocyte. *Science* **306**:1930–1933.

431. Martin, M. J., J. C. Rayner, P. Gagneux, et al. 2005. Evolution of human-chimpanzee differences in malaria susceptibility: relationship to human genetic loss of N-glycolylneuraminic acid. *Proc. Natl. Acad. Sci. U.S.A.* **102**:12819–12824.
432. Martin, R. E., R. I. Henry, J. L. Abbey, et al. 2005. The 'permeome' of the malaria parasite: an overview of the membrane transport proteins of Plasmodium falciparum. *Genome Biol.* **6**:R26.
433. Martin, R. E. and K. Kirk. 2007. Transport of the essential nutrient isoleucine in human erythrocytes infected with the malaria parasite Plasmodium falciparum. *Blood* **109**:2217–2224.
434. Mather, M. W., K. W. Henry, and A. B. Vaidya. 2007. Mitochondrial drug targets in apicomplexan parasites. *Curr. Drug Targ.* **8**:49–60.
435. Matuschewski, K., J. C. Hafalla, S. Borrmann, and J. Friesen. 2010. Arrested Plasmodium liver stages as experimental anti-malaria vaccines. *Hum. Vaccines* **7 (Suppl)**:16–21.
436. Matuschewski, K., J. Ross, S. M. Brown, et al. 2002. Infectivity-associated changes in the transcriptional repertoire of the malaria parasite sporozoite stage. *J. Biol. Chem.* **277**:41948–41953.
437. Maurice, J. 1993. Controversial vaccine shows promise. *Science* **259**: 1689–1690.
438. Mayer, D. C., L. Jiang, R. N. Achur, et al. 2006. The glycophorin C N-linked glycan is a critical component of the ligand for the Plasmodium falciparum erythrocyte receptor BAEBL. *Proc. Natl. Acad. Sci. U.S.A.* **103**:2358–2362.
439. McAllister, B. 1989. Ex-Malaria Research Aide Indicted. *Washington Post*, 18 October.
440. McCutchan, T. F., A. A. Lal, V. F. de la Cruz, et al. 1985. Sequence of the immunodominant epitope for the surface protein on sporozoites of Plasmodium vivax. *Science* **230**:1381–1383.
441. McGhee, R. B. 1988. Major animal models in malaria research: avian, pp. 1545–1568. In W. Wernsdorfer and I. McGregor (eds.), *Malaria: Principles and Practices of Malariology*, vol. 2. Churchill Livingstone, Edinburgh.
442. McGregor, I. A., H. M. Gilles, J. H. Walters, et al. 1956. Effects of heavy and repeated malarial infections on Gambian infants and children; effects of erythrocytic parasitization. *BMJ* **2**:686–692.
443. McHardy, N. 1978. *In vitro* studies on the action of menoctone and other compounds on Theileria parva and T. annulata. *Ann. Trop. Med. Parasitol.* **72**:501–511.
444. McHardy, N. and D. W. Morgan. 1985. Treatment of Theileria annulata infection in calves with parvaquone. *Res. Vet. Sci.* **39**:1–4.

445. McKee, R. W. 1951. Biochemistry of Plasmodium and influence of antimalarials, pp. 251–322. In A. Lwoff (ed.), *Biochemistry and Physiology of Protozoa*, vol. 1. Academic Press, San Diego.
446. McNeil, J. G. 2007. The Soul of a New Vaccine. *New York Times*, 11 December.
447. McRobert, L., P. Preiser, S. Sharp, *et al.* 2004. Distinct trafficking and localization of STEVOR proteins in three stages of the Plasmodium falciparum life cycle. *Infect. Immun.* **72**:6597–6602.
448. Meister, S., A. C. Koutsos, and G. K. Christophides. 2004. The Plasmodium parasite — a 'new' challenge for insect innate immunity. *Int. J. Parasitol.* **34**:1473–1482.
449. Merckx, A., G. Bouyer, S. L. Thomas, *et al.* 2009. Anion channels in Plasmodium-falciparum-infected erythrocytes and protein kinase A. *Trends Parasitol.* **25**:139–144.
450. Meshnick, S. 1998. From quinine to qinghaosu: historical perspectives, pp. 341–354. In I. W. Sherman (ed.), *Malaria: Parasite Biology, Pathogenesis and Protection*. ASM Press, Washington, DC.
451. Meshnick, S. 2001. Artemisinin and its derivatives, pp. 191–201. In P. Rosenthal (ed.), *Antimalarial Chemotherapy*. Humana Press, Totowa, NJ.
452. Meshnick, S. 2002. Artemisinin: mechanism of action, resistance and toxicity. *Int. J. Parasitol.* **32**:1655–1660.
453. Michel, K., C. Suwanchaichinda, I. Morlais, *et al.* 2006. Increased melanizing activity in Anopheles gambiae does not affect development of Plasmodium falciparum. *Proc. Natl. Acad. Sci. U.S.A.* **103**:16858–16863.
454. Miller, L. H. 1969. Distribution of mature trophozoites and schizonts of Plasmodium falciparum in the organs of Aotus trivirgatus, the night monkey. *Am. J. Trop. Med. Hyg.* **18**:860–865.
455. Miller, L. H. 1977. Hypothesis on the mechanism of erythrocyte invasion by malaria merozoites. *Bull. World Health Organ.* **55**:157–162.
456. Miller, L. H., M. Aikawa, J. G. Johnson, and T. Shiroishi. 1979. Interaction between cytochalasin B-treated malarial parasites and erythrocytes. Attachment and junction formation. *J. Exp. Med.* **149**:172–184.
457. Miller, L. H., J. D. Haynes, F. M. McAuliffe, *et al.* 1977. Evidence for differences in erythrocyte surface receptors for the malarial parasites, Plasmodium falciparum and Plasmodium knowlesi. *J. Exp. Med.* **146**:277–281.
458. Miller, L. H., D. Hudson, and J. D. Haynes. 1988. Identification of Plasmodium knowlesi erythrocyte binding proteins. *Mol. Biochem. Parasitol.* **31**:217–222.
459. Miller, L. H., S. J. Mason, D. F. Clyde, and M. H. McGinniss. 1976. The resistance factor to Plasmodium vivax in blacks. The Duffy-blood-group genotype, FyFy. *New Engl. J. Med.* **295**:302–304.

460. Miller, L. H., S. J. Mason, J. A. Dvorak, *et al.* 1975. Erythrocyte receptors for (Plasmodium knowlesi) malaria: Duffy blood group determinants. *Science* **189**:561–563.
461. Mitchell, G. H., G. A. Butcher, and S. Cohen. 1975. Merozoite vaccination against Plasmodium knowlesi malaria. *Immunology* **29**:397–407.
462. Mitchell, G. H., T. J. Hadley, M. H. McGinniss, *et al.* 1986. Invasion of erythrocytes by Plasmodium falciparum malaria parasites: evidence for receptor heterogeneity and two receptors. *Blood* **67**:1519–1521.
463. Mitchell, G. H., W. H. Richards, A. Voller, *et al.* 1979. Nor-MDP, saponin, corynebacteria, and pertussis organisms as immunological adjuvants in experimental malaria vaccination of macaques. *Bull. World Health Organ.* **57 Suppl 1**:189–197.
464. Mitchell, G. H., A. W. Thomas, G. Margos, *et al.* 2004. Apical membrane antigen 1, a major malaria vaccine candidate, mediates the close attachment of invasive merozoites to host red blood cells. *Infect. Immun.* **72**:154–158.
465. Morgan, G. J. 1998. Emile Zuckerkandl, Linus Pauling, and the molecular evolutionary clock, 1959–1965. *J. Hist. Biol.* **31**:155–178.
466. Moulder, J. 1948. The metabolism of malarial parasites. *Ann. Rev. Microbiol.* **2**:101–120.
467. Mu, J., J. Duan, K. D. Makova, *et al.* 2002. Chromosome-wide SNPs reveal an ancient origin for Plasmodium falciparum. *Nature* **418**:323–326.
468. Mu, J., D. A. Joy, J. Duan, *et al.* 2005. Host switch leads to emergence of Plasmodium vivax malaria in humans. *Mol. Biol. Evol.* **22**:1686–1693.
469. Mueller, I., M. R. Galinski, J. K. Baird, *et al.* 2009. Key gaps in the knowledge of Plasmodium vivax, a neglected human malaria parasite. *Lancet Infect. Dis.* **9**:555–566.
470. Mulligan, H. W., P. F. Russell, and B. N. Mohan. 1941. Active immunization of fowls against Plasmodium gallinaceum by injections of killed homologous sporozoites. *J. Mal. Inst. India* **4**:25–34.
471. Murphy, S. C., S. Fernandez-Pol, P. H. Chung, *et al.* 2007. Cytoplasmic remodeling of erythrocyte raft lipids during infection by the human malaria parasite Plasmodium falciparum. *Blood* **110**:2132–2139.
472. Nagao, E., K. B. Seydel, and J. A. Dvorak. 2002. Detergent-resistant erythrocyte membrane rafts are modified by a Plasmodium falciparum infection. *Exp. Parasitol.* **102**:57–59.
473. Nardin, E., R. Nussenzweig, I. McGregor, and J. Bryan. 1979. Antisporozoite antibodies. Their frequent occurrence in individuals living in an area of hyperendemic malaria. *Science* **206**:597–599.

474. Neame, K. D. and C. A. Homewood. 1975. Alterations in the permeability of mouse erythrocytes infected with the malaria parasite, Plasmodium berghei. *Int. J. Parasitol.* **5**:537–540.
475. Neibauer, M. 2010. Remembering the Blizzard of '96. *The Examiner*, 7 February.
476. Newbold, C. 2002. The history of the malaria genome project. How the international consortium came together. Available at: http://malaria.wellcome.ac.uk/print/wtd023852_print.html [Accessed 23 September 2011].
477. Nguitragool, W., A. A. Bokhari, A. D. Pillai, et al. 2011. Malaria parasite clag3 genes determine channel-mediated nutrient uptake by infected red blood cells. *Cell* **145**:665–677.
478. Niang, M., X. Y. Yam, and P. Preisier. 2009. The Plasmodium falciparum STEVOR multigene family mediates antigenic variation of the infected erythrocyte. *PLos Pathog.* **5**:e1000307.
479. Nussenzweig, R. S., J. Vanderberg, H. Most, and C. Orton. 1967. Protective immunity produced by the injection of x-irradiated sporozoites of Plasmodium berghei. *Nature* **216**:160–162.
480. Nussenzweig, R. S., J. Vanderberg, G. L. Spitalny, et al. 1972. Sporozoite-induced immunity in mammalian malaria. A review. *Am. J. Trop. Med. Hyg.* **21**:722–728.
481. O'Donnell, R. A. and M. J. Blackman. 2005. The role of malaria merozoite proteases in red blood cell invasion. *Curr. Opin. Microbiol.* **8**:422–427.
482. O'Donnell, R. A., F. Hackett, S. A. Howell, et al. 2006. Intramembrane proteolysis mediates shedding of a key adhesin during erythrocyte invasion by the malaria parasite. *J. Cell Biol.* **174**:1023–1033.
483. O'Donnell, R. A., A. Saul, A. F. Cowman, and B. S. Crabb. 2000. Functional conservation of the malaria vaccine antigen MSP-119across distantly related Plasmodium species. *Nat. Med.* **6**:91–95.
484. O'Neill, P. M., S. A. Ward, N. G. Berry, et al. 2006. A medicinal chemistry perspective on 4-aminoquinoline antimalarial drugs. *Curr. Top. Med. Chem.* **6**:479–507.
485. Ockenhouse, C. F., P. F. Sun, D. E. Lanar, et al. 1998. Phase I/IIa safety, immunogenicity, and efficacy trial of NYVAC-Pf7, a pox-vectored, multiantigen, multistage vaccine candidate for Plasmodium falciparum malaria. *J. Infect. Dis.* **177**:1664–1673.
486. Oh, S. S. and A. H. Chishti. 2005. Host receptors in malaria merozoite invasion. *Curr. Top. Microbiol. Immunol.* **295**:203–232.
487. Okie, S. 1989. 2 Indicted in Theft of $130, 000 from AID Grant. *Washington Post*, 17 September.

488. Ollliaro, P. and W. Taylor. 2004. Developing artemisinin-based drug combinations for the treatment of drug resistant falciparum malaria: a review. *J. Postgrad. Med.* **50**:40–44.
489. Olszewski, K. L., M. W. Mather, J. M. Morrisey, *et al.* 2010. Branched tricarboxylic acid metabolism in Plasmodium falciparum. *Nature* **466**:774–778.
490. Olumse, P. 2006. Guidelines for the Treatment of Malaria. World Health Organization.
491. Oransky, I. 2003. David Francis Clyde. *Lancet* **361**:439.
492. Orjih, A. U., A. H. Cochrane, and R. S. Nussenzweig. 1982. Comparative studies on the immunogenicity of infective and attenuated sporozoites of Plasmodium berghei. *Trans. R. Soc. Trop. Med. Hyg.* **76**:57–61.
493. Osta, M. A., G. K. Christophides, D. Vlachou, and F. C. Kafatos. 2004. Innate immunity in the malaria vector Anopheles gambiae: comparative and functional genomics. *J. Exp. Biol.* **207**:2551–2563.
494. Outchkourov, N. S., W. Roeffen, A. Kaan, *et al.* 2008. Correctly folded Pfs48/45 protein of Plasmodium falciparum elicits malaria transmission-blocking immunity in mice. *Proc. Natl. Acad. Sci. U.S.A.* **105**:4301–4305.
495. Overman, R. R. 1948. Reversible cellular permeability alterations in disease. *In vivo* studies on sodium, potassium and chloride concentrations in erythrocytes of malarious monkeys. *Am. J. Physiol.* **152**:113–121.
496. Overman, R. R., A. C. Bass, and T. H. Tomlinson. 1950. Ionic alterations in chickens infected with plasmodium gallinaceum. *Fed. Proc. Fedn Am. Soc. Exp. Biol.* **9**:96–97.
497. Overman, R. R., T. S. Hill, and U. T. Wong. 1949. Physiological studies in the human malaria host. I. Blood plasma, "extracellular" fluid volumes and ionic balance in therapeutic P. vivax and P. falciparum. *J. Nat. Mal. Soc.* **8**:14–31.
498. Overstreet, M. G., I. A. Cockburn, and F. Zavala. 2008. Protective CD8+ T cells against Plasmodium liver stages: immunobiology of an 'unnatural' immune response. *Immunol. Rev.* **225**:272–283.
499. Ozaki, L. S., P. Svec, R. S. Nussenzweig, *et al.* 1983. Structure of the Plasmodium knowlesi gene coding for the circumsporozoite protein. *Cell* **34**:815–822.
500. Painter, H. J., J. M. Morrisey, M. W. Mather, and A. B. Vaidya. 2007. Specific role of mitochondrial electron transport in blood-stage Plasmodium falciparum. *Nature* **446**:88–91.
501. Palmer, K. J., S. M. Holliday, and R. W. Brogden. 1973. Mefloquine. A review of its antimalarial activity, pharmacokinetic properties and therapeutic efficacy. *Drugs* **45**:470–475.

502. Palmer, K. L., G. S. Hui, W. A. Siddiqui, and E. L. Palmer. 1982. A large-scale in vitro production system for Plasmodium falciparum. *J. Parasitol.* **68**:1180–1183.
503. Pandey, K. C., S. Singh, P. Pattnaik, et al. 2002. Bacterially expressed and refolded receptor binding domain of Plasmodium falciparum EBA-175 elicits invasion inhibitory antibodies. *Mol. Biochem. Parasitol.* **123**:23–33.
504. Parker, M. D., R. J. Hyde, S. Y. Yao, et al. 2000. Identification of a nucleoside/nucleobase transporter from Plasmodium falciparum, a novel target for anti-malarial chemotherapy. *Biochem. J.* **349**:67–75.
505. Pasvol, G. 1984. Receptors on red cells for Plasmodium falciparum and their interaction with merozoites. *Philos. Trans. R. Soc. Lond.* **307**:189–200.
506. Pasvol, G., J. A. Chasis, N. Mohandas, et al. 1989. Inhibition of malarial parasite invasion by monoclonal antibodies against glycophorin A correlates with reduction in red cell membrane deformability. *Blood* **74**:1836–1843.
507. Pasvol, G., J. S. Wainscoat, and D. J. Weatherall. 1982. Erythrocytes deficiency in glycophorin resist invasion by the malarial parasite Plasmodium falciparum. *Nature* **297**:64–66.
508. Pasvol, G. and R. J. Wilson. 1982. The interaction of malaria parasites with red blood cells. *Br. Med. Bull.* **38**:133–140.
509. Patarroyo, M. E., R. Amador, P. Clavijo, et al. 1988. A synthetic vaccine protects humans against challenge with asexual blood stages of Plasmodium falciparum malaria. *Nature* **332**:158–161.
510. Patarroyo, M. E., P. Romero, M. L. Torres, et al. 1987. Induction of protective immunity against experimental infection with malaria using synthetic peptides. *Nature* **328**:629–632.
511. Pavlovic-Djuranovic, S., J. F. Kun, J. E. Schultz, and E. Beitz. 2006. Dihydroxyacetone and methylglyoxal as permeants of the Plasmodium aquaglyceroporin inhibit parasite proliferation. *Biochim. Biophys. Acta* **1758**:1012–1017.
512. Pennisi, E. 2002. Parasite genome sequenced, scrutinized. *Science* **298**:33–34.
513. Perkins, M. E. 1984. Surface proteins of Plasmodium falciparum merozoites binding to the erythrocyte receptor, glycophorin. *J. Exp. Med.* **160**:788–798.
514. Perlaza, B. L., C. Zapata, A. Z. Valencia, et al. 2003. Immunogenicity and protective efficacy of Plasmodium falciparum liver-stage Ag-3 in Aotus lemurinus griseimembra monkeys. *Eur. J. Immunol.* **33**:1321–1327.
515. Perry, A. 2010. Battling a Scourge. *Time*, 10 June.
516. Peters, W. 1987. *Chemotherapy and drug resistance in malaria*, vol. 1. Academic Press, London.

517. Peters, W. 1987. *Chemotherapy and drug resistance in malaria*, vol. 2. Academic Press, London.
518. Peterson, M. G., V. M. Marshall, J. A. Smythe, *et al.* 1989. Integral membrane protein located in the apical complex of Plasmodium falciparum. *Mol. Cell Biol.* **9**:3151–3154.
519. Pinto, S. B., F. Lombardo, A. C. Koutsos, *et al.* 2009. Discovery of Plasmodium modulators by genome-wide analysis of circulating hemocytes in Anopheles gambiae. *Proc. Natl. Acad. Sci. U.S.A.* **106**:21270–21275.
520. Pizarro, J. C., B. Vulliez-Le Normand, M. L. Chesne-Seck, *et al.* 2005. Crystal structure of the malaria vaccine candidate apical membrane antigen 1. *Science* **308**:408–411.
521. Plouffe, D., A. Brinker, C. McNamara, *et al.* 2008. In silico activity profiling reveals the mechanism of action of antimalarials discovered in a high-throughput screen. *Proc. Natl. Acad. Sci. U.S.A.* **105**:9059–9064.
522. Ponts, N., E. Y. Harris, J. Prudhomme, *et al.* 2010. Nucleosome landscape and control of transcription in the human malaria parasite. *Genome Res.* **20**:228–238.
523. Ponzi, M., I. Siden-Kiamos, L. Bertuccini, *et al.* 2009. Egress of Plasmodium berghei gametes from their host erythrocyte is mediated by the MDV-1/PEG3 protein. *Cell. Microbiol.* **11**:1272–1288.
524. Portugal, F. H. and J. S. Cohen. 1977. *A Century of DNA. A History of the Discovery of the Structure and Function of the Genetic Substance*. MIT Press, Cambridge, MA.
525. Potocnjak, P., N. Yoshida, R. S. Nussenzweig, and V. Nussenzweig. 1980. Monovalent fragments (Fab) of monoclonal antibodies to a sporozoite surface antigen (Pb44) protect mice against malarial infection. *J. Exp. Med.* **151**:1504–1513.
526. Povelones, M., R. M. Waterhouse, F. C. Kafatos, and G. K. Christophides. 2009. Leucine-rich repeat protein complex activates mosquito complement in defense against Plasmodium parasites. *Science* **324**:258–261.
527. Pradel, G. 2007. Proteins of the malaria parasite sexual stages: expression, function and potential for transmission blocking strategies. *Parasitology* **134**:1911–1929.
528. Preechapornkul, P., M. Imwong, K. Chotivanich, *et al.* 2009. Plasmodium falciparum pfmdr1 amplification, mefloquine resistance, and parasite fitness. *Antimicrob. Agents Chemother.* **53**:1509–1515.
529. Preiser, P., M. Kaviratne, S. Khan, *et al.* 2000. The apical organelles of malaria merozoites: host cell selection, invasion, host immunity and immune evasion. *Microbes Infect./Inst. Past.* **2**:1461–1477.

530. Promeneur, D., Y. Liu, J. Maciel, et al. 2007. Aquaglyceroporin PbAQP during intraerythrocytic development of the malaria parasite Plasmodium berghei. *Proc. Natl. Acad. Sci. U.S.A.* **104**:2211–2216.
531. Prudhomme, J. and I. W. Sherman. 1999. A high capacity *in vitro* assay for measuring the cytoadherence of Plasmodium falciparum infected erythrocytes. *J. Immunol. Meth.* **229**:169–176.
532. Przyborski, J. M., S. K. Miller, J. M. Pfahler, et al. 2005. Trafficking of STEVOR to the Maurer's clefts in Plasmodium falciparum-infected erythrocytes. *EMBO J.* **24**:2306–2317.
533. Pullman, T. N., L. Eichelberger, A. S. Alving, et al. 1948. The use of SN-10,275 in the prophylaxis and treatment of sporozoite-induced vivax malaria (Chesson strain). *J. Clin. Invest.* **27**:12–16.
534. Putaporntip, C., T. Hongsrimuang, S. Seethamchai, et al. 2009. Differential prevalence of Plasmodium infections and cryptic Plasmodium knowlesi malaria in humans in Thailand. *J. Infect. Dis.* **199**:1143–1150.
535. Quashie, N. B., D. Dorin-Semblat, P. G. Bray, et al. 2008. A comprehensive model of purine uptake by the malaria parasite Plasmodium falciparum: identification of four purine transport activities in intraerythrocytic parasites. *J. Biochem.* **411**:287–295.
536. Rager, N., C. B. Mamoun, N. S. Carter, et al. 2001. Localization of the Plasmodium falciparum PfNT1 nucleoside transporter to the parasite plasma membrane. *J. Biol. Chem.* **276**:41095–41099.
537. Rahman, A., M. I. Choudry, and W. J. Thomsen. 2001. *Bioassay Techniques for Drug Development.* Harward Academic Publishers, The Netherlands.
538. Ranawaka, M. B., Y. D. Munesinghe, D. M. de Silva, et al. 1988. Boosting of transmission-blocking immunity during natural Plasmodium vivax infections in humans depends upon frequent reinfection. *Infect. Immun.* **56**: 1820–1824.
539. Ranson, H., L. Rossiter, F. Ortelli, et al. 2001. Identification of a novel class of insect glutathione S-transferases involved in resistance to DDT in the malaria vector Anopheles gambiae. *Biochem. J.* **359**:295–304.
540. Ravetch, J. V., R. Feder, A. Pavlovec, and G. Blobel. 1984. Primary structure and genomic organization of the histidine-rich protein of the malaria parasite Plasmodium lophurae. *Nature* **312**:616–620.
541. Rayner, J. C., E. Vargas-Serrato, C. S. Huber, et al. 2001. A Plasmodium falciparum homologue of Plasmodium vivax reticulocyte binding protein (PvRBP1) defines a trypsin-resistant erythrocyte invasion pathway. *J. Exp. Med.* **194**:1571–1581.

542. Reed, M. B., S. R. Caruana, A. H. Batchelor, *et al.* 2000. Targeted disruption of an erythrocyte binding antigen in Plasmodium falciparum is associated with a switch toward a sialic acid-independent pathway of invasion. *Proc. Natl. Acad. Sci. U.S.A.* **97**:7509–7514.
543. Reese, R. T., W. Trager, J. B. Jensen, *et al.* 1978. Immunization against malaria with antigen from Plasmodium falciparum cultivated *in vitro*. *Proc. Natl. Acad. Sci. U.S.A.* **75**:5665–5668.
544. Remarque, E., B. Faber, C. Kocken, and A. Thomas. 2007. Apical membrane antigen 1: a malaria vaccine candidate. *Trends Parasitol.* **24**:74–84.
545. Reyes-Sandoval, A., J. T. Harty, and S. M. Todryk. 2007. Viral vector vaccines make memory T cells against malaria. *Immunology* **121**:158–165.
546. Reyes-Sandoval, A., S. Sridhar, T. Berthoud, *et al.* 2008. Single-dose immunogenicity and protective efficacy of simian adenoviral vectors against Plasmodium berghei. *Eur. J. Immunol.* **38**:732–741.
547. Rich, S. M., F. H. Leendertz, G. Xu, *et al.* 2009. The origin of malignant malaria. *Proc. Natl. Acad. Sci. U.S.A.* **106**:14902–14907.
548. Rich, S. M., M. C. Licht, R. R. Hudson, and F. J. Ayala. 1998. Malaria's Eve: evidence of a recent population bottleneck throughout the world populations of Plasmodium falciparum. *Proc. Natl. Acad. Sci. U.S.A.* **95**:4425–4430.
549. Richman, S. J. and R. T. Reese. 1988. Immunologic modeling of a 75-kDa malarial protein with carrier-free synthetic peptides. *Proc. Natl. Acad. Sci. U.S.A.* **85**:1662–1666.
550. Ricklefs, R. E. and D. C. Outlaw. 2010. A molecular clock for malaria parasites. *Science* **329**:226–229.
551. Rieckmann, K. H. 1990. Human immunization with attenuated sporozoites. *Bull. World Health Organ.* **68 (Suppl 1)**:13–16.
552. Riehle, M. A. and M. Jacobs-Lorena. 2005. Using bacteria to express and display anti-parasite molecules in mosquitoes: current and future strategies. *Insect Biochem. Mol. Biol.* **35**:699–707.
553. Riehle, M. M., K. Markianos, O. Niare, *et al.* 2006. Natural malaria infection in Anopheles gambiae is regulated by a single genomic control region. *Science* **312**:577–579.
554. Riglar, D. T., D. Richard, D. W. Wilson, *et al.* 2011. Super-resolution dissection of coordinated events during malaria parasite invasion of the human erythrocyte. *Cell Host Microbe* **9**:9–20.
555. Rogers, W. O., W. R. Weiss, A. Kumar, *et al.* 2002. Protection of rhesus macaques against lethal Plasmodium knowlesi malaria by a heterologous DNA priming and poxvirus boosting immunization regimen. *Infect. Immun.* **70**:4329–4335.

556. Roggwiller, E., M. E. Betoulle, T. Blisnick, and C. Braun-Breton. 1996. A role for erythrocyte band 3 degradation by the parasite gp76 serine protease in the formation of the parasitophorous vacuole during invasion of erythrocytes by Plasmodium falciparum. *Mol. Biochem. Parasitol.* **82**:13–24.
557. Roggwiller, E., T. Blisnick, and C. Braun-Breton. 1998. A Plasmodium falciparum hemolytic activity. *Mol. Biochem. Parasitol.* **94**:303–307.
558. Rohrbach, P., C. P. Sanchez, K. Hayton, et al. 2006. Genetic linkage of pfmdr1 with food vacuolar solute import in Plasmodium falciparum. *EMBO J.* **25**:3000–3011.
559. Roper, C., R. Pearce, S. Nair, et al. 2004. Intercontinental spread of pyrimethamine-resistant malaria. *Science* **305**:1124.
560. Ross, R. 1923. *Memoirs: With a Full Account of the Great Malaria Problem and Its Solution.* Murray, London.
561. Roth, E. F., Jr., M. C. Calvin, I. Max-Audit, et al. 1988. The enzymes of the glycolytic pathway in erythrocytes infected with Plasmodium falciparum malaria parasites. *Blood* **72**:1922–1925.
562. Rottmann, M., C. McNamara, B. K. Yeung, et al. 2010. Spiroindolones, a potent compound class for the treatment of malaria. *Science* **329**:1175–1180.
563. Rozman, R. and C. J. Canfield. 1979. New experimental antimalarial drugs. *Adv. Pharmacol. Chemother.* **16**:1–43.
564. Russell, P., H. W. Mulligan, and B. N. Mohan. 1942. Specific agglutinogenic properties. *J. Mal. Inst. India* **4**:311–319.
565. Russell, P. F. 1950. International preventive medicine. *Scientific Monthly* **71**:393–400.
566. Russell, P. F. 1955. *Man's Mastery of Malaria.* Oxford University Press, London.
567. Russell, P. F., H. W. Mulligan, and B. D. Mohan. 1941. Specific agglutinogenic properties of inactivated sporozoites of P. gallinaceum. *J. Mal. Inst. India* **4**:15–24.
568. Ryan, J. R., J. A. Stoute, J. Amon, et al. 2006. Evidence for transmission of Plasmodium vivax among a duffy antigen negative population in Western Kenya. *Am. J. Trop. Med. Hyg.* **75**:575–581.
569. Saliba, K. J., I. Ferru, and K. Kirk. 2005. Provitamin B5 (pantothenol) inhibits growth of the intraerythrocytic malaria parasite. *Antimicrob. Agents Chemother.* **49**:632–637.
570. Saliba, K. J., H. A. Horner, and K. Kirk. 1998. Transport and metabolism of the essential vitamin pantothenic acid in human erythrocytes infected with the malaria parasite Plasmodium falciparum. *J. Biol. Chem.* **273**:10190–10195.

571. Saliba, K. J., S. Krishna, and K. Kirk. 2004. Inhibition of hexose transport and abrogation of pH homeostasis in the intraerythrocytic malaria parasite by an O-3-hexose derivative. *FEBS Lett.* **570**:93–96.
572. Saliba, K. J., R. E. Martin, A. Broer, *et al.* 2006. Sodium-dependent uptake of inorganic phosphate by the intracellular malaria parasite. *Nature* **443**: 582–585.
573. Salmon, B. L., A. Oksman, and D. E. Goldberg. 2001. Malaria parasite exit from the host erythrocyte: a two-step process requiring extraerythrocytic proteolysis. *Proc. Natl. Acad. Sci. U.S.A.* **98**:271–276.
574. Salvatore, D., A. N. Hodder, W. Zeng, *et al.* 2002. Identification of antigenically active tryptic fragments of apical membrane antigen-1 (AMA1) of Plasmodium chabaudi malaria: strategies for assembly of immunologically active peptides. *Vaccine* **20**:3477–3484.
575. Sanchez, C. P., A. Dave, W. D. Stein, and M. Lanzer. 2010. Transporters as mediators of drug resistance in Plasmodium falciparum. *Int. J. Parasitol.* **40**:1109–1118.
576. Santos, J. M. and D. Soldati-Favre. 2011. Invasion factors are coupled to key signalling events leading to the establishment of infection in apicomplexan parasites. *Cell. Microbiol.* **13**:787–796.
577. Saridaki, T., K. S. Frohlich, C. Braun-Breton, and M. Lanzer. 2009. Export of PfSBP1 to the Plasmodium falciparum Maurer's clefts. *Traffic (Copenh., Denm.)* **10**:137–152.
578. Sauerwein, R. 2007. Transmission blocking vaccines: the bonus of effective malaria control. *Microbes Infect./Inst. Past.* **9**:792–795.
579. Saul, A. 2007. Mosquito stage, transmission blocking vaccines for malaria. *Curr. Opin. Inf. Dis.* **20**:476–481.
580. Saul, A., G. Lawrence, A. Allworth, *et al.* 2005. A human phase 1 vaccine clinical trial of the Plasmodium falciparum malaria vaccine candidate apical membrane antigen 1 in Montanide ISA720 adjuvant. *Vaccine* **23**:3076–3083.
581. Schellenberg, K. A. and G. R. Coatney. 1961. The influence of antimalarial drugs on nucleic acid synthesis in Plasmodium gallinaceum and Plasmodium berghei. *Biochem. Pharmacol.* **6**:143–152.
582. Schenkel, G., E. Cabrera, M. Barr, and P. Silverman. 1975. A new adjuvant for use in vaccination against malaria. *J. Parasitol.* **61**:549–550.
583. Scherf, A., J. J. Lopez-Rubio, and L. Riviere. 2008. Antigenic variation in Plasmodium falciparum. *Annu. Rev. Microbiol.* **62**:445–470.
584. Schildkraut, C. L., M. Mandel, S. Levisohn, *et al.* 1962. Deoxyribonucleic acid base composition and taxonomy of some protozoa. *Nature* **196**: 795–796.

585. Schmidt, L. H., R. Crosby, J. Rasco, and D. Vaughan. 1978. Antimalarial activities of various 4-quinolonemethanols with special attention to WR-142,490 (mefloquine). *Antimicrob. Agents Chemother.* **13**:1011–1030.
586. Sedegah, M., R. Hedstrom, P. Hobart, and S. L. Hoffman. 1994. Protection against malaria by immunization with plasmid DNA encoding circumsporozoite protein. *Proc. Natl. Acad. Sci. U.S.A.* **91**:9866–9870.
587. Sedegah, M., W. O. Rogers, M. Belmonte, *et al.* 2010. Vaxfectin enhances both antibody and *in vitro* T cell responses to each component of a 5-gene Plasmodium falciparum plasmid DNA vaccine mixture administered at low doses. *Vaccine* **28**:3055–3065.
588. Seixas, S., N. Ferrand, and J. Rocha. 2002. Microsatellite variation and evolution of the human Duffy blood group polymorphism. *Mol. Biol. Evol.* **19**:1802–1806.
589. Serazin, A. C., A. N. Dana, M. E. Hillenmeyer, *et al.* 2009. Comparative analysis of the global transcriptome of Anopheles funestus from Mali, West Africa. *PloS One* **4**:e7976.
590. Seydel, K. B., D. A. Milner, Jr., S. B. Kamiza, *et al.* 2006. The distribution and intensity of parasite sequestration in comatose Malawian children. *J. Infect. Dis.* **194**:208–205.
591. Shakespeare, P. G., P. I. Trigg, S. I. Kyd, and L. Tappenden. 1979. Glucose metabolism in the simian malaria parasite Plasmodium knowlesi: activities of the glycolytic and pentose phosphate pathways during the intraerythrocytic cycle. *Ann. Trop. Med. Parasitol.* **73**:407–415.
592. Shanks, G. D. 1994. The rise and fall of mefloquine as aan antimalarial drug in South East Asia. *Mil. Med.* **159**:275–281.
593. Sherman, I. W. 1961. Molecular heterogeneity of lactic dehydrogenase in avian malaria (Plasmodium lophurae). *J. Exp. Med.* **114**:1049–1062.
594. Sherman, I. W. 1966. Malic dehydrogenase heterogeneity in malaria (Plasmodium lophurae and P. berghei). *J. Protozool.* **13**:344–349.
595. Sherman, I. W. 1966. Heterogeneity of lactic dehydrogenase in avian malaria demonstrated by the use of coenzyme analogs, pp. 73. In A. Corradetti (ed.), *Proceedings of the First International Congress of Parasitology.* Pergamon, Oxford.
596. Sherman, I. W. 1988. The Wellcome Trust lecture. Mechanisms of molecular trafficking in malaria. *Parasitology* **96 (Suppl S)**:S57–81.
597. Sherman, I. W. 2006. *The Power of Plagues.* ASM Press, Washington, DC.
598. Sherman, I. W. 2008. Reflections on a century of malaria biochemistry. *Adv. Parasitol.* **64**:1–402.
599. Sherman, I. W. 2009. *The Elusive Malaria Vaccine. Miracle or Mirage?* ASM Press, Washington, DC.

600. Sherman, I. W., L. Mole, V. McDonald, and L. Tanigoshi. 1983. Failure to protect ducklings against malaria by vaccination with histidine-rich protein. *Trans. R. Soc. Trop. Med. Hyg.* **77**:87–90.
601. Sherman, I. W., I. Peterson, L. Tanigoshi, and I. P. Ting. 1971. The glutamate dehydrogenase of Plasmodium lophurae (avian malaria). *Exp. Parasitol.* **29**:433–439.
602. Sherman, I. W. and V. G. Sherman. 1989. *Biology, A Human Approach*, 4th edition. Oxford University Press, New York.
603. Sherman, I. W. and L. Tanigoshi. 1971. Alterations in sodium and potassium in red blood cells and plasma during the malaria infection (Plasmodium lophurae). *Comp. Biochem. Physiol. A* **40**:543–546.
604. Sherman, I. W. and L. Tanigoshi. 1974. Glucose transport in the malarial (Plasmodium lophurae) infected erythrocyte. *J. Protozool.* **21**:603–607.
605. Sherman, I. W., L. Tanigoshi, and J. B. Mudd. 1971. Incorporation of 14-C amino acids by malaria (Plasmodium lophurae). II. Migration and incorporation of amino acids. *Int. J. Biochem.* **2**:13.
606. Sherman, I. W., I. P. Ting, and L. Tanigoshi. 1970. Plasmodium lophurae: Glucose-1-14C and glucose-6-14C catabolism by free plasmodia and duckling host erythrocytes. *Comp. Biochem. Physiol.* **34**:625–639.
607. Sherman, I. W., R. A. Virkar, and J. A. Ruble. 1967. The accumulation of amino acids by Plasmodium lophurae (avian malaria). *Comp. Biochem. Physiol.* **23**:43–57.
608. Sherman, I. W. and E. Winograd. 1990. Antigens on the Plasmodium falciparum infected erythrocyte surface are not parasite derived. *Parasitol. Today* **6**:317–320.
609. Shortt, H. E. and P. C. C. Garnham. 1948. Pre-erythrocytic stage in mammalian malaria parasites. *Nature* **161**:126.
610. Shott, J. P., S. M. McGrath, M. G. Pau, et al. 2008. Adenovirus 5 and 35 vectors expressing Plasmodium falciparum circumsporozoite surface protein elicit potent antigen-specific cellular IFN-gamma and antibody responses in mice. *Vaccine* **26**:2818–2823.
611. Shuler, A. V. 1985. *Malaria. Meeting the Global Challenge*. Oelshlager, Gunn & Hain, Boston.
612. Siddiqui, W. A. 1977. An effective immunization of experimental monkeys against a human malaria parasite, Plasmodium falciparum. *Science* **197**:388–389.
613. Siddiqui, W. A. 1980. Immunization against asexual blood-inhabiting stages of plasmodia, pp. 231–262. In J. P. Kreier (ed.), *Malaria*, vol. 3. Academic Press, San Diego.

614. Siddiqui, W. A., K. Kramer, and S. M. Richard-Crum. 1978. *In vitro* cultivation and partial purification of Plasmodium falciparum antigen suitable for vaccination studies in Aotus monkeys. *J. Parasitol.* **64**:168–169.
615. Siddiqui, W. A., L. Q. Tam, K. J. Kramer, *et al.* 1987. Merozoite surface coat precursor protein completely protects Aotus monkeys against Plasmodium falciparum malaria. *Proc. Natl. Acad. Sci. U.S.A.* **84**:3014–3018.
616. Siddiqui, W. A., D. W. Taylor, S. C. Kan, *et al.* 1978. Partial protection of Plasmodium falciparum-vaccinated Aotus trivirgatus against a challenge of a heterologous strain. *Am. J. Trop. Med. Hyg.* **27**:1277–1278.
617. Siddiqui, W. A., D. W. Taylor, S. C. Kan, *et al.* 1978. Vaccination of experimental monkeys against Plasmodium falciparum: a possible safe adjuvant. *Science* **201**:1237–1239.
618. Sidhu, A. B., A. C. Uhlemann, S. G. Valderramos, *et al.* 2006. Decreasing pfmdr1 copy number in plasmodium falciparum malaria heightens susceptibility to mefloquine, lumefantrine, halofantrine, quinine, and artemisinin. *J. Infect. Dis.* **194**:528–535.
619. Silvie, O., J. F. Franetich, S. Charrin, *et al.* 2004. A role for apical membrane antigen 1 during invasion of hepatocytes by Plasmodium falciparum sporozoites. *J. Biol. Chem.* **279**:9490–9496.
620. Sim, B. K., C. E. Chitnis, K. Wasniowska, *et al.* 1994. Receptor and ligand domains for invasion of erythrocytes by Plasmodium falciparum. *Science* **264**:1941–1944.
621. Simpson, G. L., R. H. Schenkel, and P. H. Silverman. 1974. Vaccination of rhesus monkeys against malaria by use of sucrose density gradient fractions of Plasmodium knowlesi antigens. *Nature* **247**:304–305.
622. Sinden, R. E. 2007. Malaria, mosquitoes and the legacy of Ronald Ross. *Bull. World Health Org.* **85**:894–896.
623. Singh, A. P., H. Ozwara, C. H. Kocken, *et al.* 2005. Targeted deletion of Plasmodium knowlesi Duffy binding protein confirms its role in junction formation during invasion. *Mol. Microbiol.* **55**:1925–1934.
624. Singh, J., A. Ray, and C. Nair. 1953. Isolation of a new strain of Plasmodium knowlesi. *Naure* **172**:122.
625. Singh, S., M. M. Alam, I. Pal-Bhowmick, *et al.* 2010. Distinct external signals trigger sequential release of apical organelles during erythrocyte invasion by malaria parasites. *PLoS Pathog.* **6**:e1000746.
626. Singh, S., K. Miura, H. Zhou, *et al.* 2006. Immunity to recombinant Plasmodium falciparum merozoite surface protein 1 (MSP1): protection in Aotus nancymai monkeys strongly correlates with anti-MSP1 antibody titer and *in vitro* parasite-inhibitory activity. *Infect. Immun.* **74**:4573–4580.

627. Singh, S. K., R. Hora, H. Belrhali, *et al.* 2006. Structural basis for Duffy recognition by the malaria parasite Duffy-binding-like domain. *Nature* **439**:741–744.
628. Slater, L. B. 2005. Malarial birds: modeling infectious human disease in animals. *Bull. Hist. Med.* **79**:261–294.
629. Slater, L. B. 2009. *War and Disease. Biomedical Research in the Twentieth Century*. Rutgers University Press, New Brunswick.
630. Slavic, K., S. Krishna, E. T. Derbyshire, and H. M. Staines. 2011. Plasmodial sugar transporters as anti-malarial drug targets and comparisons with other protozoa. *Mal. J.* **10**:165.
631. Slavic, K., U. Straschil, L. Reininger, *et al.* 2010. Life cycle studies of the hexose transporter of Plasmodium species and genetic validation of their essentiality. *Mol. Microbiol.* **75**:1402–1413.
632. Smith, D. and L. Sanford. 1985. Laveran's germ. The reception and use of a medical discovery. *Am. J. Trop. Med. Hyg.* **34**:2–20.
633. Smith, E. A. and J. A. Erickson. 1980. The Agency of International Development Program for malaria vaccine research and development, pp. 331–334. In J. P. Kreier (ed.), *Malaria*, vol. 3. Academic Press, San Diego.
634. Smith, J. D., C. E. Chitnis, A. G. Craig, *et al.* 1995. Switches in expression of Plasmodium falciparum var genes correlate with changes in antigenic and cytoadherent phenotypes of infected erythrocytes. *Cell* **82**:101–110.
635. Smith, T. G., P. Lourenco, R. Carter, *et al.* 2000. Commitment to sexual differentiation in the human malaria parasite, Plasmodium falciparum. *Parasitology* **121 (Pt 2)**:127–133.
636. Speck, J. F. and E. A. Evans. 1945. The biochemistry of the malaria parasite. II. Glycolysis in cell-free parasite preparations of the malaria parasite. *J. Biol. Chem.* **159**:71–81.
637. Speck, J. F., J. Moulder, and E. A. Evans. 1946. The biochemsitry of the malaria parasite. V. mechanism of pyruvate oxidation in the malarial parasite. *J. Biol. Chem.* **164**:119–144.
638. Spielmann, T. and T. W. Gilberger. 2009. Protein export in malaria parasites: do multiple export motifs add up to multiple export pathways? *Trends Parasitol.* **26**:6–10.
639. Spry, C., C. L. Chai, K. Kirk, and K. J. Saliba. 2005. A class of pantothenic acid analogs inhibits Plasmodium falciparum pantothenate kinase and represses the proliferation of malaria parasites. *Antimicrob. Agents Chemother.* **49**:4649–4657.
640. Spycher, C., M. Rug, N. Klonis, *et al.* 2006. Genesis of and trafficking to the Maurer's clefts of Plasmodium falciparum-infected erythrocytes. *Mol. Cell Biol.* **26**:4074–4085.

641. Sridhar, S., A. Reyes-Sandoval, S. J. Draper, *et al.* 2008. Single-dose protection against Plasmodium berghei by a simian adenovirus vector using a human cytomegalovirus promoter containing intron A. *J. Virol.* **82**:3822–3833.

642. Staines, H. M., A. Alkhalil, R. J. Allen, *et al.* 2007. Electrophysiological studies of malaria parasite-infected erythrocytes: current status. *Int. J. Parasitol.* **37**:475–482.

643. Staines, H. M., J. C. Ellory, and K. Kirk. 2001. Perturbation of the pump-leak balance for Na(+) and K(+) in malaria-infected erythrocytes. *Am. J. Physiol. Cell Physiol.* **280**:C1576–1587.

644. Staines, H. M. and K. Kirk. 1998. Increased choline transport in erythrocytes from mice infected with the malaria parasite Plasmodium vinckei vinckei. *Biochem. J.* **334 (Pt 3)**:525–530.

645. Stanisic, D. I., I. Mueller, I. Betuela, *et al.* 2010. Robert Koch redux: malaria immunology in Papua New Guinea. *Parasite Immunol.* **32**:623–632.

646. Stanley, H. A., S. G. Langreth, and R. T. Reese. 1989. Plasmodium falciparum antigens associated with membrane structures in the host erythrocyte cytoplasm. *Mol. Biochem. Parasitol.* **36**:139–149.

647. Stanley, H. A., J. T. Mayes, N. R. Cooper, and R. T. Reese. 1984. Complement activation by the surface of Plasmodium falciparum infected erythrocytes. *Mol. Immunol.* **21**:145–150.

648. Stanley, H. A. and R. T. Reese. 1984. *In vitro* inhibition of intracellular growth of Plasmodium falciparum by immune sera. *Am. J. Trop. Med. Hyg.* **33**:12–16.

649. Stanley, H. A. and R. T. Reese. 1985. Monkey-derived monoclonal antibodies against Plasmodium falciparum. *Proc. Natl. Acad. Sci. U.S.A.* **82**:6272–6275.

650. Steenkeste, N., S. Incardona, S. Chy, *et al.* 2009. Towards high-throughput molecular detection of Plasmodium: new approaches and molecular markers. *Mal. J.* **8**:86.

651. Stewart, V. A., S. M. McGrath, P. M. Dubois, *et al.* 2007. Priming with an adenovirus 35-circumsporozoite protein (CS) vaccine followed by RTS,S/AS01B boosting significantly improves immunogenicity to Plasmodium falciparum CS compared to that with either malaria vaccine alone. *Infect. Immun.* **75**:2283–2290.

652. Stoute, J. A., J. Gombe, M. R. Withers, *et al.* 2007. Phase 1 randomized double-blind safety and immunogenicity trial of Plasmodium falciparum malaria merozoite surface protein FMP1 vaccine, adjuvanted with AS02A, in adults in western Kenya. *Vaccine* **25**:176–184.

653. Stoute, J. A., M. Slaoui, D. G. Heppner, *et al.* 1997. A preliminary evaluation of a recombinant circumsporozoite protein vaccine against Plasmodium falciparum malaria. RTS,S Malaria Vaccine Evaluation Group. *New Eng. J. Med.* **336**:86–91.
654. Stowers, A. W., V. Cioce, R. L. Shimp, *et al.* 2001. Efficacy of two alternate vaccines based on Plasmodium falciparum merozoite surface protein 1 in an Aotus challenge trial. *Infect. Immun.* **69**:1536–1546.
655. Su, X. and T. E. Wellems. 1998. Genome discovery and malaria research: current status and promise, pp. 253–266. In I. W. Sherman (ed.), *Malaria: Parasite Biology, Pathogenesis and Protection*. ASM Press, Washington, DC.
656. Su, X. Z., V. M. Heatwole, S. P. Wertheimer, *et al.* 1995. The large diverse gene family var encodes proteins involved in cytoadherence and antigenic variation of Plasmodium falciparum-infected erythrocytes. *Cell* **82**:89–100.
657. Sulston, J. and G. Ferry. 2002. *The Common Thread. The Story of Science, Politics, Ethics and the Human Genome*. Bantam, New York.
658. Summers, R. L. and R. E. Martin. 2010. Functional characteristics of the malaria parasite's "chloroquine resistance transporter": implications for chemotherapy. *Virulence* **1**:304–308.
659. Sweeney, T. 2003. *Malaria Frontline. Australian Army Research During World War II*. Melbourne University Press, Melbourne.
660. Sweeney, T. R. 1981. The present status of malaria chemotherapy: Mefloquine a novel antimalarial. *Med. Res. Rev.* **1**:281–301.
661. Tamez, P. A., S. Bhattacharjee, C. van Ooij, *et al.* 2008. An erythrocyte vesicle protein exported by the malaria parasite promotes tubovesicular lipid import from the host cell surface. *PLoS Pathog.* **4**:e1000118.
662. Tanabe, K., T. Mita, T. Jombart, *et al.* 2010. Plasmodium falciparum accompanied the human expansion out of Africa. *Curr. Biol.* **20**:1283–1289.
663. Taylor, S., J. Juliano, P. Trottman, *et al.* 2010. High-throughput pooling and real-time PCR-based strategy for malaria detection. *J. Clin. Microbiol.* **48**:512–519.
664. Taylor-Robinson, A. W. 2003. Immunity to liver stage malaria: considerations for vaccine design. *Immunol. Res.* **27**:53–70.
665. Thera, M. A., O. K. Doumbo, D. Coulibaly, *et al.* 2008. Safety and immunogenicity of an AMA-1 malaria vaccine in Malian adults: results of a phase 1 randomized controlled trial. *PloS One* **3**:e1465.
666. Thomas, A. W., J. A. Deans, G. H. Mitchell, *et al.* 1984. The Fab fragments of monoclonal IgG to a merozoite surface antigen inhibit Plasmodium knowlesi invasion of erythrocytes. *Mol. Biochem. Parasitol.* **13**:187–199.

667. Thomas, S. L. and V. L. Lew. 2004. Plasmodium falciparum and the permeation pathway of the host red blood cell. *Trends Parasitol.* **20**:122–125.
668. Thompson, F. M., D. W. Porter, S. L. Okitsu, *et al.* 2008. Evidence of blood stage efficacy with a virosomal malaria vaccine in a phase IIa clinical trial. *PloS One* **3**:e1493.
669. Thomson, K., J. Freund, H. Somer, and A. Walter. 1947. Immunization of ducks against malaria by means of killed parasites with or without adjuvants. *Am. J. Trop. Med.* **27**:79–105.
670. Tilley, L. and E. Hanssen. 2008. A 3D view of the host cell compartment in P. falciparum-infected erythrocytes. *Transf. Clin. Biol.* **15**:72–81.
671. Tilley, L., R. Sougrat, T. Lithgow, and E. Hanssen. 2008. The twists and turns of Maurer's cleft trafficking in P. falciparum-infected erythrocytes. *Traffic (Copenh., Denm.)* **9**:187–197.
672. Tomar, D., S. Biswas, V. Tripathi, and D. N. Rao. 2006. Development of diagnostic reagents: raising antibodies against synthetic peptides of PfHRP-2 and LDH using microsphere delivery. *Immunobiology* **211**:797–805.
673. Topolska, A. E., C. G. Black, and R. L. Coppel. 2004. Identification and characterisation of RAMA homologues in rodent, simian and human malaria species. *Mol. Biochem. Parasitol.* **138**:237–241.
674. Topolska, A. E., A. Lidgett, D. Truman, *et al.* 2004. Characterization of a membrane-associated rhoptry protein of Plasmodium falciparum. *J. Biol. Chem.* **279**:4648–4656.
675. Topolska, A. E., T. L. Richie, D. H. Nhan, and R. L. Coppel. 2004. Associations between responses to the rhoptry-associated membrane antigen of Plasmodium falciparum and immunity to malaria infection. *Infect. Immun.* **72**:3325–3330.
676. Topolska, A. E., C. Wang, C. G. Black, and R. Coppel. 2003. Merozoite cell biology, pp. 365–444. In A. P. Waters and C. J. Janse (eds.), *Malaria Parasites: Genomes and Molecular Biology*. Caister Academic Press, Norfolk, UK.
677. Tournamille, C., Y. Colin, J. P. Cartron, and C. Le Van Kim. 1995. Disruption of a GATA motif in the Duffy gene promoter abolishes erythroid gene expression in Duffy-negative individuals. *Nat. Genet.* **10**:224–228.
678. Tournamille, C., A. Filipe, C. Badaut, *et al.* 2005. Fine mapping of the Duffy antigen binding site for the Plasmodium vivax Duffy-binding protein. *Mol. Biochem. Parasitol.* **144**:100–103.
679. Tracy, S. M. and I. W. Sherman. 1972. Purine uptake and utilization by the avian malaria parasite Plasmodium lophurae. *J. Protozool.* **19**:541–549.
680. Trager, W. 1977. Cofactors and vitamins in the metabolism of malarial parasites. Factors other than folates. *Bull. World Health Organ.* **55**:285–289.

681. Trager, W. 1987. The cultivation of Plasmodium falciparum: applications in basic and applied research. *Ann. Trop. Med. Parasitol.* **81**:511–529.
682. Trager, W. 2002. On the release of malaria merozoites. *Trends Parasitol.* **18**:60–61.
683. Trager, W. and J. Jensen. 1980. Cultivation of erythrocytic and exoerythrocytic stages of plasmodia, pp. 271–319. In J. P. Kreier (ed.), *Malaria*, vol. 2. Academic Press, New York.
684. Trager, W. and J. Jensen. 1997. Continuous culture of Plasmodium falciparum: its impact on malaria research. *Int. J. Parasitol.* **27**:17.
685. Trager, W., M. A. Rudzinska, and P. C. Bradbury. 1966. The fine structure of Plasmodium falciparum and its host erythrocytes in natural malarial infections in man. *Bull. World Health Organ.* **35**:883–885.
686. Treeck, M., N. S. Struck, S. Haase, *et al.* 2006. A conserved region in the EBL proteins is implicated in microneme targeting of the malaria parasite Plasmodium falciparum. *J. Biol. Chem.* **281**:31995–32003.
687. Trenholme, G., R. Williams, R. E. Desjardins, *et al.* 1975. Mefloquine (WR-142,490) in the treatment of human malaria. *Science* **190**:792–794.
688. Trigg, P. I. 1989. Qinghaosu (artemisinin) as an antimalarial drug. *Econom. Med. Plant Res.* **3**:21–55.
689. Triglia, T., P. Wang, P. F. Sims, *et al.* 1998. Allelic exchange at the endogenous genomic locus in Plasmodium falciparum proves the role of dihydropteroate synthase in sulfadoxine-resistant malaria. *EMBO J.* **17**:3807–3815.
690. Tripathi, A. K., P. V. Desai, A. Pradhan, *et al.* 2004. An alpha-proteobacterial type malate dehydrogenase may complement LDH function in Plasmodium falciparum. Cloning and biochemical characterization of the enzyme. *Eur. J. Biochem.* **271**:3488–3502.
691. Turgut-Balik, D., E. Akbulut, D. K. Shoemark, *et al.* 2004. Cloning, sequence and expression of the lactate dehydrogenase gene from the human malaria parasite, Plasmodium vivax. *Biotechnol. Lett.* **26**:1051–1055.
692. Tuteja, R. 2002. DNA vaccine against malaria: a long way to go. *Crit. Rev. Biochem. Mol. Biol.* **37**:29–54.
693. Vaidya, A. B. 2004. Mitochondrial and plastid functions as antimalarial drug targets. *Curr. Drug Targ.* **4**:11–23.
694. Vaidya, A. B. 2005. The mitochondrion, pp. 234–252. In I. W. Sherman (ed.), *Molecular Approaches to Malaria*. ASM Press, Washington, DC.
695. Vaidya, A. B., R. Akella, and K. Suplick. 1989. Sequences similar to genes for two mitochondrial proteins and portions of ribosomal RNA in tandemly arrayed 6-kilobase-pair DNA of a malarial parasite. *Mol. Biochem. Parasitol.* **35**:97–107.

696. Vaidya, A. B. and P. Arasu. 1987. Tandemly arranged gene clusters of malarial parasites that are highly conserved and transcribed. *Mol. Biochem. Parasitol.* **22**:249–257.
697. Vaidya, A. B. and M. W. Mather. 2005. A post-genomic view of the mitochondrion in malaria parasites. *Curr. Top. Microbiol. Immunol.* **295**:233–250.
698. Valderramos, S. G. and D. A. Fidock. 2006. Transporters involved in resistance to antimalarial drugs. *Trends Pharmacol. Sci.* **27**:594–601.
699. Valero, M. V., L. R. Amador, C. Galindo, *et al.* 1993. Vaccination with SPf66, a chemically synthesised vaccine, against Plasmodium falciparum malaria in Colombia. *Lancet* **341**:705–710.
700. van Dooren, G., l. M. Stimmler, and G. I. McFadden. 2006. Metabolic maps and functions of the Plasmodium mitochondrion. *FEMS Microbiol. Rev.* **30**:596–630.
701. VanBuskirk, K. M., M. T. O'Neill, P. De La Vega, *et al.* 2009. Preerythrocytic, live-attenuated Plasmodium falciparum vaccine candidates by design. *Proc. Natl. Acad. Sci. U.S.A.* **106**:13004–13009.
702. VanBuskirk, K. M., E. Sevova, and J. H. Adams. 2004. Conserved residues in the Plasmodium vivax Duffy-binding protein ligand domain are critical for erythrocyte receptor recognition. *Proc. Natl. Acad. Sci. U.S.A.* **101**:15754–15759.
703. Vander Jagt, D. L., L. A. Hunsaker, M. Kibirige, and N. M. Campos. 1989. NADPH production by the malarial parasite Plasmodium falciparum. *Blood* **74**:471–474.
704. Vander Jagt, D. L., C. Intress, J. E. Heidrich, *et al.* 1982. Marker enzymes of Plasmodium falciparum and human erythrocytes as indicators of parasite purity. *J. Parasitol.* **68**:1068–1071.
705. Vanderberg, J., R. Nussenzweig, and H. Most. 1969. Protective immunity produced by the injection of x-irradiated sporozoites of Plasmodium berghei. V. *In vitro* effects of immune serum on sporozoites. *Mil. Med.* **134**:1183–1190.
706. Vanderberg, J., R. Nussenzweig, and H. Most. 1970. Protective immunity produced by the bite of x-irradiated mosquitoes infected with Plasmodium berghei. *J. Parasitol.* **56**:350–351.
707. Vanderberg, J. P. and U. Frevert. 2004. Intravital microscopy demonstrating antibody-mediated immobilization of *Plasmodium berghei* sporozoites injected into skin by mosquitoes. *Int. J. Parasitol.* **34**:991–996.
708. Vanderberg, J. P., R. S. Nussenzweig, H. Most, and C. G. Orton. 1968. Protective immunity produced by the injection of x-irradiated sporozoites of Plasmodium berghei. II. Effects of radiation on sporozoites. *J. Parasitol.* **54**:1175–1180.

709. Varadharajan, S., S. Dhanasekaran, Z. Q. Bonday, *et al.* 2002. Involvement of delta-aminolaevulinate synthase encoded by the parasite gene in de novo haem synthesis by Plasmodium falciparum. *Biochem. J.* **367**:321–327.
710. Vaughan, A. M., R. Wang, and S. H. Kappe. 2010. Genetically engineered, attenuated whole-cell vaccine approaches for malaria. *Hum. Vaccines* **6**:107–113.
711. Verhave, J. P. 1975. *Immunization with Sporozoites. An Experimental Study of Plasmodium berghei Malaria.* Catholic University Press, Nijmegen.
712. Verloo, P., C. H. Kocken, A. Van der Wel, *et al.* 2004. Plasmodium falciparum-activated chloride channels are defective in erythrocytes from cystic fibrosis patients. *J. Biol. Chem.* **279**:10316–10322.
713. Vermeulen, A. N., T. Ponnudurai, P. J. Beckers, *et al.* 1985. Sequential expression of antigens on sexual stages of Plasmodium falciparum accessible to transmission-blocking antibodies in the mosquito. *J. Exp. Med.* **162**:1460–1476.
714. Vermeulen, A. N., J. van Deursen, R. H. Brakenhoff, *et al.* 1986. Characterization of Plasmodium falciparum sexual stage antigens and their biosynthesis in synchronised gametocyte cultures. *Mol. Biochem. Parasitol.* **20**:155–163.
715. Vial, H. J., P. Eldin, D. Martin, *et al.* 1999. Transport of phospholipid synthesis precursors and lipid trafficking into malaria-infected erythrocytes. *Novartis Found. Symp.* **226**:74–83; discussion 82–78.
716. Volkman, S. K., A. E. Barry, E. J. Lyons, *et al.* 2001. Recent origin of Plasmodium falciparum from a single progenitor. *Science* **293**:482–484.
717. Wade, N. 2002. Genetic Decoding may bring Advances in Worldwide Fight against Malaria. *New York Times*, 3 October.
718. Wagner, J. T., H. Ludemann, P. M. Farber, *et al.* 1998. Glutamate dehydrogenase, the marker protein of Plasmodium falciparum — cloning, expression and characterization of the malarial enzyme. *Eur. J. Biochem.* **258**:813–819.
719. Walsh, C. J. and I. W. Sherman. 1968. Purine and pyrimidine synthesis by the avian malaria parasite, Plasmodium lophurae. *J. Protozool.* **15**:763–770.
720. Wang, P., N. Nirmalan, Q. Wang, *et al.* 2004. Genetic and metabolic analysis of folate salvage in the human malaria parasite Plasmodium falciparum. *Mol. Biochem. Parasitol.* **135**:77–87.
721. Wang, P., M. Read, P. F. Sims, and J. E. Hyde. 1997. Sulfadoxine resistance in the human malaria parasite Plasmodium falciparum is determined by mutations in dihydropteroate synthetase and an additional factor associated with folate utilization. *Mol. Microbiol.* **23**:979–986.

722. Wang, R., T. L. Richie, M. F. Baraceros, *et al.* 2005. Boosting of DNA vaccine-elicited gamma interferon responses in humans by exposure to malaria parasites. *Infect. Immun.* **73**:2863–2872.
723. Wang, R., J. D. Smith, and S. H. Kappe. 2009. Advances and challenges in malaria vaccine development. *Exp. Rev. Mol. Med.* **11**:e39.
724. Ward, G. E., L. H. Miller, and J. A. Dvorak. 1993. The origin of parasitophorous vacuole membrane lipids in malaria-infected erythrocytes. *J. Cell Sci.* **106 (Pt 1)**:237–248.
725. Waters, A. P., A. W. Thomas, J. A. Deans, *et al.* 1990. A merozoite receptor protein from Plasmodium knowlesi is highly conserved and distributed throughout Plasmodium. *J. Biol. Chem.* **265**:17974–17979.
726. Weiss, W. R., A. Kumar, G. Jiang, *et al.* 2007. Protection of rhesus monkeys by a DNA prime/poxvirus boost malaria vaccine depends on optimal DNA priming and inclusion of blood stage antigens. *PloS One* **2**:e1063.
727. Wells, T. N. C., P. Alonso, and W. E. Gutteridge. 2009. New medicines to speed the eradication of malaria. *Nat. Rev. Drug Discov.*
728. Wendel, W. B. 1946. The influence of naphthoquinones upon the respiratory and carbohydrate metabolism of malaria parasites. *Fed. Proc.* **5**:406–408.
729. Werner, C., M. T. Stubbs, R. L. Krauth-Siegel, and G. Klebe. 2005. The crystal structure of Plasmodium falciparum glutamate dehydrogenase, a putative target for novel antimalarial drugs. *J. Mol. Biol.* **349**:597–607.
730. Wertheimer, S. P. and J. W. Barnwell. 1989. Plasmodium vivax interaction with the human Duffy blood group glycoprotein: identification of a parasite receptor-like protein. *Exp. Parasitol.* **69**:340–350.
731. Westenberger, S. J., C. M. McClean, R. Chattopadhyay, *et al.* 2010. A systems-based analysis of Plasmodium vivax lifecycle transcription from human to mosquito. *PLoS Neglect. Trop. Dis.* **4**:e653.
732. White, K., U. Krzych, D. M. Gordon, *et al.* 1993. Induction of cytolytic and antibody responses using Plasmodium falciparum repeatless circumsporozoite protein encapsulated in liposomes. *Vaccine* **11**:1341–1346.
733. White, N. J. 1994. Tough test for malaria vaccine. *Lancet* **344**:1548–1549.
734. White, N. J. 2007. Plasmodium knowlesi. The fiftth human malaria parasite. *Clin. Inf. Dis.* **46**:172–173.
735. White, N. J. 2008. Qinghaosu (artemisinin): the price of success. *Science* **320**:330–334.
736. Whitten, M. M., S. H. Shiao, and E. A. Levashina. 2006. Mosquito midguts and malaria: cell biology, compartmentalization and immunology. *Parasite Immunol.* **28**:121–130.

737. Wickham, M. E., J. G. Culvenor, and A. F. Cowman. 2003. Selective inhibition of a two-step egress of malaria parasites from the host erythrocyte. *J. Biol. Chem.* **278**:37658–37663.
738. Williamson, D. H. 1998. Extrachromosomal DNA: the mitochondrion, pp. 267–275. In I. W. Sherman (ed.), *Malaria: Parasite Biology, Pathogenesis and Protection.* ASM Press, Washington, DC.
739. Williamson, K. C. 2003. Pfs 230: from malaria transmission-blocking vaccine candidate toward function. *Parasite Immunol.* **25**:351–359.
740. Windbichler, N., M. Menichell, P. Papathanos, *et al.* 2011. A synthetic homing endonuclease-based gene drive system in the human malaria mosquito. *Nature* **473**:212–215.
741. Winzeler, E. A. and R. W. Davis. 1997. Functional analysis of the yeast genome. *Curr. Opin. Genet. Dev.* **7**:771–776.
742. Wipasa, J. and E. M. Riley. 2007. The immunological challenges of malaria vaccine development. *Exp. Opin. Biol. Ther.* **7**:1841–1852.
743. Wiselogle, F. Y. (ed.). 1946. *A Survey of Antimalarial Drugs 1941–1945.* J. W. Edwards, Ann Arbor.
744. Wongsrichanalai, C., S. Prajakwong, S. R. Meshnick, *et al.* 2004. Mefloquine — its 20 years in the Thai Malaria Control Program. *Southeast Asian J. Trop. Med. Pub. Health* **35**:300–308.
745. Woodrow, C. J. and S. Krishna. 2005. Molecular approaches to malaria: glycolysis in asexual stage parasites, pp. 222–233. In I. W. Sherman (ed.), *Molecular Approaches to Malaria.* ASM Press, Washington, DC.
746. Wu, Y., X. Wang, X. Liu, and Y. Wang. 2003. Data-mining approaches reveal hidden families of proteases in the genome of malaria parasite. *Genome Res.* **13**:601–616.
747. Xu, X., Y. Dong, E. G. Abraham, *et al.* 2005. Transcriptome analysis of Anopheles stephensi-Plasmodium berghei interactions. *Mol. Biochem. Parasitol.* **142**:76–87.
748. Yamada, K. A. and I. W. Sherman. 1981. Purine metabolism by the avian malarial parasite Plasmodium lophurae. *Mol. Biochem. Parasitol.* **3**:253–264.
749. Yazdani, S. S., A. R. Shakri, P. Pattnaik, *et al.* 2006. Improvement in yield and purity of a recombinant malaria vaccine candidate based on the receptor-binding domain of Plasmodium vivax Duffy binding protein by codon optimization. *Biotechnol. Lett.* **28**:1109–1114.
750. Yeoh, S., R. A. O'Donnell, K. Koussis, *et al.* 2007. Subcellular discharge of a serine protease mediates release of invasive malaria parasites from host erythrocytes. *Cell* **131**:1072–1083.

751. Yoshida, N., R. S. Nussenzweig, P. Potocnjak, *et al.* 1980. Hybridoma produces protective antibodies directed against the sporozoite stage of malaria parasite. *Science* **207**:71–73.
752. Young, M. D., J. Porter, and C. Johnson. 1966. Plasmodium vivax transmitted from monkey to man. *Science* **153**:1006–1007.
753. Yuthavong, Y. 2002. Basis for antifolate action and resistance in malaria. *Microb. Infec./Inst. Pasteur* **4**:175–182.
754. Yuthavong, Y., J. Yuvaniyama, P. Chitnumsub, *et al.* 2005. Malarial (Plasmodium falciparum) dihydrofolate reductase-thymidylate synthase: structural basis for antifolate resistance and development of effective inhibitors. *Parasitology* **130**:249–259.
755. Yuvaniyama, J., P. Chitnumsub, S. Kamchonwongpaisan, *et al.* 2003. Insights into antifolate resistance from malarial DHFR-TS structures. *Nat. Struct. Biol.* **10**:357–365.
756. Ziffer, H., R. J. Highet, and D. Klayman. 1997. Artemisinin: An endoperoxide antimalarial from Artemisia annua L. *Fortschr. Chem. Org. Naturst.* **72**:121–214.
757. Zimmer, C. 2010. Yet-another-genome syndrome. Available at: http://blogs.discovermagazine.com/loom/2010/04/02/yet-another-genome-syndrome/ [Accessed 23 September 2011].

Index

1-deoxy-D-xylulose-5-phosphate reductoisomerase 272
2,4-diamino-5-p-chlorophenyl-6-ethyl pyrimidine 72
2,4-diaminopyrimidines 71
2-hydroxy-3-alkylnaphthoquinones 173
4,4-dinitrostilbene-2,2-disulfonic acid (DNDS) 98
4,4′di-isothiocyano-2,2′-dinitrostilbene sulfonate (DIDS) 97
4-aminoquinoline 279
4-quinoline methanol derivatives 285
4-quinoline-methanols 286
6-phosphofructokinase 177

acetylpyridine analog of NAD 179
acetylpyridine NAD 178, 188
acetylpyridine nucleotide 158
acidic base repeat antigen (ABRA) 120
acquired immunity 195, 202, 218
actomyosin 112
adenovirus vectors 263
adenovirus, AdCh63 265

adjuvant 196, 197, 208
Aedes 12
aestivo-autumnal malaria or malignant malaria 8
Africa 161, 165
  migration out of 161
Agouron Institute 213, 214
aldolase 189
Altmann, Richard 25
American Institute for Biological Sciences (AIBS) 215, 216
amino acid 86
  alanine 94
  isoleucine 88
  leucine 94
  tryptophan 88
Anders, Robin 223
anion transporter, band 3 protein 97
ankyrin 103, 108
Anopheles 12, 17, 18, 43
Anopheles funestus 145
Anopheles gambiae 59, 66, 146
Anopheles–Plasmodium-responsive leucine-rich repeat (APL1) 148
antibody-dependent cellular inhibition (ADCI) 259
anti-folate drug 281

antigenic variation 79–81, 83, 109
Aotus 207, 208, 212, 213, 215, 260
apical membrane antigen 1
(AMA-1) 112, 113, 120, 121,
217, 218, 223, 224, 262, 264, 265
apicidin 76, 77
apicoplast 44, 46, 47, 266, 270,
272, 302
apicoplast genome 47, 271
aquaglyceroporin (AQP) 96
Archaeoglobus fulgidus 66
arteether (AE) 296
artelinic acid 296
artemether (AM) 296
Artemisia 293
Artemisia annua 294
artemisinin 287, 295–298
artemisinin-based combination
therapies (ACTs) 299
artemisinins 293
artemotil (= arteether) 296
Artequin 299
artesunate (AS) 296, 297
AS02 224, 244, 227
AS02A 221
Asia
migration out of 161
Asia Pacific Conference on
Malaria 214
Atabrine 171–173, 177, 276, 277
Atovaquone 289, 292, 293
Avery, O. T. 28, 29, 30
azithromycin 271

Bacillus malariae 3
bacteriophage 36
BAEBL 118
Ball, Eric 172, 183
Ballou, W. Ripley 227, 244–246,
248, 252

band 3 protein 100, 101, 103, 119,
120, 122
Bastianelli, Giuseppe 78
Beaudoin, Richard L. 254, 257,
258
benflumetol 299
β-hematin 279
Bignami, Amico 78
bisthiazolium compound T4
271
Blackman, M. J. (Mike) 220
blood-stage vaccine 218
blood-stage vaccine
candidates 217
Bloom, William 20
bonobos (*Pan paniscus*) 160, 161
boost 262
Bovarnick, Marianna 172, 176
Bozdech, Zybnek 70
branched pathway 185, 186
Brown, Ivor 79
Brown, Neil 79
Brown, Patrick O. 67
Bueding, Ernest 177, 178
BW566C 292
BW58C 292

Ca^{2+}-ATPase 6 (adenosine
triphosphatase) 298
Cabantchik, Ioav 88
Caenorhabditis elegans 66
calcium-dependent kinase
PfCDPK4 131
calpain-1 129
calpstatin 129
Canfield, Craig 136
carbon dioxide fixation 180,
181
Carlton, Jane 58, 61
Carter, Richard 228–230

Carucci, Daniel 52
caveolae–vesicle complex 134, 141
CD4 T cells 257
CD8 T cell 255, 257, 258, 261, 266
cDNA 38
CDPK3 131
cecropin 147, 152, 153
Celera Genomics 50, 57, 59, 251
cell 41
cerebral malaria 78, 79, 103
Chargaff rule 32–34
Chargaff, Erwin 32–34
Chase, Martha 30
chimpanzee (*Pan troglodytes*) 160
chips 68, 73
chloroquine 275, 277–280, 284–287, 298
chlorproguanil 281
chlorproguanil-dapsone (Lap-Dap) 281
choline analog 74
choline ethanolamine phosphatidyl transferase (CEPT) 271
Christophers, Rickard 171
chromatin 26, 27, 75–77
 remodeling 76
chromosome 26, 36, 37
chymostatin 122, 126
chymotrypsin 114, 118, 136, 137
circumsporozoite precipitation (CSP) reaction 235, 245, 246, 249, 255–258, 262, 264, 265
circumsporozoite protein, CSP 236, 243
citric acid cycle 175, 176
clag (cytoadherence-linked antigen) 99

Clark, W. Mansfield 172
clindamycin 271
Clinton, Bill 50, 301
Clyde, David 235, 238–240, 251
Coartem 299
coenzyme A 95
Cohen, Joe 246
Cohen, Sydney 114, 136, 201, 202, 204, 205
Collins, Francis 50, 301
complementary base pairing 34
complementary DNA (cDNA) 37
compound 3361 [3-O-(undec-10-en)-yl]-D-glucose] 90, 91
Coulson, Frederick 254
Crick, Francis 33
Cross, George 218, 219
CS gene 243, 244
CSP reaction 236, 240
C-type lectin CTL4 149
C-type lectin mannose CTLMA2 149
Culex 12, 15
cycloguanil 281, 282
cytoadherence-linked asexual gene [CLAG3.1] 113
cytochalasin B 139
cytochrome b gene 173, 160, 293
cytochrome c 183, 290, 291
cytoplasm 42
cytostome 44

Dame, John 58
dapsone 285
Dave, Dilip 289
Davis, Ronald W. 67
DDT-resistant 150
Deans, Judith 222

defensin   147, 152, 153
dense granule   44, 111, 113, 128
DeRisi, Joseph   67, 68
Desowitz, Robert   136, 278
DIDS   98
dihydroartemisinin (DHA)   295, 297
dihydrofolate (DHF)   283
dihydrofolate reductase (DHFR)   72, 73, 271, 281–284
dihydro-orotate (DHO)   183, 289
dihydro-orotate dehydrogenase (DHODH)   183, 289
dihydropteroate synthase (DHPS)   282, 283
dipeptidyl peptidases (DPAPs)   128
DNA duplication   34
DNA microarrays   67, 68
DNA synthesis   72
DNA vaccines   261, 262
double helix   33
doxycycline   271
DPAP3   128
Druilhe, Pierre   252, 259
Duffy blood antigen   137–139, 164, 165
Duffy negative   137, 138, 167
Duffy positive   137

Eaton, Monroe   79
EBA-140 [BAEBL]   117
EBA-165 [PEBL] and EBL-1   117
EBA-175   116–118, 161
EBA-181 [JESEBL]   117
EE form   254, 257
EE stage   19, 21, 22
egress   127
Ehrlich, Paul   xiii, 6, 273

electron transport chain   182, 289, 290
Elion, Gertrude   71
Ellis, Joan   243
En (a–)   114
epigenetic   75
Erickson, James   215–217
erythrocyte vesicle protein 1   105
erythrocyte-binding antigen (EBA)-175]   115
erythroid transcription   138
eukaryote   42
Evans, Earl A.   173, 177
evasion   77
evolutionary history
    falciparum   159
    man and gorilla   157
    P. brazilianum   168
    P. cynomolgi   163
    P. malariae   167
    P. ovale   167, 168
    P. vivax   162–166
[exo-erythrocytic (EE)] form   19
exoneme   128
expressed sequence tags (ESTs)   49
expression library   37
Eyles, Don E.   228

FabB/F gene   266
Fairley, Neil   20
falcipain-2   129
falciparum merozoite protein 1 [FMP1]   220
Falciparum Uganda Palo Alto (FUP)   207, 208
Falciparum Vietnam Oak-Knoll (FVO)   207, 210
Fansidar   281, 288, 298

Fansimef 288
ferriprotoporphyrin IX (FP) 279
Fidock, David 280
First Law of Technology 301
Fitch, Coy 279
flagella 4
flagellated bodies 8
Flemming, Walther 26
FMP1 221
folinic acid 72
fosmidomycin 272
fowlpox strain (FP9) 264
fowlpox virus (FP9) 262, 263
FP dimerization 279, 280
Franklin, Rosalind 33
Freeman, Robert (Robbie) 218, 219
French, Sarah L. 51, 52
Freund, Jules 196, 197, 199
Freund's adjuvant 197, 204
Freund's complete adjuvant [FCA] 199, 205, 208, 212, 215
Freund's incomplete adjuvant [FIA] 204, 208, 212, 213
Friedman, Milton 211
FSV-1 244, 245
Fulton, James D. 171, 172, 177
functional genomics 67
furosemide 88, 94, 95

gametes 45
gametocyte 9, 22, 43, 45
gametocyte egress 131
g-globulin 201, 202
GAP 265, 266, 267
Gardner, Malcolm 52, 56–58, 60, 61
Garnham, P. C. C. (Cyril) 20–21, 136

Gates Foundation 248, 252
Geiman, Quentin M. 173, 206, 207, 210
gene expression 77, 82
 P. falciparum 70
gene expression libraries 37
General Accounting Office (GAO) 215, 217
genetic code 34
genome 47, 59
 P. yoelii 58
genome sequenced 162
genome sequencing 168
genomic DNA 38
genomic libraries 37
Gerrick International 216
Ginsberg, Hagai 88
GlaxoSmithKline (GSK) 224, 244–248, 269
 malaria vaccine program 246
Global Malaria Eradication Campaign 278
Global Malaria Eradication Program 201–202
glucose transport 89
glutamic dehydrogenase (GDH) 179, 180, 181, 186
glutamine 181, 185, 186
glutathione S-transferase 150
glycolysis 174, 177, 181, 185
glycolytic enzyme 181
glycophorin 114, 115, 120
glycophorin A 115, 116
glycophorin B 118
glycophorin C 117, 118
glycosylphosphatidylinositol (GPI) anchored 118
Godson, Nigel 243, 244
Golgi, Camillo 7, 34

Goodwin, Len 72
GPI anchor 119, 122, 221
Grassi, Giovanni Battista 7, 15–19
Griffith, Frederick 28, 29
GTP cyclohydrolase I 272
guanine adenine thymine adenine (GATA-1) 138
Gutteridge, William 289
Gwadz, Robert J. 229–230

Haemophilus influenzae 65
Haemoproteus 9
Hall, Neil 60
halofantrine 287
Harvard 172, 173
Heidelberger, Michael 198, 199
Hellerman, Leslie 172
heme biosynthetic pathway 184
hemoglobin
    amino acid 94
hemozoin 4, 22, 109, 296, 297
Heppner, Gray 245–248
Herrera, Socrates 225
Hershey, Alfred 30
Hertwig, Oscar 26
hexose transporter 90, 91
Hill, Adrian 263
histidine-rich protein (HRP) 63
histone 75, 76, 83
histone deacetylase 77
Hitchings, George 71
Hoffman, Stephen L. 51–54, 57, 59, 245, 250–253, 261
Hoffmann-La Roche 288
Holder, Anthony 219, 220
Holz, George G. 74
Hooke, Robert 41
Hoppe-Seyler, Felix 23–25

Hotchkiss, Rollin xiv
Huff, Clay 20, 253
Human Genome Project xii, 50, 301
hydroxynaphthoquinones 290
hypnozoite 21, 134, 135, 141
hypothetical proteins 47
hypoxanthine 92

I. G. Farben cartel 276
IFN-g 255, 257, 261
immune serum 198
incubation period 20
insecticide resistance 150
invasion 111
irradiated sporozoite 237–239, 241, 242, 249, 251, 253–257

Jacob, Walter 31
James, Sydney Pryce 20
Jensen, James B. 210–211
JESEBL 118
Johns Hopkins School of Medicine 172

KAHRP 108
Kaplan, Nathan 178, 187
Kilejian, Araxie 212
kinase 129, 131, 270, 272
kitasamycin 271
Klayman, Daniel 295
Klebs, Edwin 3
knob 79, 103, 104, 136
knob-associated histidine-rich protein (KAHRP) 106
Koch, Robert xiii, 2, 18, 195, 200, 210
Krebs cycle 180, 183, 184, 186
Krebs, Hans 175, 176

Krotowski, W.   21

lactate dehydrogenase (LDH)   157, 158, 178
lactic dehydrogenase (LDH)   179, 187–189
Langreth, Susan   212
lapinone   291
Lariam   287, 288, 289
Laveran, Chalres-Louis Alphonse   4, 5, 7, 8, 133
Le Roch, Karine   69, 74
Leeuwenhoek, Antony van
leucine-rich repeat immune protein 1 (LRIM1)   148
leupeptin   122, 126
Levene, Phoebus A.   30, 31
life cycle   21
lipid metabolism   74
LISP 1 (liver-specific protein 1)   130
liver development
    P. falciparum   46
    P. malariae   46
    P. ovale   46
    P. vivax   46
liver egress   130
liver stage   254, 257
liver-stage antigens (LSA)   256, 258
liver stage immunity   259
liver-stage parasites   266
Lofton, Susan 214
Long, Carole   220
loop-mediated isothermal amplification   193
Loucq, Christian   253
LRIM1   149
LSA-1   260, 261, 265
LSA-3   260, 261
lumefantrine   287

MacCallum, William   9
macrogamete   22
MAHRP   108
Malaria Genome Project   52–54
Malaria Genome Sequencing Meeting   57
Malaria Immunology and Vaccine Research (MIVR)   217
malaria pigment   44
malaria vaccine   195, 200, 214, 215, 238, 251, 302
Malaria Vaccine Development Program (MVDP)   217
Malaria Vaccine Initiative (MVI)   248
Malaria Vaccine Workshop   203
Malaria's Eve hypothesis   159–160
Malarone   282, 293
male development gene/protein of early gametocyte 3 (MDV/Peg3)   131
malic dehydrogenase (MDH)   179–181, 186
Malstat   158, 179, 188
Manson, Patrick   10, 11, 13, 15
Marchiafava, Ettore   3
Marmer, Julius   xiv, 157
Maurer's cleft associated histidine rich protein   108
Maurer's clefts   48, 105, 109, 121, 122
McCarty, Maclyn   29
McGregor, Ian   201, 202, 211
McLeod, Colin   29
Medea   152

mefloquine   284, 286–289
melanization   147–49
melting temperature   xiv
membrane transport   86, 101
menoctone (3-(8-cyclohexyl)-octyl-2-hydroxy-1,4-naphthoquinone)   291, 292
mepacrine   171
Mephaquine   288
Mepron   293
merogony   22
merosome   46, 47, 130
merozoite   22, 43, 44, 111, 137
merozoite antigen   213
merozoite invasion   137, 141
merozoite release   126–128, 130
merozoite surface proteins (MSPs)   21, 111
merozoite vaccination   204
messenger RNA (mRNA)   35, 37
methyl glucose   89, 90
methyl transferase   77
methylglyoxal   96
microarray   73, 74, 149, 150, 270, 271, 272, 300, 303
microgamete   22
microneme   43, 44, 111–113, 117, 120, 140
Miescher, Friedrich   23–27
Miller, Louis H.   115, 135, 136
Mitchell, Graham   114
mitochondrial DNA   158, 160
Mitochondrial Eve   158, 159
mitochondrial genome   47, 182
mitochondrion   185
modified vaccinia Ankara (MVA) virus   262
molecular clock   155, 157–159
monoclonal antibody   219, 222, 230

Montanide   260
Montanide ISA720   223
Mosquirix   252
Mosquito immunity   147
Most, Harry   234–236, 240, 277
Moulder, James W.   174, 176
MSP   264
MSP-1   119, 121, 217, 218, 220, 221, 262, 265
MSP-3   120, 259
MSP-6   120
MSP-7   120
MSP-9   120
Mulligan, H. W.   233
multidrug resistance (MDR)   284, 287
Mupirocin   271
muramyl dipeptide (MDP)   208
MuStDo (MultiStage DNA Vaccine Operation)   261
Mycoplasma genitalium   66

N-acetylmuramyl-L-alanyl-D-isoglutamine   213
N-glycolylneuraminic acid (Neu5Gc)   161
National Institutes of Health in Bogota, Colombia   215
Naval Medical Research Center [NMRC]   51
Neu5Ac (sialic acid)   161
neuraminidase   114, 116–118, 136
new permeability pathway (NPP)   88
New York University Medical Center (NYU)   234–236, 244, 251, 255
Night owl (*Aotus*) monkeys   206

NMRI/USAID/WHO Workshop on the Immunology of Malaria 206
normocyte-binding proteins (NBPs) 140
Novartis Institute for Tropical Disease 270
NPP 93–96, 98, 100, 101
nuclear genome 47, 141
nuclear membrane 42
nucleic acid 25, 27
nucleic acid-based tests 189
    LAMP 193
    PCR assay 189
    real-time PCR 190–192
nuclein 24–27
nucleoside (purine) transporter of P. falciparum 91
nucleoside transport 91–93
nucleotide 31
nucleus 25–27, 35, 36
Nussenzweig, Ruth 235–237, 243, 244, 251

odor receptors 153
oocyst 13, 14, 22, 45
ookinete 9, 14, 22, 45
Opie, Eugene 9
OptiMal 179, 188, 189, 192
organelle 42
orotidine 5′-monophosphate phosphorylase 72
Oscillaria malariae 5
Osler, William 8

P. berghei 43, 177, 179
P. cathemerium 196
P. chabaudi 43
P. cynomolgi 21, 43, 237
P. elongatum 254
P. falciparum 7, 8, 36, 66, 160
    genes 39
    nuclear DNA 39
    nucleosome 76
    repetitive interspersed family (rif) 83
    sub-telomeric open reading frame (stevor) 83
    transport 88
    var (variant) genes 81, 82
P. falciparum CQ resistance transporter 280
P. falciparum erythrocyte surface protein 1 (PfEMP1) 80
P. falciparum mitochondrion 182
P. falciparum normocyte-binding protein 1 (PfNBP1) 140
P. falciparum reticulocyte homology 4 (PfRH4) protein 116
P. falciparum skeleton-binding protein 1 (PfSBP1) 122
P. falciparum-specific HRP-2 189
P. gallinaceum 43, 72, 176, 228–230, 232, 254
P. inui 43
P. knowlesi 7, 43, 79, 137, 139, 197, 198, 204–206, 237
P. knowlesi merozoites 222
P. kochi 21
P. lophurae xiii, xiv, 43, 72, 179, 196
P. malariae 7, 168
P. ovale 7, 21
P. reichenowi 160, 161
P. relictum 43
P. simium 167
P. vinckei 43

P. vivax   7, 19, 21, 133–135, 137, 138, 140, 142, 198, 199
   malaria therapy   20
P. yoelii   43, 66, 218, 255, 258
P. yoelii merozoite-specific protein   220
P36   266
P52   266
pantothenate   95, 96
para-aminobenzoic acid (pABA)   282
parasitophorous vacuolar membrane (PVM)   85, 113
parasitophorous vacuole (PV)   44
parvaquone   292
passive transfer   202
Pasteur, Louis   2
Patarroyo, Manuel Elkin   224–227
PATH MVI   253
Pauling, Linus   155, 156
perforin-like protein   127
permeome   90
Petri dish cultures   211
Petri, Julius   210
PEV3A   263
Pf195   220
Pf195 = MSP-1   215
Pf41   120
PfADET1   92
PfADET2   92
PfATPase6   298
PfCG2   109
PfCRT   281, 283
PfEMP1   81, 82, 106, 108
PfEMP3   108
PfENT1   92
Pfg377   131
PfHT   90
PfHT1   91
PfLAAT   92
pfmdr   287
PfMDR1   287, 288
PfNT1   91, 92, 93
PfNT2   93
PfNT3   93
PfNT4   93
PfPgh-1   287
PfRH2b   113
PfRH4   117
Pfs230   230, 231
Pfs48/45   230, 231
PfSPZ vaccine   251
Pg26   232, 233
P-glycoprotein   287
Pgs25   231
Pgs28   231
phage   37
phloretin   98
phosphatidylcholine   74, 95
phospholipase C   112
PkDBP   138
placental malaria   103
plasma membrane   42
plasmepsin   109
plasmepsin II   129
plasmepsin V   107
plasmid   36, 37
Plasmodium   6,
   discovery   2
   hiding   19
   transmitted   10
Plasmodium export element (PEXEL)   106–109
Plasmodium surface anion channel (PSAC)   99
Pneumocystis carinii   293
pneumonia   27, 28
polar ring   43

polymerase chain reaction (PCR)   38, 192
population replacement   152
porin   129
pre-blood stage   21
pre-blood-stage immunity   258
pre-blood-stage vaccine   242
precursor to major merozoite surface antigens   220
prevention   18, 19
prime-boost   264, 265
Program for Appropriate Technology in Health (PATH)   248
proguanil   281, 293
prokaryotes   42
protamine   25, 75
protein export   104, 107
Ps230   232
Ps25   233
Ps28   233
Ps48/45   232
PSAC   100, 101
P-type cation transporter ATPase 4   270
pulsed-field gel electrophoresis   36
purified protein factor (PPF)   204
purine salvage   91–92
   adenine   93
   adenosine   93
   guanine   93
   hypoxanthine   93
pus cell   23, 24, 26
PvDBP   138, 139
PVM   86, 95, 104, 105, 107, 109, 121, 126–129, 131, 266
PvRBP-2   140
Py235   220

pyrimethamine (Daraprim)   71, 72, 281, 282, 284, 285
   resistance   73
pyrimidine pathway   72
pyrimidine synthesis   48, 289

qinghaosu   294, 295
quinacrine   171
quinine   171, 173, 176, 285, 287, 290

rapid diagnostic tests [RDTs]   187–189, 192
Read, Clark P.   86
recombinant DNA   36
recrudescence   21
red blood cell receptor   114, 115
Reese, Robert   212–214
relapse   21, 142, 199
replication   34
residual body   44, 45
resistance   195, 273–275, 280–282, 284, 288, 298, 303
Resochin   276, 277
reticulocyte-binding proteins, PvRBP-1   140
reticulocytes   140, 142
reverse transcriptase   37
rhomboid protease   113, 121, 122
RhopH   121
rhoptry   43, 44, 111–113, 121, 122
rhoptry-associated membrane antigen (RAMA)   121
rhoptry-associated protein 1 (RAP1)   122
Riamet   299
ribosomal RNA   160
ribosome   35
Richards, W. H. G. (Harry)   289

Rieckmann, Karl   205, 238, 251
Riley, Eleanor   220
ring stage   44
ring-exported proteins, REX 1 and REX 2   108
RNA polymerase   35
Rockefeller Institute   28, 30, 31
Roman Campagna   17
Ross, Ronald   11–15, 18, 19, 46, 145–147
RTS, S   242, 246, 247, 248, 252, 255
RTS, S/AS02D   248
RTS, S/AS02   249
Russell, Paul F.   233, 234

Salvador I strain   162
Sanaria   250, 251, 253
Sanger Institute   55, 56
Sanger sequencing   39
Sanger, Frederick   38, 64
Schaudinn, Fritz   19
schizogony   22
schizont   43, 44
schizont-infected cell agglutination reaction (SICA)   79
Schüffner, William   134
Schüffner's dots   134, 142
Scripps Research Institute   213
SERA 8   131
serine-rich antigen (SERA) protease   120, 128
serpin   148
Shortt, Henry E.   21, 136
shotgun sequencing   64
sialic acid (N-acetylneuraminic acid [Neu5Ac])   114
SICA antigens   80

sickle hemoglobin   156
Siddiqui, Wasim A.   207–210, 214, 215
Silverman, Paul   203–206, 250
single nucleotide polymorphisms (SNPs)   158, 283
skeleton-binding protein 1 (SBP1)   108
Sontochin   276, 277
Speck, John F.   174, 176, 177
spectrin   103, 108
SPf66   224–227
spiroindolones   270
sporozoite   19, 20, 22, 46, 235
sporozoite asparagine-rich protein 1 (SAP1)   266
sporozoite surface protein (CSP)   63
sporozoite vaccine   233–236, 239
squirrel (*Saimiri*) monkeys   215
S-s-U (–)   114
Stanley, Harold   212
stearoyl-MDP   208
sterile immunity   235, 237, 258, 260
sterile insect technique   150–151
sterilizing immunity   249
Sternberg, George   3, 7
STEVOR   109
Streptococcus pneumoniae   27, 28
SUB1   119, 122, 127, 128
SUB2   121, 122, 127
SUB3   127
subtilisin-like protease SUB2   112, 119
sulfadiazine   285
sulfadoxine   281, 282
sulfamethaxozole   282

sulfathiazole   282
Sulston, John   54, 55

Taliaferro, William H.   254, 263
taste receptors   153
Tate, Parr   20
TBV   227, 228, 232, 304
TDR (*see* WHO-TDR)
telomeres   48, 81
TEP1   148, 149
tetrahydrofolate   72
tetranucleotide hypothesis
   31, 32
Tettelin, Herve   57, 58
thioester-containing protein 1
   148
thrombospondin-related
   antigenic protein (TRAP)   258
thymidine   72
thymidylate synthetase   72
Tiggert, William   285
tight junction   112
TIGR   49, 50, 52–58, 60, 65
Tommasi-Crudeli, Corrado   3
Trager, William   xiii, 125, 207,
   210, 211
transcription   35, 75–77, 271
transcriptome   73
transfer RNA (tRNA)   35
transforming principle   28, 29
transgenic mosquito   151, 152
translation   35
transmission   15, 18, 149, 304
transmission-blocking
   antibodies   231, 232
transmission-blocking
   immunity   229, 230
transmission-blocking vaccines
   (TBVs)   130, 227

transport   48, 85, 88, 94, 96, 97
transporter
   adenine   92, 93
   adenosine   91, 92, 93
   anion   99
   carrier   87
   channel   87
   fructose   90
TRAP   261, 262, 264, 265
triplet code   34
trophozoite   43, 44
trypan red   273
trypsin   114, 118, 136
tubulovesicular network
   (TVN)   48, 104, 105

ubiquinone   290, 291, 293
UIS3   266
UIS4   266
University of Chicago   173, 174
upregulated in infectious
   sporozoites (UISs)   266
U.S. Agency for International
   Development (USAID) malaria
   vaccine program   202, 203,
   205–207, 209–212, 214, 215,
   217, 244
U.S. Army Research Program on
   Malaria   285

vaccination   196–200, 204, 205,
   207–209, 212, 213, 221–223, 237,
   240, 241
vaccine   212
Vaidya, Akhil B.   181, 182
Vanderberg, Jerome   234–237,
   239–241, 251
var   48
var gene   83

Venter, Craig   49, 50, 52, 53–55, 58, 301
vir genes   141

Walter and Eliza Hall Institute   223
Walter Reed Army Institute of Research (*see* WRAIR)
Warhurst, David   279
Waters, Andy   223
Watson–Crick model   34
Watson, James   33
Wellcome Foundation   218
Wellcome Research Laboratories   289, 291
Wellcome Trust Sanger Institute   50
Wellems, Thomas   280
White, Nick   227
WHO   200–202, 278, 279, 296, 299
   candidate vaccines   233, 264
WHO Scientific Group on the Immunology of Malaria   205
WHO-TDR (the Special Programme for Research and Training in Tropical Diseases)   146

Wilkins, Maurice   33
Wilson, Allan   157
Wilson, E. B.   27
Winthrop Chemical Company   276, 277
Winzeler, Elizabeth   67–69
WR142,490 (a-(2-piperidyl)-2,8-bis-(trifluoromethyl)-4-quinolinemethanol)   286
WR99210   73, 271, 283, 284
WR99210 (4,6-diamino-1,2-dihydro-2,2-dimethyl-1-(2,4,5-trichlorophenoxy) propyloxy)-1,3,5 triazine hydrobromide)   281
WRAIR   285, 296

xanthurenic acid   131
X-ray crystallography   33

Yoeli, Meier   235

zanzarone   17
Zavala, Fidel   255–258
Zuckerandl, Emile   155, 156